Contemporary Knowledge and Systems Science

W. B. Lee
Hong Kong Polytechnic University, China

Farzad Sabetzadeh
Hong Kong Polytechnic University, China

A volume in the Advances in Knowledge Acquisition, Transfer, and Management (AKATM) Book Series

IGI Global
DISSEMINATOR OF KNOWLEDGE

Published in the United States of America by
IGI Global
Information Science Reference (an imprint of IGI Global)
701 E. Chocolate Avenue
Hershey PA, USA 17033
Tel: 717-533-8845
Fax: 717-533-8661
E-mail: cust@igi-global.com
Web site: http://www.igi-global.com

Library of Congress Cataloging-in-Publication Data

Names: Lee, W. B., editor. | Sabetzadeh, Farzad, 1981- editor.
Title: Contemporary knowledge and systems science / W.B. Lee and Farzad Sabetzadeh, editors.
Description: Hershey, PA : Information Science Reference, [2018]
Identifiers: LCCN 2017049731| ISBN 9781522556558 (h/c) | ISBN 9781522556565 (eISBN)
Subjects: LCSH: Knowledge management.
Classification: LCC HD30.2 .C65454 2018 | DDC 658.4/038011--dc23 LC record available at https://lccn.loc.gov/2017049731

British Cataloguing in Publication Data
A Cataloguing in Publication record for this book is available from the British Library.

Advances in Knowledge Acquisition, Transfer, and Management (AKATM) Book Series

ISSN:2326-7607
EISSN:2326-7615

Editor-in-Chief: Murray E. Jennex, San Diego State University, USA

MISSION

Organizations and businesses continue to utilize knowledge management practices in order to streamline processes and procedures. The emergence of web technologies has provided new methods of information usage and knowledge sharing.

The **Advances in Knowledge Acquisition, Transfer, and Management (AKATM) Book Series** brings together research on emerging technologies and their effect on information systems as well as the knowledge society. **AKATM** will provide researchers, students, practitioners, and industry leaders with research highlights surrounding the knowledge management discipline, including technology support issues and knowledge representation.

COVERAGE

- Cognitive Theories
- Cultural Impacts
- Information and Communication Systems
- Knowledge Acquisition and Transfer Processes
- Knowledge Management Strategy
- Knowledge Sharing
- Organizational Learning
- Organizational Memory
- Small and Medium Enterprises
- Virtual Communities

IGI Global is currently accepting manuscripts for publication within this series. To submit a proposal for a volume in this series, please contact our Acquisition Editors at Acquisitions@igi-global.com or visit: http://www.igi-global.com/publish/.

The Advances in Knowledge Acquisition, Transfer, and Management (AKATM) Book Series (ISSN 2326-7607) is published by IGI Global, 701 E. Chocolate Avenue, Hershey, PA 17033-1240, USA, www.igi-global.com. This series is composed of titles available for purchase individually; each title is edited to be contextually exclusive from any other title within the series. For pricing and ordering information please visit http://www.igi-global.com/book-series/advances-knowledge-acquisition-transfer-management/37159. Postmaster: Send all address changes to above address.

Titles in this Series

For a list of additional titles in this series, please visit:
https://www.igi-global.com/book-series/advances-knowledge-acquisition-transfer-management/37159

Maximizing Social Science Research Through Publicly Accessible Data Sets
S. Marshall Perry (Saint Mary's College of California, USA)
Information Science Reference • ©2018 • 328pp • H/C (ISBN: 9781522536161) • US $195.00

Promoting Interdisciplinarity in Knowledge Generation and Problem Solving
Mohammed Nasser Al-Suqri (Sultan Qaboos University, Oman) Abdullah Khamis Al-Kindi (Sultan Qaboos University, Oman) Salim Said AlKindi (Sultan Qaboos University, Oman) and Naifa Eid Saleem (Sultan Qaboos University, Oman)
Information Science Reference • ©2018 • 324pp • H/C (ISBN: 9781522538783) • US $195.00

Techniques for Coding Imagery and Multimedia Emerging Research and Opportunities
Shalin Hai-Jew (Kansas State University, USA)
Information Science Reference • ©2018 • 483pp • H/C (ISBN: 9781522526797) • US $235.00

Cross-Disciplinary Approaches to Action Research and Action Learning
Tome Awshar Mapotse (University of South Africa, South Africa)
Information Science Reference • ©2018 • 339pp • H/C (ISBN: 9781522526421) • US $215.00

Cross-Cultural Analysis of Image-Based Assessments Emerging Research and Opportunities
Lisa Keller (University of Massachusetts Amherst, USA) Robert Keller (Measured Progress, USA) and Michael Nering (Measured Progress, USA)
Information Science Reference • ©2018 • 166pp • H/C (ISBN: 9781522526919) • US $130.00

Social Media for Knowledge Management Applications in Modern Organizations
Francesca Di Virgilio (University of Molise, Italy)
Business Science Reference • ©2018 • 384pp • H/C (ISBN: 9781522528975) • US $215.00

Philosophical Perceptions on Logic and Order
Jeremy Horne (International Institute of Informatics and Systemics, USA)
Information Science Reference • ©2018 • 402pp • H/C (ISBN: 9781522524434) • US $225.00

For an entire list of titles in this series, please visit:
https://www.igi-global.com/book-series/advances-knowledge-acquisition-transfer-management/37159

IGI Global
DISSEMINATOR OF KNOWLEDGE

701 East Chocolate Avenue, Hershey, PA 17033, USA
Tel: 717-533-8845 x100 • Fax: 717-533-8661
E-Mail: cust@igi-global.com • www.igi-global.com

Table of Contents

Preface xiii

Chapter 1
A Systems Theory of Organizational Information 1
Tuan Nguyen Manh, HCMC University of Technology, Vietnam

Chapter 2
Theorizing Sociomaterial Practices: Gilbert Simondon's Theory of Individuation 38
Alexander Styhre, University of Gothenburg, Sweden

Chapter 3
Organizational Culture and Organizational Performance 50
Anas M. Bashayreh, Al-Zahra'a College for Women, Oman

Chapter 4
A Knowledge Creation Model for Graduate Research 70
Yoshiteru Nakamori, Japan Advanced Institute of Science and Technology, Japan
Jing Sun, Japan Advanced Institute of Science and Technology, Japan
Jianguo Wu, Wuhan University of Technology, China
Jing Tian, Wuhan University of Technology, China
Van-Nam Huynh, Japan Advanced Institute of Science and Technology, Japan

Chapter 5
Knowledge Retention in Smaller Firms 100
Susanne Durst, University of Skövde, Sweden
Guido Bruns, Incipere, Germany
Ingi Runar Edvardsson, University of Iceland, Iceland

Chapter 6
Knowledge-Scientific Evaluation of Social Service Systems 120
Yoshiteru Nakamori, Japan Advanced Institute of Science and Technology, Japan
Fei Meng, Japan Advanced Institute of Science and Technology, Japan
Van-Nam Huynh, Japan Advanced Institute of Science and Technology, Japan

Chapter 7
Exploring the Determinants Affecting Academics' Knowledge-Sharing Behavior in United Arab Emirates Public Universities 151
Huda Skaik, Independent Researcher, Malaysia
Roslina Othman, International Islamic University Malaysia, Malaysia

Chapter 8
Knowledge-Intensive Evolutionary Algorithms for Solving a Healthcare Fleet Optimization Problem: An Ontological Approach 192
Carlos Adrian Catania, National University of Cuyo, Argentina
Cecilia Zanni-Merk, INSA Rouen Normandie, France
François de Bertrand de Beuvron, INSA Strasbourg, France
Pierre Collet, Strasbourg University, France

Chapter 9
Assessing Knowledge Management Performance in Organisations Based on the Criteria of Total Quality Management...224
Kit Fai Pun, University of the West Indies, Trinidad and Tobago
Man Yin Rebecca Yiu, University of the West Indies, Trinidad and Tobago

Chapter 10
Integration Between Knowledge Management and Total Quality Management in Jordanian Universities: Empirical Study...256
Issa Mahmoud Shehabat, Yarmouk University, Jordan
Mohammad Berrish, Shaqra University, Saudi Arabia

Compilation of References 296

About the Contributors 334

Index 339

Detailed Table of Contents

Preface xiii

Chapter 1
A Systems Theory of Organizational Information 1
Tuan Nguyen Manh, HCMC University of Technology, Vietnam

Standing on both the Peircean pragmatism semiotics and the Churchmanian systems thinking, this chapter is an empirically grounded conceptualization of the phenomena of organizational information in terms of the nature and the formulation process of information in organizations. By the author's systemic conception, organizational information would be a unity that comprises six nonexclusive aspects: structure, function, process, context, time, and epistemology. From the relational perspective, organizational information would manifest itself as a dynamically triadic process that comprises three states of mind (i.e., surprise, doubt, and belief) and three human activities (i.e., experience, abduction, and inquiry). The author's system of organizational information introduces a foundational framework for both information and organization domains, which offers that information and organization constitute each other. The author also posits that the model of organizational information would imply an information paradigm for and hence a theory native to the information systems and knowledge management field.

Chapter 2
Theorizing Sociomaterial Practices: Gilbert Simondon's Theory of Individuation 38
Alexander Styhre, University of Gothenburg, Sweden

The recent interest in sociomaterial practices and forms of imbrication of social and material resources in, for example, studies of the use of information technology and information systems in organizations, has called for new theoretical developments to enact and fully understand materiality. This chapter examines the work of French philosopher Gilbert Simondon as an important, yet to date little explored resource in organization studies. Following Orlikowski's call for a relational ontology apprehending the enfolding of materiality and social resources, Simondon's analytical framework, which includes key concepts such as individual-tion, transduction, and relations, is presented. The chapter contributes to the recent debates regarding how to theorize and examine material resources such as information and communication technologies used in organizations.

Chapter 3
Organizational Culture and Organizational Performance....................................50
Anas M. Bashayreh, Al-Zahra'a College for Women, Oman

Organizational cultures can have varying impacts on organization performance. Oftentimes, employees work harder to achieve organizational goals if they consider themselves to be part of the corporate culture. Different cultures operating in one company can also impact organization performance. Organizational culture is an important part to be considered by dynamic organizations in order to develop some competitive advantage to ensure enhanced organizational performance. This chapter aims to close the gap between understanding the role of organizational culture and the effect on organizational performance among the employees in insurance companies.

Chapter 4
A Knowledge Creation Model for Graduate Research...70
Yoshiteru Nakamori, Japan Advanced Institute of Science and Technology, Japan
Jing Sun, Japan Advanced Institute of Science and Technology, Japan
Jianguo Wu, Wuhan University of Technology, China
Jing Tian, Wuhan University of Technology, China
Van-Nam Huynh, Japan Advanced Institute of Science and Technology, Japan

This chapter proposes a model of knowledge creation and knowledge validation for the research of postgraduate students, and then introduces an evaluation method on research ability and research environment based on this model. The chapter reports the results of statistical analysis, using the data obtained by a questionnaire survey at a graduate school of computer science and technology. Among several discoveries, the great differences in the evaluation of research environment between teachers and students, and in the self-evaluation of research ability between female students and male students are noteworthy. It was confirmed that the proposed model and the evaluation method are effective for supporting research of graduate students.

Chapter 5
Knowledge Retention in Smaller Firms 100
Susanne Durst, University of Skövde, Sweden
Guido Bruns, Incipere, Germany
Ingi Runar Edvardsson, University of Iceland, Iceland

In environments that are characterized by skill shortage, knowledge retention should be a key activity. Our understanding of this knowledge practice, however, is underdeveloped. The purpose of this chapter is to contribute to the further development of this field. The authors are interested in studying how smaller firms approach the retention of critical knowledge. Semi-structured interviews with organization members of five smaller Austrian firms operating in the building and construction industry were conducted. The findings presented in this chapter advance the limited understanding of knowledge retention in general and regarding smaller businesses. They also contribute to the further development of the study of knowledge management in the building and construction sector. Based on the study's findings, suitable measures to better manage the process of knowledge retention in smaller firms are derived. These measures may be useful for smaller firms operating in similar industries as well.

Chapter 6
Knowledge-Scientific Evaluation of Social Service Systems 120
Yoshiteru Nakamori, Japan Advanced Institute of Science and Technology, Japan
Fei Meng, Japan Advanced Institute of Science and Technology, Japan
Van-Nam Huynh, Japan Advanced Institute of Science and Technology, Japan

This chapter proposes a new evaluation method for social service systems in order to support participants to create new knowledge and value. The main proposal is a rating scale method that can evaluate the current level (as-is) and the future desire (to-be) of participants, and with which we can analyze achievements of individuals. By using a concrete example of an education program, which is a worth living discovery seminar for retired men, the chapter reports the results of hypothetical tests on the relationships between items of evaluation and an interesting finding related to the reason of differences in attitude of participants.

Chapter 7
Exploring the Determinants Affecting Academics' Knowledge-Sharing
Behavior in United Arab Emirates Public Universities 151
Huda Skaik, Independent Researcher, Malaysia
Roslina Othman, International Islamic University Malaysia, Malaysia

This chapter investigates knowledge-sharing practice among academics, examines the relationship between knowledge-sharing behavior and its predictors based on theory of planned behavior, and identifies the motivators and barriers affecting these predictors. Data were collected using a web-based questionnaire. Using SPSS and PLS-SEM, the analysis process involved (1) measuring the extent of knowledge sharing practice, (2) assessing the measurement model, (3) assessing the structural model, (4) testing the hypotheses, (5) validating research model fit. Results showed (1) great extent of knowledge-sharing practice, (2) knowledge-sharing behavior is significantly affected by intention, which is affected by attitude, subjective norms, self-efficacy, but not affected by controllability, (3) attitude is positively affected by trust and reputation as motivators of knowledge sharing, while controllability is negatively affected by lack of time and poor communication as barriers of knowledge sharing. This is the first research on knowledge-sharing behavior in higher education in UAE.

Chapter 8
Knowledge-Intensive Evolutionary Algorithms for Solving a Healthcare Fleet
Optimization Problem: An Ontological Approach 192
Carlos Adrian Catania, National University of Cuyo, Argentina
Cecilia Zanni-Merk, INSA Rouen Normandie, France
François de Bertrand de Beuvron, INSA Strasbourg, France
Pierre Collet, Strasbourg University, France

In this chapter, the authors show how knowledge engineering techniques can be used to guide the definition of evolutionary algorithms (EA) for problems involving a large amount of structured data, through the resolution of a real problem. Various representations of the fitness functions, the genome, and mutation/crossover operators adapted to different types of problems (routing, scheduling, etc.) have been proposed in the literature. However, real problems including specific constraints (legal restrictions, specific usages, etc.) are often overlooked by the proposed generic models. To ensure that these constraints are effectively considered, the authors propose a methodology based on the structuring of the conceptual model underlying the problem, as a labelled domain ontology suitable for optimization by EA. The authors show that a precise definition of the knowledge model with a labelled domain ontology can be used to describe the chromosome, the evaluation functions, and the crossover and mutation operators. The authors show the details for a real implementation and some experimental results.

Chapter 9
Assessing Knowledge Management Performance in Organisations Based on the Criteria of Total Quality Management..224
Kit Fai Pun, University of the West Indies, Trinidad and Tobago
Man Yin Rebecca Yiu, University of the West Indies, Trinidad and Tobago

Knowledge management (KM) has been gaining momentum as the means toward organisational growth. Recent literature also gives much prominence to emerging performance measurement (PM) systems for assessing performance. This chapter comprehends the concepts of integrating KM with PM, and in short, KM performance measures in organisations. It reviews the issues that surround KM/PM initiatives, and discusses the assessment criteria that integrate the philosophy and the principles of total quality management (TQM) with KM performance measures. Five KM/PM enablers/criteria are identified, namely 1) senior management leadership, 2) management by processes, 3) people development, 4) continuous improvement, and 5) results orientation. These concepts are constituted the development of an integrated knowledge management (IKM) model for assessing organisational KM performance. Future research could evaluate the efficacy of the IKM model, validate the key KM/PM criteria, and develop an implementation framework to foster the KM/PM efforts in organisations in different sectors.

Chapter 10
Integration Between Knowledge Management and Total Quality Management in Jordanian Universities: Empirical Study..256
Issa Mahmoud Shehabat, Yarmouk University, Jordan
Mohammad Berrish, Shaqra University, Saudi Arabia

Resulting from the development of management theory in the past few years, knowledge management has emerged, which is identifying information of value and how to use it well, as well as understanding cognitive assets of an organization and how to exploit them. Knowledge management is important and necessary for the survival of an organization and its superiority. It can be integrated into the philosophy of total quality management to play a role in the development of the basis of performance of modern organizations. The total quality management philosophy seeks to achieve customer satisfaction through the commitment of the leadership of an organization and its workers by a process of continuous improvement of the quality of performance in various aspects of an organization. This chapter aims to identify the integration between knowledge management and total quality management, when applied to public Jordanian universities as a sample for the study. The chapter found the presence of integrity and a strong correlation between knowledge management and total quality management.

Compilation of References 296

About the Contributors 334

Index 339

Preface

I am glad to introduce this special edition of *Contemporary Knowledge and Systems Science*.

Over the years, we have witnessed an unprecedented convergence and intersection among various academic disciplines not only in physical science, but also in all other scientific domains such as social science, economics and technological developments. Nevertheless, this does not mean our knowledge frontier is expanding towards generalization and abstraction. On the contrary, we are indeed observing a more organic synergy of different branches of knowledge, spiraling and spouting in various new directions. Nowadays, it is almost impossible for people to communicate effectively if they are only knowledgeable about their own specialized fields. Knowledge science provides a good platform to better understand and position different types of new knowledge from the systems perspective.

In this book's collection, we have a mix of theoretical and empirical chapters on some of the current topics in knowledge and systems science. Organizations are always the pinnacles of human systems. We have gathered various topics such as use of systems theory to analyze the interplay between technology and information in organizations, theorizing sociometrical practices, evaluating organizational culture and performance, and retention of knowledge in small firms.

There are also valuable research findings in specific applications. These include evolutionary algorithms for solving health care fleet optimization problem, creation of a knowledge model to evaluate ability in research environment, evaluation methods in social service systems, knowledge sharing behaviors in public universities, assessing knowledge management performance and integration of knowledge and total quality management.

I hope, reading the above topics throughout this book provide some fruits for thought for both researchers and practitioners in knowledge and systems science (and beyond!).

I would like to thank all the contributing authors to make this book possible and the co-editor, Dr. Farzad Sabetzadeh for his efforts to compile this book. If you have any ideas and suggestions about the theme of our next book series, you are welcome to contact us and the publishers.

W. B. Lee
The Hong Kong Polytechnic University, China

Chapter 1
A Systems Theory of Organizational Information

Tuan Nguyen Manh
HCMC University of Technology, Vietnam

ABSTRACT

Standing on both the Peircean pragmatism semiotics and the Churchmanian systems thinking, this chapter is an empirically grounded conceptualization of the phenomena of organizational information in terms of the nature and the formulation process of information in organizations. By the author's systemic conception, organizational information would be a unity that comprises six nonexclusive aspects: structure, function, process, context, time, and epistemology. From the relational perspective, organizational information would manifest itself as a dynamically triadic process that comprises three states of mind (i.e., surprise, doubt, and belief) and three human activities (i.e., experience, abduction, and inquiry). The author's system of organizational information introduces a foundational framework for both information and organization domains, which offers that information and organization constitute each other. The author also posits that the model of organizational information would imply an information paradigm for and hence a theory native to the information systems and knowledge management field.

INTRODUCTION

The influential management guru Drucker (1999) reminded us that the biggest challenge to the contemporary age is the very information, which an executive needs for his/her specific knowledge work. It was warned that the key point may be the

DOI: 10.4018/978-1-5225-5655-8.ch001

own way the knowledge worker works on data to formulate his/her information that, emphatically, needs to be specified contextually.

Given contemporary organizations as information-bond systems (Gharajedaghi 2011), information has been the central object in several information related fields, including information systems (IS), knowledge management (KM), and management science (Mingers 2006), to name a few. Unfortunately, although knowledge of the nature of information was non-trivial, there was still no agreement on what information is (Mingers 2006) and it is further acknowledged that IS field is lacking of an information paradigm (McKinney and Yoos 2010).

Several studies have recently attempted to construct conceptual models of information. For example, Mingers (2006) proposed a comprehensive theory of semantic and pragmatic information, in which information, an object of information system, is associated closely with meaning, an object of human cognition. Despite those contributions, the extant studies have neither modeled completely the nature of organizational information nor developed satisfactorily the formulation process of organizational information. First, there exist several contradictory views with regard specifically to the information and knowledge hierarchy (e.g. Tuomi 1999). For example, most researchers simply take one sort of hierarchy either conventional or reversed (see e.g. Ackoff 1989). But few contextually recognize the value of both hierarchies (e.g. Bach and Belardo 2003), or suggest a conception that could be emerged beyond the two hierarchies (e.g. Stenmark 2002). Second, the contradictory views in the information and knowledge hierarchy could be in turn caused by a lack of a satisfactory distinction among the notions of information, knowledge, and data. This shortcoming has not resolved yet in the literature, although it has been long time (Gourlay 2006), and considerable to the body of knowledge of IS and KM in particular (Mingers 2008). More fundamentally, Mingers (2008) emphasized that it was just the weakness of all approaches to the definitions of information and knowledge as well. Third, the formulation process of organizational information or knowledge in almost extant studies suffers from ill theoretical (i.e. logical, psychological, and sociological) foundations because the underlying notions (e.g. tacit knowledge) are not well defined conceptually. For example, Nonaka's (1994) classic model of organizational knowledge creation, which has ever achieved the paradigmatic status since the mid-1990's (Gourlay 2006), has been criticized as "seriously incomplete and selectively blind" (Zhu 2006, p.109), or as cracked in its conceptual framework of knowledge conversion process (Gourlay 2006). Additionally, Nonaka's model could hardly explain how to generate new ideas and how to maintain collaborative work (Bereiter 2002), which are inherently crucial to knowledge creation in general and under the community view of knowledge in specific. Fourth, the formulation process is inadequate because the process is developed onto several underlying concepts taken for granted, or the entity-centered view is mostly assumed and it does not show the linking between the information or knowledge types. In fact, very few studies start with the process view of information or knowledge (e.g. Polanyi's (1962) tacit knowing). Last but not least, it is widely recognized that the relevant literature is lacking an empirical understanding of information in organizations (Gourlay 2006). It also fails to take a systemic perspective that would be more adequate for an understanding of inherently dynamic and complex socio-cultural phenomena (Gharajedaghi 2011).

To rectify these issues, our new conceptualization focusing on the nature of information and its formulation process is offered within the following boundary. First, organizational information should be a social construct that is inherently one of symbolic structures, and of uncertainty reduction (Newman 2001). Second, the notion 'information' is standing in the non-trivial yet not well defined relation to those of 'knowledge' and 'data' (e.g. Stenmark 2002). Third, organizational information should be viewed as, on the one hand, both entity and process, on the other hand, both individual and collective perspectives to reflect the locus view of knowledge in organizations (e.g. von Krogh 2009). At this point, several confusing typologies and types of information and knowledge in the relevant literature such as individual knowledge vs. organizational knowledge, tacit knowledge vs. explicit knowledge, subjective knowledge vs. objective knowledge, subjective information vs. objective information, and so on should be also made satisfactorily clear. Fourth, organizational information should be justified to some extent to validate itself as scientific information, truth, and so on (e.g. Mingers 2008). Finally, due to its dynamic and complex nature, organizational information should be empirically investigated from a systems thinking perspective.

The paper is structured as follows. The next section provides background for the whole study, starting with a brief of organizational information as the object of study, and then presenting some sensitizing concepts for our theorizing later. The third section is on methodology, which is both building the iterative procedure of and conducting data gathering and analysis to reach the grounded constructs and themes of organizational information. We report the empirical findings in the fourth section that much elaborates on sample case evidence and relevant literature to satisfactorily gain a comprehensively explanatory model of organizational information. The fifth section starts with a proposal of semiotics based foundation for both information and organization studies, and then, discusses the contribution of this research. Finally, our middle range theory is summarized in the sixth section.

BACKGROUND

Organizational Information

In this paper, organizational information is simply understood as information in organizational life, which is, however, the unique object of IS studies, and could feature a purposefully social product of symbolic structures and of pragmatic use (Newman 2001). In this sense, organizational information could be "facts given and taken, and inferences drawn and established by participants within an organizational situation" (Swanson 1978, p.238).

To formally position organizational information as the object of study, we looked to Gorry and Scott Morton's (1989) classic framework of managerial activities or management information systems, and to von Krogh's (2009) locus of knowledge in organizations. As a result, we came to the following scope of study: (i) organizational business activities as instances of organizational information, and (ii) locus of organizational information covering both individualist and collectivist perspectives.

Theoretical Perspectives

Assuming that organizational information is purposefully enacted reality (Newman 2001), at Orlikowski and Baroudi's (1991) advice of the compatibility of phenomenon of interest and research approach, we argued that the nature of organizational information should be investigated with the research paradigm of systems philosophy, a sort of systemic pragmatism (e.g. Britton and McCallion 1994). The paradigm was useful here because it primarily aims to describe the historical emergence of social structures and contemporary contexts for social action and human freedom (e.g. Gephart 2004).

With Churchman's (1971) philosophy of systems as a research approach of our choice, several of theoretical perspectives could be used throughout our study in terms of sensitizing concepts. These were primarily Gharajedaghi's (2011) systems thinking, and Peirce's (1931) pragmatism semiotics. The former would be used to investigate multiple-perspective social constructs as found in the information phenomena in various organizations (e.g. Gharajedaghi 2011). The latter would fully show complicated relations in real world phenomena (Hausman 1993), and specifically organizational phenomena of symbolic structures as information (e.g. Newman 2001).

Systems Thinking

To explore organizations as purposefully social systems, Gharajedaghi's (2011) design based systems thinking, which is in turn rooted from Churchman's (1971) inquiring systems, is preferred (Tuan and Huy 2008). According to this school of thought, the organizational phenomena should be investigated via the plurality of structure, function, process and their interactions in a specific context.

Pragmatism Semiotics

Peirce's semiotics, the formal doctrine of signs, plays a crucial role in the Peircean pragmatism as it is really a foundation of his theories of perception, cognition, and inquiry (e.g. Peirce 1931). The semiotics centers on a general but parsimonious triad of sign, object and interpretant with Peirce's well-known expression "A sign is something that stands to somebody for something in some respect or capacity" (CP 2.228)[1]. Philosophically, the semiotics is based on three foundational categories, which are necessary and sufficient for all human experience and designated as firstness, secondness and thirdness (CP 1.417-418), and thus, can be also noted as a pragmatism semiotics to emphasize its general, triadic, and pragmatic nature (Tuan and Huy 2008). Therefore, it could be adequately employed as the underlying principle to clarify ideas, or a method to determine the meaning of words and concepts (Peirce 1931). In specific, the combination between the Peircean pragmatism and the Churchmanian systems thinking would be appropriate to base the study (e.g. Tuan and Huy 2008).

METHODOLOGY

We employed both Glaser and Strauss' (1967) grounded theory methodology and Yin's (2003) case study strategy. The former is for a method of data collection and analysis and the latter for a research design. There were two key reasons for our choice. First, grounded theory research is trans-disciplinary and commonly accepted to be inductive, processual and contextual (Orlikowski 1993), and attentive to issues of interpretation and process and does not bind one too closely to long-standing assumptions (Suddaby 2006). Second, case research was effective for problem formulation and grounded theory building (van de Ven 2007).

Research Setting and Data Sources

Our design was an embedded multiple case study (e.g. Yin 2003). For this, the main unit of analysis was organization but sub-units of analysis could be also individuals, groups, departments or projects in organizations.

Although organizational information was very generic, the research setting in this study was the consulting industry, which is a very emerging economic sector in Vietnam. The industry was playing the central role in the on-going economic restructure of HCM City, the economic hub of the country. The study involved four professional service firms ranging from 20 to 200 employees (see Table 1). Of the four companies, two were privately held, and two were shared holding. Three of the four firms were in the developmental stage of profit making, and one was in product development.

Field data were collected from the consulting organizations in the following manner. For HY-ICT as the first organizational sample, we identified four embedded units of analysis within the company case. With TP-DTF as the second company sample, we relied on two embedded units of analysis, and finally, one unit of analysis was selected for each of the two remaining firms.

It should be also noted that with grounded theory, data collection and analysis were simultaneous in accompany with purposeful, or theoretical sampling (Glaser and Strauss 1967). After working on data, we identified four sets of data in four organizational focal cases that suggested a satisfactory theoretical saturation (e.g. Glaser and Strauss 1967). Our number of cases and sets of data were also adequate to Martin and Turner's (1986) notes respectively. Table 1 summarizes how data were collected.

Data Gathering

Employing Gorry and Scott Morton's classic framework, we considered all varieties of organizational business activities as instances of organizational information. As a result, we were able to approach the phenomenon of organizational information in a generic sense.

Next, we used the interview protocol that comprises open-ended and in-depth questions to look at adequate business actions and events. An interview typically started with an introductory background of organization and individual who was respondent at the moment. The interview flow was conducted around descriptions of organizational businesses that could range from the level of unit, department, project or the whole organization. As soon as an organizational unit was determined, the interview focused on the respective business activities as well as the roles and involvement of the respondent. The inquiries were guided into time frames (e.g.

Table 1. Organizations, cases and embedded units of analysis

Organization	Number of Cases and Units of Analysis
HY-ICT	*Case:* Enterprise package application implementation
	Embedded units: Project sales Training activities Client business process restructuring Customization in implementation
TP-DTF	*Case:* Cluster of projects of designing coastal ports to 2020
	Embedded units: Coastal port designing R&D in consultancy
NN-TCD	*Case:* Intra-business technological consulting
HL-POM	*Case:* Lean production solutions
Total: 4 organizations	Total: 4 organizational cases and 8 embedded units of analysis

Gorry and Scott Morton 1989), representational structures (e.g. Newman 2001), processual and contextual boundaries (e.g. Gorry and Scott Morton 1989), human actor roles (e.g. Newman 2001), and more in journalist's basic interrogative pronouns (e.g. van de Ven 2007). For data triangulation, we attempted to use more than one participant for evidence account as well as company web sites and documents.

Data Analysis

We transcribed all interviews following the "24-hour rule" (Eisenhardt 1989) that required all interview notes be finished completely within the same day. Some interviews needed to be clarified via follow-up e-mails and phones. Private names were disguised due to their sensitivity. Under Glaser and Strauss' (1967) grounded theory spirit, we elaborated our own guidelines of data collection and analysis, drawing mainly on Lincoln and Guba's (1985) methodological foundations of naturalistic inquiry, Turner's (1983), and Martin and Turner's (1986) practical directions of grounded organization studies rather than on Strauss and Corbin's (1998) rigorous frame of procedures. Our procedure finally included five iterative and intertwined steps as follows.

We started with the organizational case of HY-ICT, a big corporation in the field of software solution in Vietnam. We conducted interviews, gathered more organizational documents, then visited the company web sites to select and finally analyze the company business activities to look into organizational information. During the process, we made clear the nested units of analysis (Miles and Huberman 1994) based on the company's specialized services such as ERP Application Implementation,

Custom Application Development, Technical Support Services, and so on. These services centered on enterprise package application implementation, and hence, we were able to identify four organizational units to work with on the same business of ERP implementation, which is first sketched as in Table 2.

We entered the first step of the unitizing by making initial codes. Our inputs were mainly field data of 8 interviews and the company's 20 web pages. All data were coded in the manner of sentence-by-sentence or paragraph-by-paragraph basing on the interview protocol.

In the second step of concept categorizing, we first determined provisional rules of category properties or inclusion of incidents for each conceptual category to be established. Sensitized with Gharajedaghi's social system covering four ingredients (i.e. structure, function, process, and context), our four rules of category properties respectively included structure as means, function as ends, process as behavior, and context as threshold (e.g. Gharajedaghi 2011). Then we transformed the initial codes into tentative conceptual properties according to the four rules, and hence, came up with four piles of incidents. For such incidents to be tentatively theoretical categories, we needed to get more data to fit with categorical rules above. Following theoretical sampling (Glaser and Strauss 1967), we continued the process with the second case named TP-DTF, a big company in the field of consultancy of planning and designing transportation facilities. We identified two organizational units centered on cluster of projects of designing coastal ports up to 2020. We were able to reach the convergence of conceptual categories after the second case when its codes and conceptual properties were quite similar to those of the first case.

Table 2. HY-ICT's Business of ERP Implementation: An Overview

HY-ICT's mission is to consult implementation of ERP packages. Hence, several of its basic business activities are, for example, sales of ERP packages and training of ERP implementation. HY-ICT's standardized sales procedure of off-the-shelf package implementation is relying on: (i) historical project profiles and original product provider's training and support; (ii) pre-sales activities including visits of customer sites, exploration of customer software practices and so on; (iii) implementation activities including user training and product installation as well as customizing application modules; starting with signed off agreement of specifications of application packages to be installed and detailed planning for user training and final product installation. Meanwhile, training in ERP implementation projects is based on: (i) consultant training about functionalities and usage of original products like Oracle ERP; (ii) training of key users/division heads about main uses of business applications on the existing business processes such as accounting practices; (iii) client end-user training about deployment and operation of the new systems or applications customized. After all, ERP implementation projects are based on: (i) existing business operations, and systems, with new business requirements for high performance or competitive advantages as well as off-the-shelf powerful ERP products offered; (ii) exploration of existing business processes; defining "future process"; selection of ERP products; training key users/department heads, changing current practices, identifying some customization/ development of ERP modules/subsystems; and doing customization or developing ERP modules/subsystems; (iii) making "future process" official, end-user training about operations of the new systems/applications customized/developed, and transferring the new system to customer sites.

For the third step of categorizing and category development, we took Peirce's (1931) semiotics to dimension conceptual properties and then narrowed them down into six piles of incidents, each comprising three and only three concepts. We also fine-tuned dimensions of conceptual properties adequately and then labeled the dimensions theoretically. Along the flow of abstraction from data to theoretical constructs, we advanced the triad of dimensions for each of the systemic properties of organizational information as follows: (source; content; effect) for structure, (uncertainty; equivocality; disorganization) for function, (resource; development; product) for process, and (action; inquiry; habit) for context. The labels of the dimensions for each triad were acquired when we linked conceptual categories and their dimensional data to the relevant literature as illustratively provided in Appendix. Then in the similar manner, we came up with the following three-value states of the remaining properties: (past, present, future) for time, and (objectivity, subjectivity, inter-subjectivity) for epistemology. A greater detail discussion on relations among the triads for each property will be provided in a later section.

The established triads helped to re-assert, on the one hand, the relations among the organizational information properties that were emerged, on the other hand, the contextual aspects of organizational information under investigation. The former implied a context-sensitive aggregation among organizational means (i.e. structure), ends (i.e. functions), and behaviors (i.e. process), which was neither a causal nor a producer-product link (e.g. Gharajedaghi 2011), of generic business activities modeling as organizational information. The latter was to position organizational information in the interaction with organizational actors that experienced their real worlds by a continuously activity chain of business action (e.g. decision making), inquiry (e.g. questioning and identification of business problem), and habit (e.g. forming business rules and policies). Such an interaction would be a selective extension of Checkland and Holwell's (1998) notion of information in support of inquiry and action. The chain was also supported by the evidence of the time endlessly ranging from the past, the present, and to the future, and by the incidents of epistemological status continuously flowing from the material world (i.e. objectivity), to the mental (i.e. subjectivity), and to the social one (i.e. inter-subjectivity).

More data were required to confirm the conceptual properties emerged. We continued sampling two more organizational cases, one (NN-TCD) with business activities of intra-organizational supports and on-site implementation of enterprise applications, and the other (HL-POM) with business activities of client consulting on clean production solutions. The status of theoretical saturation of concepts and relations was in fact reached at the eighth embedded case in HL-POM. At this point, we observed the critical state of data analysis when we could not gain any new insight while adding more field data into examination.

Our findings provided six properties for organizational information: epistemology, time, structure, function, process, and context. Each property in turn comprised a triadic relation among its three dimensions. In regard to organizational information formulation, two properties of process and context were dedicatedly discovered. Table 3 shows the preliminary findings.

In the fourth step of the analysis, we proceeded with varieties of dimensions of six properties established and found that many patterns of organizational information were possible. But note that each of our properties had three dimensions organized in a Peircean triad, which were firstness, secondness, and thirdness, and using the epistemological property as the key identifier, we ended up with only three extreme but substantial patterns of organizational information. Pattern1 was formed when all dimensions were firstness only. This pattern of organizational information was more factual and objective. In a similar manner, pattern2 was formed when both firstness and secondness were included to point out the opposite pattern of organizational information, which was more interpretive and subjective. Finally, pattern3 comprised all dimensions simultaneously firstness, secondness, and thirdness to represent the communal pattern of organizational information, which was more critical and intersubjective. Thus, pattern3 could be specified as a semiotic triad of pattern1, pattern2, and pattern3 itself. Alternatively, pattern3 presented itself as an intermediate point or mediation between pattern1 and pattern2. Table 4 presents the three fundamentally conceptual patterns.

For the fifth and final step, we continued coding theoretical concepts emerged to reach some conceptual categories and respective relationships by extension of Gharajedaghi's (2011) 'system' as sensitized at the outset. Its abstract extensions were the properties of time, of epistemology, and especially, triadic essence of the social phenomena (Peirce 1931). Regrouping the properties, we proposed organizational information as a 'system' of ontological (which in turn include the ones of structure, function, process, and context), epistemological, and time ingredients.

At a different logic of pattern or process rather than entity, the core category 'system' and its ingredients were necessarily triadic relations, which were conditions for both the emergent property and the boundary of organizational information as

Table 3. Conceptual properties and dimensions

Property	Dimension1	Dimension2	Dimension3
Structure	*source*	*content*	*effect*
Function	*uncertainty*	*equivocality*	*disorganization*
Process	*resource*	*development*	*product*
Context	*inquiry*	*action*	*habit*
Time	*past*	*present*	*future*
Epistemology	*objective*	*subjective*	*inter-subjective*

Table 4. Three emerging fundamental patterns of organizational information

Pattern3: Information (Dimension1, 2 & 3 – Intersubjectivity)			
Pattern2: Knowledge (Dimension1 & 2 – Subjectivity)			
Pattern1: Data (Dimension1 - Objectivity)			
Property	**Dimension1**	**Dimension2**	**Dimension3**
Structure	*source*	*content*	*effect*
Function	*uncertainty*	*equivocality*	*disorganization*
Process	*resource*	*development*	*product*
Context	*inquiry*	*action*	*habit*
Time	*past*	*present*	*future*
Epistemology	*objective*	*subjective*	*inter-subjective*

system. The former indicated that information (pattern3) was the emergent property of the organizational information as system, which turned out to be an interaction between data (pattern1) and knowledge (pattern2). This is important because it suggested that information could be necessarily distinguished from knowledge and data. The latter informed the interaction between organizational information and organizational actors those who created (i.e. the role of knowledge) and used it (i.e. the role of data) in their organizational settings. With Peirce's (1931) semiosis, an evolutionary stream of sign production and transmission, the categorical triad suggested could be sufficient for positioning such a boundary. Table 5 shows the central conceptual category of the emerging grounded theory.

At this point, when drawing on our emerging grounded patterns, we naturally came up with two central theoretical themes for the organizational information formulation process and for the nature of organizational information as a whole. The themes were emerged out of the thirdness of the respective triad, which as a law of being (Peirce 1931) is recognized as Peirce's key contribution to phenomenology (e.g. Hausman 1993). Thus by focusing on the thirdness, our theoretical themes were expected to accommodate and explain various phenomena of organizational information such as environmental data in organizational scanning (Daft and Weick 1984), or rich representation of understanding in supporting distributed cognition in organizations

Table 5. The grounded central category of organizational information

Central Category	*System*		
Sub-Categories	*Data*	*Knowledge*	*Information*
Relations	*Triadic*		

(Boland et al. 1994). Explicitly, the former theme was that organizational information formulation as habit production, and the latter, organizational information as social system.

EMPIRICAL FINDINGS

Our endeavour was discovery of theory from field data (Glaser and Strauss 1967), or a bit more formally, a middle-range theorizing (van de Ven 2007).

Nature of Organizational Information or Organizational Information as System

Our key finding was organizational information as system, which would involve six perspectives: structure, function, process, context, time (indicator), and epistemological (status). Further, our data also proposed that each perspective would in turn spread over three dimensions. Any single measure would lead one to a partial understanding of the essence of organizational information.

Structural Aspect of Organizational Information

Organizational actors could always point out the means of organizational information through organizational business activities in forms of symbolic actions (Zott and Huy 2007). In this sense, our findings proposed the structural property to consist of three nuances, which were source, content, and effect. We used 'source' to mainly refer to media, collection, body, or bases from which actors could extract or build some next artifacts. For example, a higher chief officer of TP-DTF described that, his company was distinguished from other consulting firms by its data collection that was "much older, much more valuable". Its data sets were accumulated results from past projects of investigations of natural and geographical properties of territories and environments. Those were valuable assets and deployed over time, the CEO admitted. In a sense, those are essential bases for formulation and implementation of the current cluster of projects of designing coastal ports up to 2020.

We used 'content' to refer to an abstraction from meanings, senses, or soul from which some like art masterpieces or work projects could start. One should not confuse between 'source' and 'content' if the perspective of epistemology is well taken. In specific, the 'source' is more objective and the 'content' is more subjective. Given the same 'source', depending on different analyses and insights, there might be some different 'contents' that could result in some very different organizational outcomes. For example in TP-DTF, we found that the same record of past projects of natural investigations could be used to form a few of master planning and long range visions of coastal transportation facilities, which definitely resulted in different work projects of consulting. For the 'content' of project cluster of consulting and designing coastal ports up to 2020, the local and central governments were the subjects to make final decisions, the Chairman of TP-DTF told us so.

Ultimately, we used 'effect' to refer to a pragmatic measurement, or justification of business transactions. If 'source' acts as physical body or material, and 'content' plays the role of mental mind, soul, or individual, then 'effect' created the value added or the significance figured out by putting itself as mediation between 'source' and 'content'.

In this triadic relation, both 'content' and 'source' were required to determine 'effect'. However, many 'content's could be drawn from single 'source', and some 'content's should be commonly shared or agreed to result in 'effect'. No 'effect' could be determined for a specific 'source', if shared 'content' could not be reached. More, 'effect' could affect the next 'source' in the subsequent structural intervention. In our case of TP-DTF, 'source' was a long time collection of investigations of natural and geographical conditions of territories and environments. Drawing on such a 'source', local and central governments would create 'content's that were long range planning and visions, some of which might be feasible and become true, but some might turn out unfeasible. 'Effect's were determined when the designs of coastal ports, which were both starting from such governmental long range plans and relying on physical investigations of territories and environments, were shared and agreed. However, some resultant designs were qualified, and some were not during the next implementation that would interfere with results of originally physical investigations. TP-DTF's senior director exclamatorily claimed that "moving coastal ports from SG to VT was not imaginative 10 years ago", which had a great socio-economic value. The latter had its form in advanced designs of coastal ports but was apparently beyond the physical blueprints. Moreover, coastal ports to be built would affect next investigations of natural and physical conditions of the relevant regions.

At this point, we employed the Peircean semiotic triad to conceptualize the structural property of organizational information in a triadic relation among 'source', 'content', and 'effect' as depicted in Figure 1. Also in this figure, one may find similar triads for function, process, context, time, and epistemology, which will be discussed next. It should be noted that, because of the semiotics relation of such each triad comprising three generic categories as firstness, secondness, and thirdness, the different dimensions (e.g. source, uncertainty) positioning at the same vertex (e.g. firstness) could be of equal status in organizational information expressions.

Figure 1. Six systemic properties of organizational information

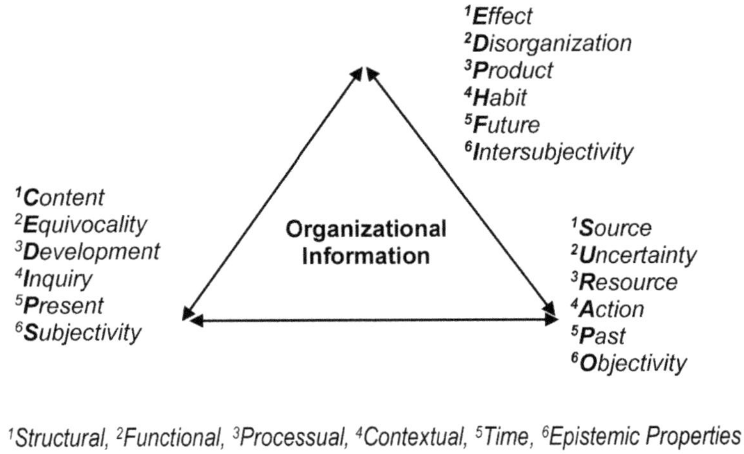

Functional Aspect of Organizational Information

Our informants indicated that their business activities were important to fulfill their business mission. The interesting finding here is that business activities were acted to perform three different nuances of the function of organizational information: (reduction of) uncertainty, equivocality, and disorganization as the ends of their business activities (Gharajedaghi 2011). For example, training activities in HY-ICT corporation could be classified into three sorts: (i) training for teams of business consultants, who directly built and provided solutions for clients; (ii) training for client key users or division heads, who made decisions about their business process improvements or redesigning; and (iii) training for client end-users, who directly run business applications in their operations. One can easily tell that the three sorts of training resolved the three different sorts of problems, and thus match well the three functions of organizational information.

First, the on-going training of business consultants those who act as solution providers for customers is to make certainty or cover most ground for projects of ERP implementation. This is the core activity for HY-ICT consultants to be able to master enterprise application packages that are delivered by the original software developers such as Oracle, SAP, and so on. Second, the dedicated training classes for key users or division heads to understand and identify core processes and issues is to make clear key business processes and main uses of business applications in the corporation. This is to specify the scope and scale of customization of off-the-shelf products, and hence, how big and how much the project of ERP implementation is. The third type is to make up or organize "future" business processes and respective uses through end-uses training of operations of new systems or application packages. It is the critical success factor of the whole ERP project.

The attribute of uncertainty centered on facts, or sense data which organizational participants often received from their environments, yet appeared usually as output of internal communication channels (e.g. official company documents). On the contrary, the aspect of equivocality was almost implied. Our data showed that it emerged across different interpretations of organizational actors, under individual human meanings, understandings, or judgments. As in HY-ICT setting, division heads were occupied with the ambiguity of legacy as well as future processes but they could not assess it until those processes went live. Finally, the attribute of disorganization was found in the description of the integration or coordination of organizational business activities in general as a featured indicator of the coupling and the coherence of the system as both a whole and part in a greater whole. In short, training activities in HY-ICT were organized to make synergy (i.e. new organization) among three types of training for: (i) mastering enterprise package (as an external entity), (ii) understanding and specifying legacy and future business processes (as business internal objects), and (iii) implementing and applying future processes onto enterprise applications (as integration) for end-user operations. Figure 1 depicts the functional property by means of the Peircean triad.

Processual Aspect of Organizational Information

Our data pointed out that organizational business activity as process could reveal three self-explanatory nuances: resource, development, and product.

A resource is a source or supply from which an organization gains profit. If a source does not produce benefit, it won't be selected for the process. Resource may be consumed, thus it may be limited availability. In HY-ICT's ERP implementation projects, resource mentioned were: (i) state-of-the-art enterprise products from some big names in the field as SAP, Oracle, and so on; (ii) proven experience of consultants teams of implementation services; and (iii) client legacy business processes and transactions as well as new ones and requirements of implementation performance if any. Hence, in a sense, the 'resource' of ERP implementation project was really given and selected.

'Development' is concerned with the selectively feasible method for applying the resources before a product goes into production. 'Development' is carried out typically after resources are selected, functions are known in a specific context. For example, the director first emphasized specially about pre-sales activities, which were designed for each client, including careful exploration of existing business processes, identifying and defining "future process" through training client key users or client department heads. Then there would be sales activities along the project life, including selection of implementation methodologies and tools, position of consultant teams, selection of ERP products, identifying some customization or development of ERP modules or subsystems, and real world implementation activities. Obviously, 'development' put itself in a transformative relation with 'resource'.

Finally, we used 'product' here to refer to things produced by development effort. It is the result of a process to meet organizational objectives. In HY-ICT case, the 'product's were customized ERP packages, new business processes made officially, end-user training well conducted, and new system transferred to client sites. When 'product' meets an intended function, it creates an effect which keeps environment, business actors in a self-control circle. In this sense, the 'product' of the process could backward affect the 'resource' (e.g. much more experience of consultant teams) and the 'development' (e.g. improved methods and tools of implementation) in the next round of the process.

As a result, we proposed that the process property of organizational information could be the Peircean triadic relation (see also Figure 1).

Contextual Aspect of Organizational Information

We used 'context' to refer to the business' actual workplace where organizational actors conducted business activities to resolve their business problems. We proposed that the context property of organizational information could be sliced into three nuances as a result of our field data analysis of organizational business activities. One, it was the context of 'inquiry' that created the conditions or needs for an activation of these activities. Two, it was the context of 'action' that used organizational information for business work. In a relation, 'inquiry' and 'action' act as pre-conditions and post-conditions for organizational information. While the former could be referred to as organizational problems or business questions, the latter could be organizational solutions or business activities.

In this contextual relation, the business 'inquiry', on the one hand, specified the business 'habit' or the business rule of conduct, on the other hand, was limited by the business 'action' that directly responded to client requirements. The 'habit' in turn ruled the 'action' and provided the client with allowed or possible services for the 'inquiry'. Moreover, the 'habit' is a "janus-faced" state of organizational information which has both autonomy (as habitual activity) and goal of inquiry. It should be noted that the 'habit' was neither viewed as a concept (i.e. in inquiry) nor an activity (i.e. in action) but a justification for organization to provide links between problems and solutions. In this view, the organization was seen as a stream of its organizational habits.

The contextual slices were evidenced in the R&D in consultancy services as described by the CEO of TP-DTF. For example, the 'inquiry' context emerged when a demanding requirement of some qualified survey works for port projects and maritime structures was initiated. TP-DTF tended to move to the 'action' context when they considered making more investments for hi-tech in-situ geological instruments and laboratory. The 'action' context was formed when the hi-tech laboratory was firstly established for demanding surveys to cope with the 'inquiry' context. Over time, the accumulation of technical people, instruments, procedures and technologies made the hi-tech laboratory the R&D department in TP-DTF consulting firm, which would create a 'habit' context. For example, since then some geomatics, topography surveys, GIS studies and remote sensing were conducted as R&D activities and much contributed to further qualified surveys. In parallel, also as a result of R&D activities, application of Real Time Kinematic technology, a very new technology in the field, to the bathymetric mapping work was confirmed preliminarily. In consequence, the R&D department has acquired a competence in topography survey, GIS, remote sensing and Real Time Kinematic, which is known as TP-DTF's 'habit' of consultancy services. The habited competence has benefited port projects and maritime structures much more then.

Whilst the context of inquiry was the business situations where port projects increasingly requested qualified surveys, the context of action referred to the establishment of the hi-tech geological instruments and laboratory to serve the survey requests that was not justified completely at that time. In such a situation, the 'habit' context was necessarily emerged and they all suggested a triadic relation. The contextual property is also captured in the semiotic triad (Figure 1).

Time Aspect of Organizational Information

The time aspect was not only a boundary condition but also a substantive ingredient of organizational activities (George and Jones 2000). Our data analysis suggested that organization business activities has spread from past to future. In TP-DTF, the time indicator was embedded in the cluster of projects of designing coastal ports up to 2020. Based on the past investigations of natural environments and territories, the projects started with long range vision and planning of coastal transportation facilities in the region to 2020, which played the role of present legal foundation at that time, to look forward to project achievements that were defined as designing new coastal ports and moving ports from SG to VT.

Without the time factor, the dynamics of processual property of organizational information could not be obtained. For example, resource is past, development is present, and product is future. But when the product is accepted, it naturally becomes resources. The time property is also represented in Figure 1.

Epistemological Aspect of Organizational Information

Epistemology is concerned with the creation and dissemination of knowledge in particular areas of inquiry. The data analysis suggested that the organizational information could be viewed from one or more of three nuances of epistemological status that was significantly present in organizational phenomena modeled as information (e.g. Orlikowski and Robey 1991). This was evidenced in HY-ICT director's discussion of training activities of consultancy services of implementation projects.

The training for business consultants and analysts was an instance of objectivity dimension of implementation consultancy services of HY-ICT. The training content was about the functionalities and usage of the original products like SAP ERP, Oracle ERP, and so on for consultant teams to master enterprise application packages, off-the-shelf products, which were more external and objective as one thought. This activity is objective because it depends neither on the businesses nor consultants. However,

the training for client's key users those who were division heads and made decisions about their business process improvements or redesigning at client sites for a better alignment with enterprise applications to be implemented next appeared to be subjective. That is because the activity was associated with (i) single subject (e.g. division heads) in the projects of multiple stakeholders (e.g. software vendors, service providers, consultants, client decision makers, client end-users, and so on) for a concrete situation of specifications of business processes to be implemented in software; and (ii) decision making on business processes could not be evaluated until the later stages of business processes implemented. Finally, the training of client end-users about the new system to be transferred is neither objective nor subjective. The activity was continuously adjusted to reflect the delivery - user acceptance process. The system was in the status of inter-subjectivity. That is because the final off-the-shelf product was more objective, but it was also more subjective when being customized by the teams of business analysts and programmers on demands of client decision makers on business processes improvement and application modules customization. The term 'inter-subjectivity' was particularly preferred for our analysis because its identity is two fold (e.g. Wiley, 1988): interaction between two or more subjects, and an effect of joint subjectivity emerged during the process.

The epistemological property is also depicted in Figure 1. In the same manner as the time property, its essence was clearly demonstrated in data analysis and pattern building above. For example, pattern1 referred to an objective, external entity, pattern2 to subjective, internal object, and pattern3 to an inter-subjective object. In TP-DTF case, pattern1 was the collection of past natural investigations of the company, pattern2 was the long range planning and visions of governments, and pattern3 was the corporation designs of coastal ports.

Systemic Nature of Organizational Information

At this point, we specified six triadic relations (i.e. four ones on ontology, one on temporality, and one on epistemology) for six properties of organizational information respectively (i.e. structure, function, process, context, time, and epistemology). The first four ingredients (structure, function, process, and context) from Gharajedaghi (2011) were necessary to define organizational information as a whole. These ingredients interacted to form a complementary set for a comprehensive explanation of the diversity of organizational information in the following manner. While 'structure' indicates the components and their relations, and 'function' defines the outcomes, 'process' concentrates the activity sequence and how to produce the outcome, and 'content' specializes the environment of the system under investigation (Gharajedaghi 2011). In addition, the ingredient of time set by Geogre and Jones (2000), and of epistemology suggested by Orlikowski and Robey (1991) were to emphasize the dynamically social nature of organizational information, and hence, embodied Peirce's principle of continuity of universe, which analytically comprises an inner world, an outer world, and a logic or inbetweeness world (7.438), and indicates contiguity in time or space (Mingers 2006). More, the three different dimensions of each triad within any ingredient could clarify the role of the 'choice' as requested in purposeful system of organizational information. In specific, 'structure' displayed the choice of means, 'function' the choice of ends, 'process' the choice of behavior, 'context' the choice of environment, 'time' the choice of timing, and 'epistemology' the choice of the world where the 'system' was situated. As a result, our systemic notation goes beyond Gharajedaghi's (2011) purposeful system to reach a Peirce's generic meaning system of semiosis (e.g. Mingers 2006). Table 6 is a descriptive model of organizational information as system.

For a more explanatory model of organizational information, we needed to get back to the three emerging extreme patterns in Table 4. The goal of the grounded model developed here was to seek a consistently theoretical pattern, which is able to describe, predict and explain as well as to improve our understanding of the phenomenon of organizational information. However, the pattern constructed presents an acceptable fit, not always a perfect one (e.g. Glaser and Strauss 1967).

In this finding, it is important to stress that the third extreme pattern was the largest semiotic triad named pattern3 because it comprised pattern2 that in turn comprised pattern1. We found that the representation of organizational information as the semiotic triad of pattern1 or data, pattern2 or knowledge, and pattern3 or information itself is interesting. That is because such embedded model obviously

Table 6. A descriptive model of organizational information as system

Core Category	Properties	Dimensions
Organizational information as system	Structure	Source
		Content
		Effect
	Function	Uncertainty
		Equivocality
		Disorganization
	Process	Resource
		Development
		Product
	Context	Inquiry
		Action
		Habit
	Time	Past
		Present
		Future
	Epistemology	Objective
		Subjective
		Inter-subjective

would help to distinguish among the long standing elusive concepts such as knowledge, information, and even data but in an inclusive relation. In short, data was monadic, knowledge dyadic, and information was triadic. One could find out data (pattern1) in knowledge (pattern2) that could in turn be found in organizational information (pattern3).

Our three patterns could be illustrated in terms of Ackoff's (1981) strategies of problem solving. According to Ackoff's framework, to solve a problem (i.e. pattern1) is to resolve a cause (i.e. pattern2) through dissolving a treatment (i.e. pattern3). In other words, the problem (pattern1) could be identified and solved by mediation between the cause (pattern2) and the treatment (pattern3). There may be no direct attack to and no direct identification of that original problem (pattern1).

At this point, in reference to pattern1 on data (D) already mentioned as a difference of objective perspective on organizational information, pattern2 on knowledge (K) readily maintained as a difference of subjective perspective, we suggested that pattern3 on information (I) necessarily indicates a difference of inter-subjective perspective. Shortly, we called pattern1 data (D) or an objective difference of type 'D' ('data'), pattern2 knowledge (K) or a subjective difference of type 'K' ('knowledge'), and pattern3 information (I) or an inter-subjective difference of type 'I' ('information').

Finally, it is noted that the three patterns in total is more complex than pattern3 (I) alone. This is because whilst pattern3 (I) maintains a system ontologically, the three patterns as a whole promotes a system not only ontologically but also epistemologically. In fact, the ontological view focuses more on the system as specific parts and relations, whereas the epistemological view stresses much on either the boundary of the system (e.g. Metcalfe 2006) or the process of enquiry into the world (e.g. Checkland and Holwell 1998). Because all of three patterns represent the likely organization or socio-cultural system (Gharajedaghi 2011), we referred it to the systemic view to reflect simultaneously the objective, subjective, and inter-subjective aspects of organizational information as system. This view also supports the recursive structure of the three patterns as a whole as compared with pattern3 (I) to accommodate the endlessly continuous evolution of organizational information.

Figure 2 presents our three-pattern model of organizational information, which is triadic, holistic and recursive. It should be noted that from the model in Table 6 to the one in Figure 2, the theoretical level was raised from a more descriptive, data-driven to a more explanatory, theoretical sensitive one. And Table 7 provides a greater detail distinction among three key patterns of organizational information, which also stand for the phenomena of data, knowledge and information respectively.

At this point, in contrast between the descriptive model (Table 6) and the patterned model (Figure 2), we found some connections between the dimensions and the patterns of organizational information. First, it was very clear that the patterns through their properties comprised the dimensions. Second, based on the distinction among the information-related theoretical categories just mentioned, the dimensions could in turn be the references to the patterns: data, knowledge, or information. From this, we advanced both our models as follows. For our descriptive model, we made a

Figure 2. A patterned framework of organizational information

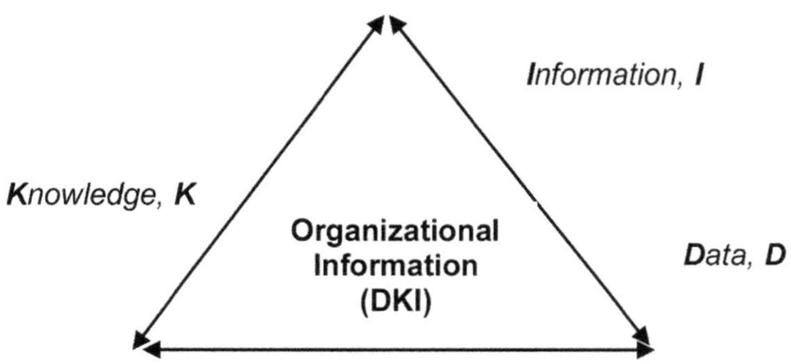

Table 7. The distinction among three basic patterns of organizational information

	Data	Knowledge	Information
Structure	*Source*	*Content*	*Effect*
Function	*Uncertainty*	*Equivocality*	*Disorganization*
Process	*Resource*	*Development*	*Product*
Context	*Inquiry*	*Action*	*Habit*
Time	*Past*	*Present*	*Future*
Epistemology	*Objective*	*Subjective*	*Inter-subjective*
Universal Categories	*Firstness*	*Secondness*	*Thirdness*

complement to it by filling out the dimensions by the patterns. For our patterned model, we described it by coupling the patterns with the original epistemological status of the dimensions. Further, to better reconcile such recursive relations between the whole (pattern) and the ingredients (dimensions), we lent our support to Mitroff and Linstone's (1993) multiple perspective concept that involves three types of concerns, which are technical, organizational, and personal ones, in considering three 'messy' organizing problems in organizational life: environmentalization, humanization, and self-control (Ackoff and Emery 1972). In doing so, we equated data or objectivity to technical, thing, or environmentalization, knowledge or subjectivity to personal, people, or humanization, and information or inter-subjectivity to organizational, organization, or self-control.

Last but not least, with the defining characteristics mentioned, we could afford to posit our holistic model of three basic patterns of organizational information in a distinct manner as follows. The first pattern 'data' could be best designated as 'objective data', the second pattern 'knowledge' as 'personal knowledge', and the third pattern 'information' as 'organizational information'. Further, from this perspective, other ways of labeling such as organizational knowledge, subjective information (e.g. Checkland and Holwell 1998), or even some concurrent or overlapping sorts of objective information and subjective information (Mingers 2006) are undesirable. We believed that there should have only three categorical notions in the relevant literature of IS, KM, organization and management, and so on: objective data, personal knowledge, and organizational information. Our explanatory model is depicted in Figure 3.

Figure 3. An explanatory model of organizational information as system (DKI model)

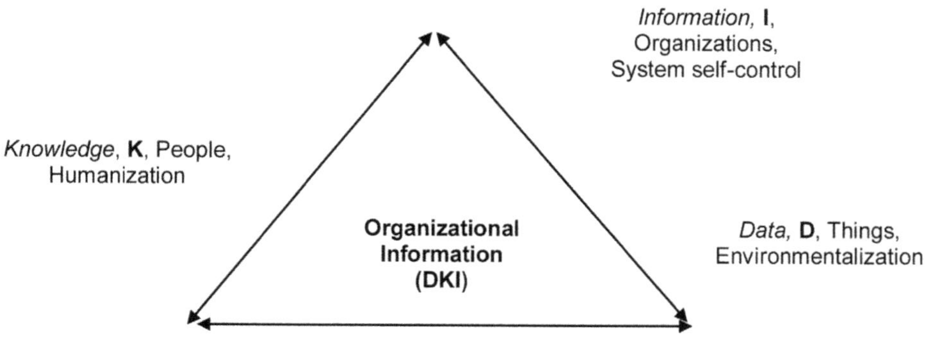

Organizational Information Formulation Process or Organizational Information Formulation as Habit Production

In the pattern finding process above, we identified that organizational information (pattern3) was formulated as mediation between data (i.e. pattern1) and knowledge (i.e. pattern2). Alternatively, data are transformed into information through knowledge.

We inferred that such a process of transformation might start with a difference of type 'D' (data that would be a more objective perception, for example a business symptom in an organizational consultancy case), through a difference of type 'K' (knowledge that would be a more subjective cognition, for example a business diagnosis in the consultancy case), and to end in a difference of type 'I' (information that would be a more inter-subjective behavior, for example a business treatment in the same consulting). Next round of the process only started again when some new sense-data (new 'D') happened as an experiential result of application of information ('I') onto the real world. From here, it should be also noted that both the property of process and of context got involved in the formulation process because the formulation was evolutionary endlessly. In fact, without the contextual property, we could not address the problem as to how data was created. In this sense, the information formulation process was also more or less linked with the information use process.

The triad of resource, development, and product could represent the triadic relation of the processual property of organizational information but mainly limiting to the logic of categories, and unable to provide a detailed logic of the processes of the phenomenon under investigation (e.g. Ball 1979). To fully develop the organizational information formulation, we borrowed Peirce's theory of knowledge including his models of inquiry and perception.

The Peircean model of perception indicates that the perceptual fact that is a representative of reality (CP 2.143), or evidence of the senses (CP 2.143), and constitutes human experience (CP 2.142) including both human action and perception on real word (CP 1.324). Alternatively, it is the factual surprise (CP 7.36), or the sensible effects or the practical bearings we have access to in a practical way (CP 5.400). The Peircean model of inquiry suggests that the inquiry process is just the struggle that starts with the irritation of doubt and ends with the cessation of doubt, which is to attain a state of belief (CP 5.374-5). Doubt is an uneasy and dissatisfied state, starting from the facts, and an occurrence, which is individual, here and now (CP 5.372; 1.419). And belief is a calm and satisfactory state (CP 5.372), a matter of community rather than of individual (CP 5.378), and asking the methods of belief fixation, which consists of four: method of tenacity, authority, a priori and science (CP 5.377-87). It is emphasized that the methods of belief fixation play a key role in shaping a community or, in a coherent integration of a community as a unity. Without the methods, no belief fixed, and thus, no habit established, which indicates that neither individual nor community could be identified.

Further, according to Peirce, with abduction possessing both insight and inference (CP 5.181), the process of perceptual judgment and the one of inquiry were likely seamlessly integrative because (i) the perceptual judgment is the extreme case of abductive judgment (CP 5.185, 5.181.3), and (ii) the full inquiry process consists of three consecutive stages in which the first stage is just the abduction process (CP 6.469). However, it should be noted that these two processes featured two very different capabilities. The inquiry process started with a factual surprise and ended with a satisfactory belief, mainly showing inference capability. In contrast, the perceptual judgment process, by human experience, started with belief and produced perceptual fact, mainly presenting action capability.

From here, we linked our organization information formulation process with both of the Peircean perceptual judgment and belief fixation process and found a good fit between D (data) with Peirce's factual surprise, K (knowledge) with Peirce's genuine doubt, and I (information) with Peirce's firm belief. Then, the Peircean belief fixation strategies were employed in our formulation model in the following manner. The process starts by human experiencing (i.e. both action and perception) real world for perceptual facts, or factual surprise, which turns into data (D). Next, one enters her inquiry process beginning with abduction to come into a mind state of genuine doubt in the form of interrogative expressions or explanatory hypotheses, which matches knowledge (K). One then, in her community of inquiry, employs a method of belief fixation, for example, method of science that uses both deduction and induction to reach a conclusion of inquiry as a satisfactory belief which becomes information (I). Alternatively, organizational information formulation process could be also expressed in terms of human activities. By this, it is an integration of experience (i.e. actions, observations, or interventions), abduction (i.e. insight and inference), as well as deduction and induction (i.e. inference). In particular, such inference indicated that our emerging formulation process was to perfectly meet Minger's (2008) premise of justifiability of some sorts of organizational information, which was set forth at the outset.

An additional point we would like to introduce here is, for methods of belief fixation, our data suggested something beyond Peirce's four original strategies. For example, in TP-DTF case, several decisions in the corporation board were simply unexplainable, as the CEO tried to describe to us: "The board atmosphere helps!". The manner of belief fixation in this case was possibly Weick's (1998) improvisation, which stressed the unforeseen and worked with the unexpected without a prior stipulation. It was interesting that, in a sense, improvisation by itself was helpful but not a method at all.

It should be noted that, in Figure 4, the information formulation process covered only one half of the closed path, from D through K to I, and matching the Peircean inquiry process, whereas the other half was for information use process (which is beyond our scope of study), from I to D, and matching the Peircean perceptual judgment process. In reference to the contextual property (Figure 1), it is also noted that one half of formulation process was for context of 'inquiry', the other half of use process was for context of 'action', and context of 'habit' could not be shown explicitly along the path.

At this point, with the fact that Figure 3 (DKI model) heavily indicated the logic of categories and Figure 4 (SDB model) much reflected the logic of processes, for the sake of simplicity and parsimony, we integrated both to reach a finally systems based model of organizational information as shown in Figure 5.

Figure 4. A triadic model of organizational information process (SDB model)

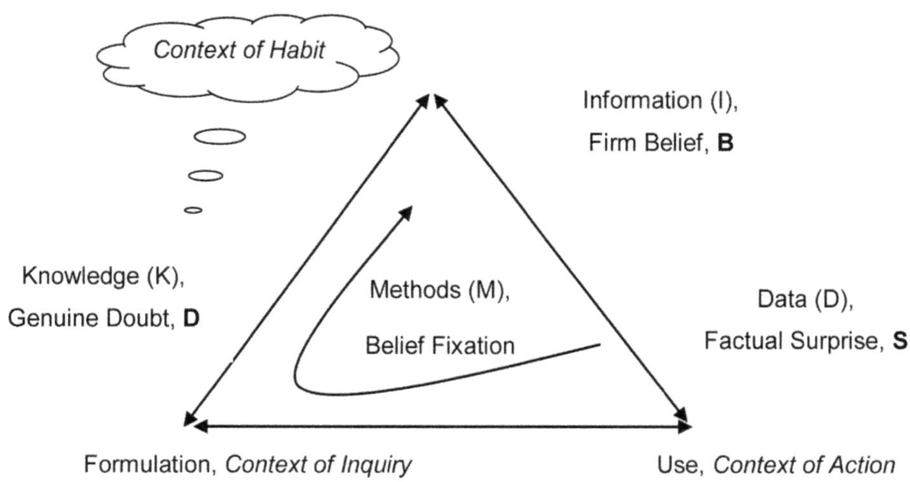

Figure 5. A systems based model of organizational information (systems model)

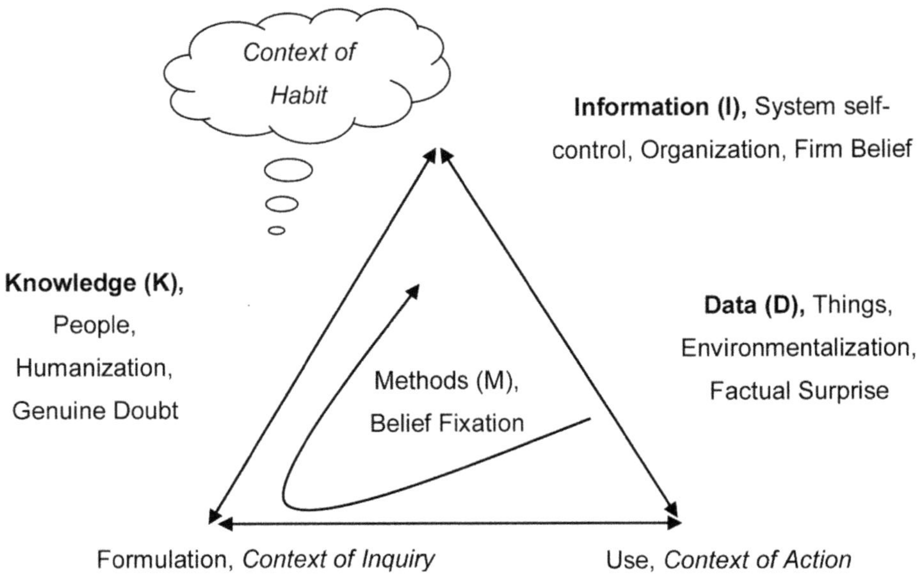

DISCUSSION AND CONTRIBUTION

In this section, we extend our findings into a semiotics based framework for both information and organization domains, and then discuss the research contribution.

First, we posited that our grounded model of organizational information would provide a new theoretical perspective on organization. The implication is emerged in the following comparison to Weick's (1979) sense-making theory of organization. Weick's model defined organization as organizing, or snapshots of ongoing cognitive processes. Such sense-making processes are to interpret the environment through continuous sequences of enactment, selection, and retention. Further, the sense-making constitute organization and vice versa (Weick 1979). Meanwhile, our process model (Figure 4), in the form of the Peircean semiosis (i.e. a continuous stream of sign production and transmission), viewed organization as habit production. That is, organization is to continuously form stream of habitual behaviors and its performance is evaluated by its making and conducting habits. To use the Weickean sense-making analogy, we argued that such information formulation and organization also constitute each other. As a result, the contextual designs of organization could be justified against the information formulation or belief fixation methods of communities of organizational actors because organization is enacted in the same way as information is formulated. Thus, one could consider, for example, the decision making tactic of authority for administrative bodies, of science for government funded research institutions, and of improvisation for firms of technological innovation, and so on.

The main contribution of our paper is an empirically grounded systems theory of organizational information on the backdrop of semiotics lens. The contributive effects are threefold. First, it could comprehensively capture the phenomena of information in organizations. Our processual model of information-as-system could explain inclusively both common sides of information at the same time, information as entity (e.g. Buckland's (1991) information as thing) and information as process (e.g. Callaos and Callaos' (2002) dialectic view of information). In addition, our model could completely incorporate the locus view of information (e.g. Swanson 1978), which was transparently embodied in the dynamic balance in our information formulation process, between the individual perspective (i.e. especially personal knowledge) and the community perspective (i.e. especially methods of belief fixation). This is possible because our entity view focuses more on the role of the information user via the informational 'effect', and our process view concentrates more on the role of some relevant community and its methods of belief fixation. By showing simultaneously the three views mentioned, our single systems model could address satisfactorily the construct of information in organizations. In the relevant literature, no such an integration of the three views has been proposed (e.g. von Krogh 2009). More, with the methods of belief fixation, our formulation model could explain the justifiability of organization information, which has rarely been examined in IS studies (e.g. Mingers 2008).

Second, our model could provide taxonomic relations among data, knowledge, and information, and hence, suggest a comprehensive distinction among the three fundamental information categories. Comprising a dynamic triad of the three categories, our framework could settle contradictory hierarchies of information and knowledge (e.g. Bach and Belardo 2003). For example, by their own epistemological features (e.g. objective data, subjective knowledge, intersubjective information), they would satisfactorily make all varieties of information related common categories redundant (e.g. explicit knowledge, tacit knowledge, objective and subjective information, etc). Such a triadic schema could make completely clear the confusing and overlapping among the information categories, which has been long time (Gourlay 2006) and considerable to IS literature (Mingers 2008).

Third and finally, our systems model of pragmatic information could clarify the role of the context of habit in information formulation by bridging the gap between the two common contexts of information in the literature: the context of inquiry (e.g. Churchman 1971) and the context of action (e.g. Ackoff 1989). We designated the context of habit as the one to keep information in contrast to the context of inquiry, which enacts information, and the context of action, which uses information. Few studies have revealed the third context like ours. In terms of pragmatism products (e.g. Hickman 1994), our model considered organizational information as a contextual triad of action (i.e. practice), inquiry (i.e. ideas), and habit (i.e. competence).

In organization and management studies, our theory suggests that organization would be also justified by its habit production led by its organizational members, or alternatively, organization and organizational information-as-system constitute each other. The three segments in organizational contexts in our study are unique because each of which dedicates itself to organizational life and requires its own notion of information (i.e. data, knowledge or information) as well as respectively organizational actors (e.g. information technology personnel, new product developer, or quality control consultant). We posit that the business context of habit would be the emergent property of organization as social system in the same manner as that organizational information as habit would be the emergent property of organizational information-as-system. This potentiality in turn helps to theoretically solve March's (1991) tension between 'exploration' for new capabilities and 'exploitation' of existing capabilities, to come up an effective relationship between researchers and managers at Churchman and Schainbatt's (1965) request of a dialectic implementation, or to be evidently in line with Knott's (2002) empirical studies on knowledge exploration and knowledge exploitation as complements.

CONCLUSION

Information plays a critical role in contemporary organizations (Gharajedaghi 2011), but the understanding information is largely unreflective (Mingers 2008).

Our theory-building endeavor was to suggest a middle range theory of pragmatic information, which is also a response to McKinney and Yoos' (2010) call for a unified theory of information in business domain. By our systems based conception, organizational information would manifest itself well as a triadic unity that comprises three ingredients (i.e. data/surprise, knowledge/doubt, and information/belief), three transformations (i.e. experience, abduction, and traditional inquiry), and three respective loci (i.e. material, individual, and collective). Each ingredient, transformation, or locus could be in turn identified by its own fundamental specification of ontology, epistemology, relation, and time. Further, our systems model would also show the decisive roles of some community and its methods of belief fixation or agreement for the organizational information formulation, which were foundationally requested by von Krogh (2009) for the former (i.e. community), and by Mingers (2008) for the latter (i.e. justifiability).

Finally, with its elements as general but parsimonious semiotic triads, our proposed model is expected to be a promising starting point for future research centered on pragmatic information and naturally open for later empirical and even theorizing work. We posit that our grounded systems model propose a fundamentally theoretical framework about the nature and the process of information, which may be also a theory native to the information systems field. In specific, our triadic model also implies a semiotics based foundation for both information and organization studies: an information framework of organization, and an organization model of information. Further, this is reflexive in the sense of helping at the same time to better conduct the inquiries on organization and management in general as information is in essence inter-subjective, in community, by community, and for community.

REFERENCES

Ackoff, R. L. (1981). *Creating the corporate future*. New York: John Wiley & Son.

Ackoff, R. L. (1989). From data to wisdom. *Journal of Applied Systems Analysis*, *16*, 3–9.

Ackoff, R. L., & Emery, F. E. (1972). *On purposeful systems*. Chicago: Aldine-Atherton.

Bach, C., & Belardo, S. (2003). Scientific and philosophical aspects of information and the relationships among data, information, and knowledge. *Proceedings of the 9th ACIS*, 2629-2639.

Ball, R. (1979). The dialectical method: Its application to social theory. *Social Forces*, *57*(3), 785–798. doi:10.1093/sf/57.3.785

Bereiter, C. (2002). *Education and Mind in the Knowledge Age*. Mahwah, NJ: Lawrence Erlbaum Associates.

Britton, G., & McCallion, H. (1994). An overview of the Singer/Churchman/Ackoff school of thought. *Systemic Practice and Action Research*, *5*(5), 487–521.

Buckland, M. (1991). Information as thing. *Journal of the American Society for Information Science*, *42*(5), 351–360. doi:10.1002/(SICI)1097-4571(199106)42:5<351::AID-ASI5>3.0.CO;2-3

Callaos, N., & Callaos, B. (2002). Toward a systemic notion of information: Practical consequences. *Informing Science*, *5*(1), 1–11. doi:10.28945/532

Checkland, P., & Holwell, S. (1998). *Information, systems and information systems*. New York: John Wiley & Sons.

Churchman, C. (1971). *Design of inquiring systems*. New York: Basic Books.

Churchman, C., & Schainblatt, A. (1965). The researcher and the manager: A dialectic of implementation. *Management Science*, *11*(4), B69–B87. doi:10.1287/mnsc.11.4.B69

Daft, R., & Weick, K. (1984). Toward a model of organizations as interpretation systems. *Academy of Management Review*, *9*(2), 284–295.

Drucker, P. (1999). *Management challenges for the 21st century*. New York: HarperBusiness.

Eisenhardt, K. (1989). Making fast strategic decisions in high velocity environments. *Academy of Management Journal*, *32*(3), 543–576. doi:10.2307/256434

George, J. M., & Jones, G. R. (2000). The role of time in theory and theory building. *Journal of Management*, *26*(4), 657–684. doi:10.1177/014920630002600404

Gephart, R. (2004). Qualitative research and the Academy of Management journal. *Academy of Management Journal*, *47*(4), 454–462. doi:10.5465/AMJ.2004.14438580

Gharajedaghi, J. (2011). *Systems thinking – managing chaos and complexity: a platform for designing business architecture* (3rd ed.). Burlington, MA: Morgan Kaufmann.

Glaser, B., & Strauss, A. (1967). *The discovery of grounded theory: strategies for qualitative research.* Chicago: Aldine.

Gorry, G. & Morton, M. (1989). A framework for management information systems. *Sloan Management Review*, 49–61.

Gourlay, S. (2006). Conceptualizing knowledge creation: A critique of Nonaka's theory. *Journal of Management Studies*, *43*(7), 1415–1436. doi:10.1111/j.1467-6486.2006.00637.x

Hausman, C. (1993). *Charles S Peirce's Evolutionary Philosophy*. Cambridge, UK: Cambridge University Press. doi:10.1017/CBO9780511570773

Hickman, L. (1994). The products of pragmatism. In G. Debrock & M. Hulswit (Eds.), *Living Doubt* (pp. 13–24). Kluwer Academic Publishers. doi:10.1007/978-94-015-8252-0_2

Knott, A. M. (2002). Exploration and exploitation as complements. In C. W. Choo & N. Botis (Eds.), *The strategic management of intellectual capital and organizational knowledge* (pp. 339–358). New York: Oxford University Press.

Lauer, T. (2001). Questions and information: Contrasting metaphors. *Information Systems Frontiers*, *3*(1), 41–48. doi:10.1023/A:1011497503953

Lincoln, Y., & Guba, E. (1985). *Naturalistic inquiry*. Beverly Hills, CA: Sage.

March, J. (1991). Exploration and exploitation in organizational learning. *Organization Science*, *2*(1), 71–87. doi:10.1287/orsc.2.1.71

Markus, M., & Robey, D. (1988). Information technology and organizational change: Causal structure in theory and research. *Management Science*, *34*(5), 583–598. doi:10.1287/mnsc.34.5.583

Martin, P., & Turner, B. (1986). Grounded theory and organizational research. *The Journal of Applied Behavioral Science*, *22*(2), 141–157. doi:10.1177/002188638602200207

Mason, R. (1978). Measuring information output: A communications systems approach. *Information & Management, 1*(4), 219–234. doi:10.1016/0378-7206(78)90028-9

McKinney, E. H., & Yoos, C. J. (2010). Information about information: A taxonomy of views. *Management Information Systems Quarterly, 34*(2), 329–344.

Metcalfe, M. (2006). Seeking Churchman's system for guaranteeing knowledge claims. *Systems Research and Behavioral Science, 23*(1), 51–59. doi:10.1002/sres.676

Miles, M., & Huberman, A. (1994). *Qualitative data analysis*. Thousand Oaks, CA: Sage.

Mingers, J. (2006). Realising systems thinking: knowledge and action in management science. New York: Springer Science + Business Media.

Mingers, J. (2008). Management knowledge and knowledge management: Realism and forms of truth. *Knowledge Management Research and Practice, 6*(1), 62–76. doi:10.1057/palgrave.kmrp.8500161

Mitroff, I. (1973). Systems, inquiry, and the meanings of falsification. *Philosophy of Science, 40*(2), 255–276. doi:10.1086/288520

Mitroff, I., & Linstone, H. A. (1993). *The unbounded mind: breaking the chains of traditional business thinking*. New York: Oxford University Press.

Newman, J. (2001). Some observations on the semantics of "information. *Information Systems Frontiers, 3*(2), 155–167. doi:10.1023/A:1011439305567

Nonaka, I. (1994). A dynamic theory of organizational knowledge creation. *Organization Science, 5*(1), 14–37. doi:10.1287/orsc.5.1.14

Orlikowski, W. (1993). CASE tools as organizational change: Investigating incremental and radical changes in systems development. *Management Information Systems Quarterly, 17*(3), 309–340. doi:10.2307/249774

Orlikowski, W., & Baroudi, J. (1991). Studying information technology in organizations: Research approaches and assumptions. *Information Systems Research, 2*(1), 1–28. doi:10.1287/isre.2.1.1

Orlikowski, W., & Robey, D. (1991). Information technology and the structuring of organizations. *Information Systems Research, 2*(2), 143–169. doi:10.1287/isre.2.2.143

Peirce, C. (1931-1958). The collected papers of C.S. Peirce (Vols. 1-8). Cambridge, MA: Harvard University Press.

Polanyi, M. (1962). *Personal knowledge - Towards a post critical philosophy*. Chicago: University of Chicago Press.

Simon, H. (1997). The future of information systems. *Annals of Operations Research, 71*, 3–14. doi:10.1023/A:1018975616482

Stenmark, D. (2002). Information vs. knowledge: the role of intranets in knowledge management. *Proceedings of HICSS*. IEEE Computer Society Press. doi:10.1109/HICSS.2002.994043

Strauss, A., & Corbin, J. (1998). *Basics of qualitative research: Techniques and procedures for developing grounded theory* (2nd ed.). Thousand Oaks, CA: Sage.

Suddaby, R. (2006). What grounded theory is not? *Academy of Management Journal, 49*(4), 633–642. doi:10.5465/AMJ.2006.22083020

Swanson, K. (1978). The two faces of organizational information. *Accounting, Organizations and Society, 3*(2), 237–246. doi:10.1016/0361-3682(78)90015-6

Tuan, N., & Huy, V. (2008). Information as system in IS: A systems thinking perspective. *International Journal of Information Technologies and Systems Approach, 1*(2), 1–19. doi:10.4018/jitsa.2008070101

Tuomi, I. (1999). Data is more than knowledge: Implications of the reversed knowledge hierarchy for knowledge management and organizational memory. *Journal of Management Information Systems, 16*(3), 107–121. doi:10.1080/07421222.1999.11518258

Turner, B. (1983). The use of grounded theory for the qualitative analysis of organizational behavior. *Journal of Management Studies, 20*(3), 333–348. doi:10.1111/j.1467-6486.1983.tb00211.x

van de Ven, A. H. (2007). *Engaged scholarship: A guide for organizational and social research*. London: Oxford University Press.

von Hayek, F. (1962). The use of knowledge in society. *The American Economic Review, 35*(4), 519–530.

von Krogh, G. (2009). Individualist and collectivist perspectives on knowledge in organizations: Implications for IS research. *The Journal of Strategic Information Systems, 18*, 119–129. doi:10.1016/j.jsis.2009.08.001

Weick, K. (1979). *The social psychology of organizing*. Reading, MA: Addison-Wesley.

Weick, K. (1998). Improvisation as a mindset for organizational analysis. *Organization Science*, *9*(5), 543–555. doi:10.1287/orsc.9.5.543

Wiley, N. (1988). The micro-macro problem in social theory. *Sociological Theory*, *6*(2), 254–261. doi:10.2307/202119

Yin, R. (2003). *Case study research: design and methods* (3rd ed.). Thousand Oaks, CA: Sage.

Zhu, Z. (2006). Nonaka meets Giddens: A critique. *Knowledge Management Research and Practice*, *4*(2), 106–115. doi:10.1057/palgrave.kmrp.8500091

Zott, C., & Huy, N. (2007). How entrepreneurs use symbolic management to acquire resources. *Administrative Science Quarterly*, *52*(1), 70–105. doi:10.2189/asqu.52.1.70

APPENDIX

Table 8. Labeling dimensions of conceptual properties

(a) Labeling dimensions of conceptual property 'structure': An illustration at TP-DTF			
Dimension	*Empirical data*	*Linking to extant literature*	*Label emerged*
Dimension1	Long time, long ago, expensive investigation of natural territories.	It is the physical manipulation of data and media (i.e. records), or the technical level of the communication problem (e.g. the transmission of signals from a source to a destination over a channel) (Mason, 1978; Newman, 2001).	SOURCE
Dimension2	Planning and vision of coastal ports delivered by governments.	It is (units of) logical (and psychological) meaning (of the business activities under investigation), or the semantic level of the communication problem (Mason, 1978; Newman, 2001).	CONTENT
Dimension3	Designing and then moving coastal ports from SG to VT.	The changes in behavior, attitude, or cognition (of someone involved in the business activities under investigation), or the effectiveness or influence level of the communication problem (Mason, 1978; Newman, 2001).	EFFECT

(b) Labeling dimensions of conceptual property 'function': An illustration at TP-DTF			
Dimension	*Empirical data*	*Linking to extant literature*	*Label emerged*
Dimension1	(Grounding projects of designing new coastal ports on) long time, long ago, expensive investigation of natural territories.	The uncertainty-reduction as a defining characteristic of information (Newman, 2001), or a result of the communication technical problem (Mason, 1978).	UNCERTAINTY
Dimension2	(Making out or making clear) a plan and vision for future development of transportation facilities.	The reduction of ambiguity or equivocality to escape from multiple or conflicting interpretations of an organizational event as a way of gaining new understanding of the organizational phenomena (Daft and Lewin, 1990; Daft and Weick, 1984)	EQUIVOCALITY

continued on following page

Table 8. Continued

Dimension3	(Making up or organizing and) designing new ports as planned.	It is the pragmatic impact on information receivers, a result of the communication effectiveness problem (Mason, 1978), information behaviors for organizational life (Ellis et al, 1999), or the information needs of organizational actors, and information processes in organizations (Newman, 2001)	DISORGANIZATION
(c) Labeling dimensions of conceptual property 'process': An illustration at TP-DTF			
Dimension	*Empirical data*	*Linking to extant literature*	*Label emerged*
Dimension1	(Accumulation of) long time, long ago, as well as recent and expensive investigations of natural and geographical properties of territories and environments.	Information involves the means by which one (producing) system (a mind or device) affects another (receiving). Producing system receives inputs of data and media as all resources consumed in processing and transmitting signs (Mason, 1978). Information as resource, and knowledge workers in a firm as information consumers (Lauer, 2001).	RESOURCE
Dimension2	Long range planning and vision of transportation facilities in the region and delivered by local and central governments; and managerial capabilities of the corporation.	Developmental process or development as core concept of the systems view of the socio-cultural systems that are information-bonded (in contrast to efficiency and growth of the mechanistic and biological views respectively). Development features purposeful systems (i.e. organizations) and assumes plurality or choices of means, ends, and behaviors (Gharajedaghi, 2011).	DEVELOPMENT
Dimension3	Designing and then moving coastal ports from SG to VT.	Information as social product, reflected in cognitive processes and symbolic structures (Newman, 2001), and perceiver-concerns (Metcalfe and Powell, 1995).	PRODUCT
(d) Labeling dimensions of conceptual property 'context': An illustration at TP-DTF			
Dimension	*Empirical data*	*Linking to extant literature*	*Label emerged*
Dimension1	Port projects and maritime structures increasingly requested some qualified survey works.	Close relation between question and information (Lauer, 2001), information inseparable from inquiry (Mitroff, 1973).	INQUIRY
Dimension2	Hi-tech geological instruments and laboratory are firstly established for qualified surveys.	Information is used in organizations to help make decision. Information isn't the scarce resource; human time and attention is (Simon, 1997).	ACTION

continued on following page

Table 8. Continued

Dimension3	Then, over time, as a result of accumulation of technical people, instruments, procedures and technologies, hi-tech lab becomes R&D department. R&D department was for geomatics, and specializing in topography survey, GIS study and remote sensing, and next, these activities are contributed to further qualified surveys.	The pragmatic impact on the information receiver, or the influence level of the communications problem includes the personal, psychological, organizational, and social context of the receiver and the receiver's habit or disposition to respond to output in a predictable way (Mason, 1978, p.225). Measuring human development and accomplishment neither in terms of activities nor in terms of ideas, but in terms of habits produced (Hickman, 1994).	HABIT

[1] Citations of Peirce's work in this paper follow the form: "CP" (the abbreviation for Collected Papers) followed by *volume* and *paragraph* numbers, with a period between.

Chapter 2
Theorizing Sociomaterial Practices:
Gilbert Simondon's Theory of Individuation

Alexander Styhre
University of Gothenburg, Sweden

ABSTRACT

The recent interest in sociomaterial practices and forms of imbrication of social and material resources in, for example, studies of the use of information technology and information systems in organizations, has called for new theoretical developments to enact and fully understand materiality. This chapter examines the work of French philosopher Gilbert Simondon as an important, yet to date little explored resource in organization studies. Following Orlikowski's call for a relational ontology apprehending the enfolding of materiality and social resources, Simondon's analytical framework, which includes key concepts such as individual-tion, transduction, and relations, is presented. The chapter contributes to the recent debates regarding how to theorize and examine material resources such as information and communication technologies used in organizations.

INTRODUCTION

It seems that 'constructivism' is passé, the linguistic turn has reached a dead end and a rhetoric of materiality is almost obligatory. -Nikolas Rose (2013: 4)

This article aims to introduce the work of Gilbert Simondon as an important and still to date largely unexplored resource in the recent interest for materiality and

DOI: 10.4018/978-1-5225-5655-8.ch002

its relations in organization studies. The works of Simondon are only translated in parts into English, and the secondary literature is dispersed, but Simondon offers a genuinely original theoretical framework for the understanding of both technological artifacts and biological organisms. Simondon's contribution to a materialist theory would be useful for organization researchers being concerned about the role of e.g., information technology and information systems.

The study of information technology and information systems has from the very outset been based on an understanding of the distinction between the artifact, the hardware (Orlikowski and Iacono, 2001), and abstract, intangible resources such as data (Frandsen, 2009; Thacker, 2003), information (Kallinikos, 2006; Ciborra, 2002), code (Galloway, 2006; Ullman, 1997), and protocols (Galloway, 2004; Galloway and Thacker, 2007), and how the two resources constitute and mutually stabilize one another. In addition, Kallinikos, Aaltonen, and Marton (2013: 358) suggest that what they refer to as "digital artifacts" such as homepages, computer games, mp3-files, etc. are "ontologically ambivalent" inasmuch as they are (1) *editable*, i.e., they are "pliable" and always possible, at least in principle, "to modify and update continuously and systematically," (2) *interactive,* providing "alternative pathways along which human agents can activate functions embedded in the object, or explore the arrangement of underlying information forms" (Kallinikos, Aaltonen, and Marton, 2013: 358), (3) accessible through the means of *other digital objects*, and (4) *distributed* and are thus "seldom contained within a single source or institution." (Kallinikos, Aaltonen, and Marton, 2013: 360). Under normal conditions, the composite nature of IT/IS is concealed for the user as the technologies serve the role of what Star (1999) calls *infrastructures* (see e.g., Halavais, 2009; Galloway and Thacker, 2007), that is, as long as they work as intended, the technological systems are "invisible to the user" (Ribes and Bowker, 2009: 204).

In order to understand this "ontological ambivalence" of information technology, scholars need to abandon social constructivist and constructivist theories, Kallinikos (2006) suggests, echoing Rose's (2013) recent claim that constructionism is becoming unfashionable. Such theories deliver, in Kallinikos (2006: 144) formulation, "a rather trivial message." This rejection of constructivist framework does however not imply that anything that can be treated as being "social" in nature should be abandoned. Instead, Kallinikos (2006: 144) suggests, "social agents are not disembodied sprits; instead, they are complex ensembles of skills, proclivities, and roles, some of which are brought into being by technology itself." In addition, the technology these social agents employ is not just an "[e]xterior force that encroaches upon local, technologically 'unspoiled' contexts"; instead, most of the time, "technology partakes in the constitution of local contexts and agents," Kallinikos (2006: 144) says (see also Kallinikos, 2009; Cousins and Robey, 2005). Elsewhere, Kallinikos

(2005) suggests that technology is best understood as "a structural form that supports human action in a world beset with contingencies of every sort."

The question remains, however, on what theoretical grounds the materiality of information technology and its practical use should be understood and examined? On the one hand, materiality is enacted as being intransient and immutable, existing regardless of our analytical vocabularies; on the other hand, agency and social conditions and context influence how materiality are put to work by social agents. This paper contributes to the formulation of a relational ontology by introducing and discussing the philosophy of the French philosopher Gilbert Simondon (1989, 1992, 2009a,b, 2012), a theorists who addresses how both technologies and biological organisms are constituted through what Simondon refers to as *individuation* in a process of transduction. Simondon's writings, first published in the late 1950s and 1960s, on the ontology of both mechanical and biological systems are helpful in advancing an analytical framework that can shed further light on the relations between material and intangible resources embodied in information technology and information systems. Importantly, if we take Kallinikos's (2006) farewell to constructivism seriously, Simondon offers an analytical framework that "suggests a different entry-point upstream of social construction" and that "backs away from the social as an explanatory substratum of technologies and technical practices" (Mackenzie, 2005: 394). In this perspective, Simondon opens up for a materialism that avoids both a naïve realist ontology and ontologies that refer back to "the social" and "society" as the basis for materiality. Simondon's writing is therefore of great interest for organization scholars being concerned with sociomaterial practices. In addition, Simondon offers an analytical framework that avoids reducing the one resource to the other. "Simondon's philosophy is down to earth," Chabot (2005: 103) says. "It is an empirical philosophy, sensitive to details, with lot of description of technical mechanisms, physical processes, and the working of psychology." Simondon is in short contributing to what Orlikowski (2010) more recently has named a *relational ontology*.

This remainder of this paper is structured accordingly: First, the newly awakened interest in materiality in the social science and organization studies more specifically is addressed. Second, the Simondon's analytical framework is examined. Third, the theoretical and practical implications of Simondon's work are discussed, with special attention being given to the connections between Simondon's thinking and the interests for materiality and sociomaterial practices and imbrications between material and social resources in organization theory.

Gilbert Simondon: Materiality and Individuation

Simondon's writing, published in the late 1950s and early 1960s is little recognized in the English-speaking world due to the lack of translations of his work, but recently there are more articles and monographs published paying attention to his writing on the individuation of biological and technological systems (e.g., De Boever et al., 2012; Manning, 2010; Styhre, 2008). Simondon's own texts and the commentaries on his work is filled with technical and philosophical terms and definitions (see e.g., Bartélémy, 2012) that makes both categories of texts challenging to read and understand, not to mention how to apply the analytical framework in empirical studies.

Simondon's Ph.D. thesis was supervised by Maurice Merleau-Ponty at École Normale Supérieur and while there are, Beistegui (2005) claims, few traces of Merleau-Ponty's work in Simondon's texts, Hansen (2004: 130) still refers to Simondon as a "bio-phenomenologist." Bernard Stiegler (2011: 97), possibly Simondon's most widely recognized reader besides Gilles Deleuze, speak of Simondon as a "descendant of Bergson" inasmuch as both Simondon and Bergson were preoccupied with understanding change and stability in both technological and biological systems. Both Simondon and Bergson can be best described as process philosophers.

The Concept of Individuation: Breaking With Hylomorphism

Using the concept *ontogenesis*, Simondon wants to theorize how individuals, a term used in the widest sense of the term denoting all kinds of entities that we perceive as freestanding and integrated units, are constituted. Aristotelian hylomorphism, assuming that every individual already contains its own *potentiality*—e.g., the acorn destined to once become an oak—is rejected in any form by Simondon. Instead, individuals are the outcome from a ceaseless process of *individuating*, of being gradually stabilized as the individual encounters new forms of information in its environment: One must, Simondon (1992: 300. Original emphasis omitted) writes, "[u]nderstand the individual from the perspective of the process of individuation rather than the process of individuation by means of the individual," stresses. Simondon's concept of information derives from cybernetics (see e.g., Wiener, 1948; Ashby, 1956), wherein information is a key term denoting an "operation" (Chabot, 2013: 79). Rather than speaking of cybernetics, Simondon uses the term *the allagmatic,* which he defines as "the theory of operations" (Combes, 213: 14); operation is in turn defined as "a conversation of a structure into another structure." Information is therefore what triggers change, what "causes mutations," and what "sets in motion the process of individuation" (Chabot, 2013: 80). Simondon thus shares with cybernetics the ambition to formulate a unified and general theory of being on basis of the concept of information.

In other words, philosophers have made the unwarranted assumption that individuals exists from the outset, but Simondon wants to undo this proposition and regards the individual as the *effect* of individuation: "Living matter is far from pure indetermination or pure passivity. Neither is it a blind tendency; it is, rather, the vehicle of informed energy," Simondon (1980: 66) states. This proposition calls for some clarification. Matter becomes life under certain conditions enabling the individuation of the "living entity":

The living entity is both the agent and the theater of individuation: its becoming represents a permanent individuation or rather a series of approaches to individuation progressing from one state of metastability to another. The individual is thus no longer either a substance or a simple part of the collectivity. The collective unit provides the resolution of the individual problematic, which means that the basis of the collective reality already forms a part of the individual in the form of the preindividual reality, which remains associated with the individual reality. (Simondon, 1992: 307)

Individuation is in Simondon's understanding a generic process—a form of a "subjectivity without subject" (Boucher, 2012: 106). Unlike the hylomorphic scheme, assuming already the presence of form and matter, Simondon traces the ontogenesis of the individual, its individuation, back to the preindividual phase.

The Transduction of the Individual

In addition to the concept of individuation, the process wherein forces, energies, and information are temporarily reconciled in the metastable entity of the individual, Simondon introduces the concept of *transduction* to capture how the individuation process emerges as the individuating entity responds to stimuli in its environment. In the sciences, the term *transduction* or *signal transduction* denotes the process wherein energy (e.g., chemical energy) is translated into another form of energy (e.g., electric energy). In molecular biology, *signal transduction* denotes the specific event in which a virus carries new genetic material over into the DNA of bacteria (Mackenzie, 2002: 17). The term transduction thus denotes a *transition* or *transformation* where information produces material effects. "The term [transduction] denotes a process—be it physical, biological, mental or social—in which an activity gradually sets itself in motion, propagating within a given area, through the structuration of the different zones of the area over which it operates," Simondon (1992: 313) writes. He continues:

The transductive process is thus an individuation in progress. Physically, it might be said to occur at its simplest in the forms of progressive iteration . . . Transduction occurs when there is activity, both structural and functional, which begins at a center of the being and extends itself in various directions from this center, as if multiple dimensions of being were expanding around the central point. (Simondon, 1992: 313).

Simondon's favorite example of transduction is the growth of a crystal, starting from a tiny grain and then growing outwards as the crystalline structure responds to the energies in its close environment. The crystal is thus individuating as it responds to energies as information, making it grow from the center to its margins:

A crystal that, from a very small seed, grows and expands in all directions in its supersaturated mother liquid provides the most simple image of the transductive operation: each already constituted molecular layer serves as an organizing basis for the layer currently being formed. The result is an amplifying reticular structure. (Simondon, 2009: 11)

The pre-individual chaos is thus transformed into the metastable structure of the crystal. "A being does not possess a unity in its identity, which is that of the stable state within which no transformation is possible; rather, a being has a *transductive unity*, that is, it can pass out of phase with itself," Simondon (1992: 311) writes. In the process of transduction, Massumi (2012: 31) argues, matter is defined in term of "[a] form-taking activity immanent to the event of taking-form." In Simondon's understanding, matter is never passive and inert but is what is in a constant state of being modified as it responds to energies qua information in its environment.

Simondon's analytical framework enacts biological and technological systems as being part of the same ontogenic process of individuation. Individuation in vital, biological system and in inert, technological system is still qualitatively separated, unfolding on basis of different sources of information. Still, the strength of Simondon's thinking is the ability to render one single resource, that of information, as what triggers processes of individuation. While biological systems are commonly understood as inherently changing and adapting to their milieu, the concept of individuation is perhaps most innovative for the study of technology, in many cases portrayed as enclosed and immutable. "The machine with superior technicality is an open machine," Simondon (1980: 4) writes, adding that "the technical object is a unit of becoming" (Simondon, 1980: 12). Similar to the individuating of the biological organism, "the technical being evolves by convergence and by adaption to itself; it is unified from within according to a principle of internal resonance" (Simondon, 1980: 13). Just like the distinction between matter and life is a *matter of degree* rather than *in kind*, the matter of the machine or "technical being" is similarly transducing into a metastable arrangement. Simondon's philosophy thus holds the potential to provide new views of how biological and technological systems integrate and meld, as in e.g., stem cells or nanotechnologies being recent and heavily debated "technologies" (see e.g., Bensaude-Vincent, 2007; Franklin, 2005); In this view, life is matter individuating under certain conditions but so too are technical beings, and combining the two would not in any way violate any boundaries between mutually excluding categories of matter as biological and technological matter both individuates in accordance with the same principles.

Individuation as a Relational Ontology

Expressed differently, relations are both the cause and the effect in Simondon's model of individuation; on the one hand, it is what is constitutive and generative of the individual qua transductive unity; on the other hand, it is what is produced as individuals actively relates to itself and its interior relations and to its external environment. Simondon's analytical framework advances a relational ontology that does not succumb to the assumptions of the Aristotelian hylomorphism that form and matter are always already in place, ready to mingle and be mixed up. Simondon's escapes this tradition of thinking as he stresses relations, dynamic and changing conditions, as integral to individuation. The form of the ontology of hylomorphism is an end-point, a terminal position, not what is in-the-the-making and ceaselessly becoming.

To summarize, Simondon invites us to rethink the hylomorphic model assuming that any entities always already contain all its potentiality and instead stresses that processes of individuating are always open-ended and creative, as the individual entity is a dynamic response to energies and information in its milieu. The individual is therefore what is derived from pre-individual potentialities and being in a constant process of change. Rather than seeking to sort out all the theoretical and philosophical intricacies of Simondon's *oeuvre*, the empirically oriented researcher may ponder how Simondon's work can be used to inform studies of digital media and information technology; it offers the possibility to overcome the Plato-Descartes tradition of thinking carefully separating materialism and idealism. In addition, the Aristotelian vocabulary separating form and matter is rejected by Simondon, thereby opening up for and recognizing the need for a process-based view of how social and material resources may be brought together and individuate into a metastable unity.

CONCLUSION

Simondon's philosophy represents a non-hylomorphic tradition of thinking that enables new ways of theorizing and understanding digital media and information technology. Simondon rejects both the determinist proposition of the hylomorphic scheme where individuals always already contain all their potentiality where and the environment barely matters in the process inasmuch as form and matter are distinct ontological categories. Simondon's analytical framework examines how biological and technological systems individuate, i.e., how they emerge as what humans in everyday life regard and treats as unified and discrete entities, no longer subject to much reflection, in an environment characterized by an abundance of information. In this view, processes of individuation are neither a pure realization of inherent potentialities as in the case of the hylomorphic scheme, nor is the individual simply the product of its environment. In Simondon's analytical framework, individuation is the operation wherein a structure transforms into a new structure on basis of external information; individuation is a process of dynamic relations between what individuates and its environment. That is, when contrasting e.g. Orlikowski's (2010, 2007) concept of sociomaterial practices, Simondon regards the process of individuation as what co-aligns material and abstract resources as the process of individuation is continuous and ongoing. Simondon does not want to collapse the material and the social categories into one another, but is more concerned with how materiality—biological, mechanical, biomechanical, or biochemical matter—individuates.

Simondon (1980) uses the term *concretization* to examine technological artifacts, and distinguishes between three levels of analysis: (1) That of *elements*, individual components or aggregates of components; (2) that of *the individual*, the specific technological artifact; and (3) that of *the ensemble*, integrated systems of technological artifacts. For Simondon (1980: 12) "the technical object is a unit of becoming," it moves toward concretization as it is stabilized and further developed. Studies of information technologies should thus recognize that technological objects are moving toward concretization:

The technical object exists . . . as a specific type that is arrived at the end of a convergent series. This series goes from the abstract mode of to the concrete mode; it tends towards a state at which the technical being becomes a system that is entirely coherent with itself and entirely unified. (Simondon, 1980: 16)

At the same time, the process of concretization, is not a self-enclosed or secluded process, but is open to external influences—"superior technicality" demands what Simondon (1980) speak of as "open machines," technologies that move toward concretization by being informed from its milieu:

A purely automatic machine completely closed in on itself in predetermined operation could only give summary results. The machine with superior technicality is an open machine, and the ensemble of open machines assumes man as permanent organizer and as a living interpreter of the interrelationships of machines. Far from being the supervisor of a squad of slaves, man is the permanent organized of a society of technical objects which need him as much as musicians in an orchestra needs a conductor. (Simondon, 1980: 4)

In other words, Simondon invites scholars to think of information technology as what is never wholly reducible to its sheer materiality, nor invoke "the social" as some ultimate ground for the of these technological artifacts, but instead treat technology as what is material, yet in a process of becoming.

Orlikowski and Scott (2008) suggests that one reason for technology and others forms of materiality being a relatively marginal concern in the field of organization studies is that scholars trained in the social sciences and humanities lack a vocabulary and an analytical framework to address materiality properly. Rennstam (2012: 1072) in turn suggests that there is an "over-reliance on symbolism" in organization studies, especially in the study of knowledge-intensive work and innovation work, two fields where various types of objects are developed and acted upon. In his own research work in the telecommunication industry, Rennstam (2012) found that objects are not fruitfully understood as passive entities being acted upon but instead the objects are inscribed with a certain agency and capacity to "act back" (Rennstam, 2012: 1073). Similarly, Ribes et al. (2013) argue that organizations are increasingly populated by technologies such as computers and information systems that actively serve to eliminate or minimize human-to-human communication. Today, when studying the role of materiality and its relationship to abstract resources such as code, protocols, regimes of computation and calculation, in organizational settings, there is a need for constructing a materialist theory of organizing. In this intellectual pursuit, the writings of Gilbert Simondon, addressing materiality at an ontogenic level, serve as a resource and inspiration.

REFERENCES

Ashby, W. R. (1956). *An introduction to cybernetics*. London: Chapman & Hall. doi:10.5962/bhl.title.5851

Bartélémy, J.-H. (2012). Fifty key terms in the works of Gilbert Simondon. In Gilbert Simondon: Being and technology. Edinburgh, UK: Edinburgh University Press.

Bensaude-Vincent, B. (2007). Nanobots and nanotubes: Two alternative biomimetic paradigms of nanotechnology. In Genesis redux: Essays in the history and philosophy of artificial life. Chicago: University of Chicago Press.

Bergson, H. (1998). *Creative evolution*. Dover Publishers.

Boucher, M.-P. (2012), Infra-psychic individualization: Transductive connections and the genesis of living techniques. In Gilbert Simondon: Being and technology. Edinburgh, UK: Edinburgh University Press.

Chabot, P. (2005). The philosophical August 4th: Simondon as a reader of Bergson. *Angelaki, 10*(3), 103–108. doi:10.1080/0969725050041729

Combes, M. (2012). *Gilbert Simondon and the philosophy of the transindividual.* Cambridge, MA: The MIT Press.

Coole, D., & Frost, S. (Eds.). (2010). *New materialisms: Ontology, agency, and politics*. Durham, NC: Duke University Press. doi:10.1215/9780822392996

Cousins, K. C., & Robey, D. (2005). Human agency in a wireless world: Patterns of technology use in neomadic computational environments. *Information and Organization, 15*(2), 151–180. doi:10.1016/j.infoandorg.2005.02.008

de Beistegui, M. (2005). From Merleau-Ponty's "reduction" to Simondon's "transduction". *Angelaki, 10*(3), 109–122. doi:10.1080/09697250500417316

De Boever, A., Murray, A., Roffe, J., & Woodward, A. (Eds.). (2012). *Gilbert Simondon: Being and technology*. Edinburgh, UK: Edinburgh University Press.

Frandsen, A.-C. (2009). From psoriasis to a number and back. *Information and Organization, 19*(2), 103–128. doi:10.1016/j.infoandorg.2009.02.001

Franklin, S. (2005). Stem Cell R Us: Emergent life forms and the global biological. In Global assemblages: Technology, politics, and ethics as anthropological problems. Malden, MA: Blackwell.

Galloway, A. R. (2004). *Protocol: How power controls after decentralization.* Cambridge, MA: The MIT Press.

Galloway, A. R. (2006). Language wants to be overlooked: On software and ideology. *Journal of Visual Culture, 5*(3), 315–331. doi:10.1177/1470412906070519

Galloway, A. R., & Thacker, E. (2007). *The exploit: A theory of networks*. Minneapolis, MN: The University of Minnesota Press.

Halavais, A. (2009). *Search engine society*. Cambridge, UK: Polity.

Hansen, M. B. N. (2004). *New philosophy for new media.* Cambridge, MA: The MIT Press.

Kallinikos, J. (2005). The order of technology: Complexity and control in a connected world. *Information and Organization*, *15*(3), 185–202. doi:10.1016/j.infoandorg.2005.02.001

Kallinikos, J. (2006), Farewell to constructivism. Technology and context-embedded action. In The social study of information and communication technology: Innovations, actors, and contexts. Oxford, UK: Oxford University Press.

Kallinikos, J. (2006). *The consequences of information: Informational implications of technological change*. Cheltenham, UK: Edward Elgar. doi:10.4337/9781847204301

Kallinikos, J. (2009). On the computational rendition of reality: Artefacts and human agency. *Organization*, *16*(2), 183–202. doi:10.1177/1350508408100474

Kallinikos, J., Aaltonen, A., & Marton, A. (2013). The ontological ambivalence of digital objects. *Management Information Systems Quarterly*, *37*(2), 357–370. doi:10.25300/MISQ/2013/37.2.02

Mackenzie, A. (2002). *Transductions: Bodies and machines at speed*. London: Continuum.

Mackenzie, D. (2009). *Material markets: How economic agents are constructed*. Oxford, UK: Oxford University Press.

Manning, E. (2010). Always more than one: The collectivity of a life. *Body & Society*, *16*(1), 117–127. doi:10.1177/1357034X09354128

Massumi, B. (2012). Technical mentality revisited: Brian Massumi on Gilbert Simondon. In Gilbert Simondon: Being and technology. Edinburgh, UK: Edinburgh University Press.

Orlikowski, W. J. (2007). Sociomaterial practices: Exploring technology at work. *Organization Studies*, *28*(9), 1435–1448. doi:10.1177/0170840607081138

Orlikowski, W. J. (2010). The sociomateriality of organizational life: Considering technology in management research. *Cambridge Journal of Economics*, *34*(1), 125–141. doi:10.1093/cje/bep058

Orlikowski, W. J., & Iacono, C. S. (2001). Research commentary: Desperately seeking the 'IT' in IT research—A call to theorizing the IT artifact. *Information Systems Research*, *12*(2), 121–134. doi:10.1287/isre.12.2.121.9700

Orlikowski, W. J., & Scott, S. V. (2008). Sociomateriality: Challenging the separation of technology, work, and organization. *The Academy of Management Annals*, *2*(1), 433–474. doi:10.1080/19416520802211644

Rennstam, J. (2012). Object-control: A study of technologically dense knowledge work. *Organization Studies*, *33*(8), 1071–1090. doi:10.1177/0170840612453527

Ribes, D., & Bowker, G. C. (2009). 'Between meaning and machine: Learning to represent the knowledge of communities. *Information and Organization*, *19*(4), 199–217. doi:10.1016/j.infoandorg.2009.04.001

Ribes, D., Jackson, S., Geiger, S., Burton, M., & Finholt, T. (2013). Artifacts that organize: Delegation in the distributed organization. *Information and Organization*, *23*(1), 1–14. doi:10.1016/j.infoandorg.2012.08.001

Rose, N. (2013). The human sciences in a biological age. *Theory, Culture & Society*, *30*(1), 3–34. doi:10.1177/0263276412456569

Simondon, G. (1964/1992). The genesis of the individual. In Incorporations. New York: Zone Books.

Simondon, G. (1980). *On the mode of existence of technical objects* (B. N. Mallahphy, Trans.). London: University of Western Ontario. (Original work published 1958)

Simondon, G. (2009a). The position of the problem of ontogenesis. *Parrhesia*, *7*, 4–16.

Simondon, G. (2009b). Technical mentality. *Parrhesia*, *7*, 17–21.

Simondon, G. (2011). *Two lessons on animal and man* (D. S. Burk, Trans.). Minneapolis, MN: Univocal.

Star, S. L. (1999). The ethnography of infrastructure. *The American Behavioral Scientist*, *43*(3), 377–391. doi:10.1177/00027649921955326

Stiegler, B. (2011). *Technics and time, 3: Cinematic time and the question of malaise (S. Barker, Trans.)*. Stanford, CA: Stanford University Press.

Styhre, A. (2008). Transduction and entrepreneurship: A biophilosophical image of the entrepreneur. *Scandinavian Journal of Management*, *24*(2), 103–112. doi:10.1016/j.scaman.2008.03.004

Thacker, E. (2003). Data made flesh: Biotechnology and the discourse of the posthuman. *Cultural Critique*, *53*(1), 72–97. doi:10.1353/cul.2003.0029

Ullman, E. (1997). *Close to the machine: Technophilia and its discontents*. San Francisco: City Lights Book.

Wiener, N. (1948). *Cybernetics, or control and communication in the animal machine*. New York: John Wiley.

Chapter 3
Organizational Culture and Organizational Performance

Anas M. Bashayreh
Al-Zahra'a College for Women, Oman

ABSTRACT

Organizational cultures can have varying impacts on organization performance. Oftentimes, employees work harder to achieve organizational goals if they consider themselves to be part of the corporate culture. Different cultures operating in one company can also impact organization performance. Organizational culture is an important part to be considered by dynamic organizations in order to develop some competitive advantage to ensure enhanced organizational performance. This chapter aims to close the gap between understanding the role of organizational culture and the effect on organizational performance among the employees in insurance companies.

INTRODUCTION

Recent literature proved that management of human resource in company has become an increasingly important for organizational performance and business vision attainment. Employees are considered as valuable assets to an organization, which require effective management of these employees in firms. A lot of research in organizational theory has focused on developed countries; 95% and whereas only 5% of the studies testing organizational theories are found to be done in developing countries (Farashahi et al., 2005) in spite of the highly dynamic environment.

Organizational culture has been an important factor that powerfully influences the performance of any organization. Within the past decades, an analysis has been done on the relationship between the organizational culture and performance (Anne

DOI: 10.4018/978-1-5225-5655-8.ch003

& Maaja, 2007; Ogbonna & Harris, 2000; Murugan, 2010; Rousseau, 1991). The organizational culture is a complex phenomenon, which is characterized by many dimensions (Marcoulides & Heck, 1993).

Organizational culture is that the assortment of operational principles that verify how people behave among the context of the company (Khan, 2005). Organizational culture is considered a vital theme in management and business research for the past few decades as results of its potential impact on the preferred and expected outcomes organizationally and individually such as commitment and loyalty (Chow et al., 2001). Organizational culture research can be divided into three groups. The first group embraces scientific works, in which culture is only described (Schein, 1992; Nahav & Malekzadeh, 1993; and Spivak, 2001). The second one includes works, in which dimensions are picked out (Van der et al., 1997; Veiga et al., 2000; and Sarros et al., 2005). Their influence on success is validated theoretically or empirically in the third group of works (Ogbonna & Harris, 2000; Child, 1974; Olie, 1994; Denison and Mishra, 1995; Larson and Finkelstein, 1999; Kangis et al., 2000; Howitz et al., 2002; and Nahm et al., 2004). The further dimensions of the analyzed phenomenon will be disclosed, the more clearly it will be perceived (Ginevicius & Vaitkunaite, 2006).

Scholars, who analyze organizational culture, present a different number of dimensions. Furthermore, some of them are very similar by their content (meaning), i.e. they are components of each other. Not all researchers examine the link between dimensions and organizational outcomes (Ginevicius & Vaitkunaite, 2006).

People are the most vital resource of an organization, and that the management of people makes a difference to firm performance (Patterson, West, Lawthom & Nickell, 1997). Many studies have been brought up to study the effectiveness of people management and business performance, but very few studies were conducted on the cultural impact on performance management. Every nation has a varying effect of cultural dimensions in organizations and the employees working over there. It is important to consider the cultural dimensions before creating strategies and objectives in organizations as this helps he management in effective performance (Patterson et al., 1997).

Thus, this research tries to focus on this issue and studies the impact of culture and cultural dimensions on organizational performance especially in Jordanian insurance companies. This research intends to bring about awareness in the readers and to make them learn the importance of focusing on cultural dimensions. Insurance in the Middle East region, especially in Jordanian case has traditionally lagged in growth and development relative to other elements of the region's financial services sector. This is evidenced by the low level of demand as measured by penetration and density levels, undercapitalized supply, and generally underdeveloped legal and regulatory environments (Vayanous & Hammoud, 2007).

Therefore the goal of this paper is to realize the conceptual of organizational culture dimensions, and measurement the link of organizational culture dimensions and organizational performance. The research methods, used in this work, are systematic analysis of scientific literature, content analysis, method of descriptive, correlation and regression.

Organizational Performance

Most studies on organizational performance use a variety of success measures each financial and non financial. Some specialists have examined the organizational performance theoretically and through empirical observation and the results show different measurement (Muafi, 2009; and Gary, 2009).

Performance indicates to the degree of accomplishment of the mission at the work place that builds up an employee job (Cascio, 2006). Different researchers have different thoughts concerning performance. Mostly researcher's used the term performance to express vary of measurements of transactional efficiency and input and output efficiency (Stannack, 1996). According to Barney (1991) performance is a continuous process to controversial issue between organizational researchers. Organizational performance does not solely mean to outline the problem, but it also for resolution of problem (Heffernan & Flood, 2000). Daft (2000), mention that organizational performance is the organization's capability to accomplish its goals effectively and with efficiency using resources. As similar to Daft (2000), Ricardo (2001) said that achieving the goals of organization and objectives is known as organizational performance. Ricardo (2001) prompts that organization's success shows high returning on equity and this become possible due to establishment of better employee's performance management system.

The purpose of performance management is to transform the raw potential of human resource into performance by removing intermediate barriers as well as motivating and refreshing the human resource (Kandula, 2006). Competitive capacity of organization can be increased by building strong people and effectively managing and developing people, which is in essence performance management (Shakil, 2012).

DeWaal (2002) said the process that enables a corporation to deliver a certain contribution to sustained value creation. A world-class performance management method consists of excellent strategy development, budgeting and target setting, performance measuring, performance review, and incentive compensation sub-processes.

Consequently, the management principle underpinning performance management is that performance management could be a management 'tool' that integrates the business operations with the strategic purpose of the organization. The success of any organization is alleged to depend upon its ability to execute its functions against its strategies and as Flapper (1996) argues that strategies within the organization should make sure that "all noses are pointing in the same direction", and that every person at intervals the organization is contributing to the organization's objectives via their daily activities in support of measurements as contained within the performance management system.

Undoubtedly, the organization's performance is a function of the potential return to the inculcation of strong culture into the organization's systems enabling it to execute its routines.

Organizational Culture

Organizational culture is the set of important assumptions-often unstated-that members of an organization share in common. There are two major assumptions in common; *beliefs and values. Beliefs* are assumptions about reality and are derived and reinforced by experience. *Values* are assumptions about ideals that are desirable and worth striving for (Muya & Wesonga, 2012).

Culture is an arrangement of various attributes that express a corporation and differentiate the firm from other one (Forehand & Von Gilmer, 1964). According to Kotter and Heskett (1992), culture means fairly established set of beliefs, behaviors and values of society contains generally. In simple words, we can perceive that culture is gaining knowledge, explanations, values, beliefs, communication and behaviors of a huge group of people, at the same time and same place.

There are many ways to visualize the concept of organizational culture. One popular conceptualization is the onion model. If you cut an onion in half and look at it, you will see many layers. An organization's culture can be visually represented in this way as illustrated in Figure 1.

Culture idea should be learned and shared within the organizations (Titiey, 1959). Pettigrew (1979) argues that cultures of organization based on cognitive systems that facilitate to elucidate how employees think and make decisions. According to Tichy (1982) organizational culture is understood as "normative glue" means that to hold the overall organization along. The concept of OC else makes available a base for determining the differentiation which will survive in-between the organizations that are doing business within the same national culture (Schein, 1990). Culture outlined as system of common values which might be estimated that individuals describe the similar OC even with completely different background at different levels within the organization (Robbins & Sanghi, 2007). As stated by Kotter (2008), that organization's norms and values have a powerful affect on all of those who are attached to the organization. It is thought about by him that norms are invisible, but if the organizations want to enhance the performance of the employees and profitability, norms are placed first to look.

Figure 1. The Onion Model of Organizational Culture
Source: Johnson and Scholes (1999, p. 74)

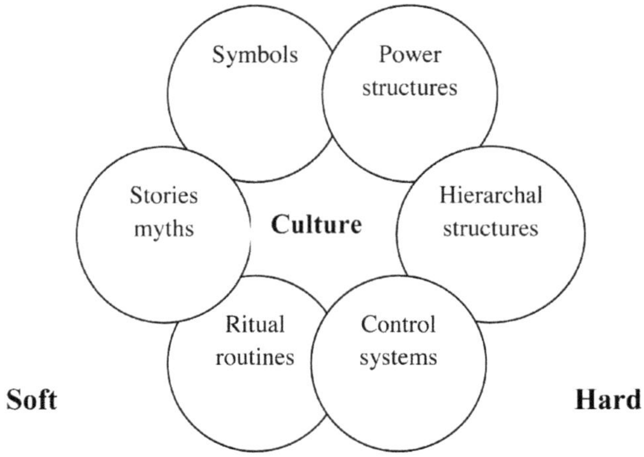

According to Handy's (1993) theory, culture has been divided into four main classes, particularly power culture, role culture, task culture, and person culture. Culture has also been termed as a link between employee and employer. The first component of culture comes from power that can be either from the top with few people or from the inner circle. The role is performed by a number of the individuals implementing rules who can be termed as bureaucracy, whereas task culture is concentrated on the accomplishment of objectives with cost-effective use of resources. Power culture relates to the degree of influence that comes from a central power or an internal circle. This means that the influence of a number of people determines the actions of others and is common in smaller organizations. Person's culture is a reflection of the central role of a person in an organization.

Organizational culture varies from business to business and forms country to country. It shows its different characteristics like diversity in the workforce. It can create both positive and negative effect on the employees and the workplace. In fact culture determines the survival of an organization over the long term. Cultures can be a responsibility to an organization if it creates barriers to change, create barriers to diversity or barriers to mergers and acquisitions (Robbins, 2009).

It is extensively accepted in the literature that the organizational culture is a complex concept. It may be analyzed from various perspectives. After a detailed investigation of the literature, the best fitting culture dimensions (in accordance with our research area) were selected from the study of Okashah (2008) cultural dimensions that focus on the five aspects of OC. These dimensions were used in the telecommunications sector within the same region of the researcher, and the researcher will re-applied in the insurance sector. The dimensions can be viewed as follow:

Stability and Communication

Communication generally emphasizes the flexibility and willingness to interact in open and critical communication. This dimension also assumed to be positively associated with the firm's ability to create and maintain long-term relationships with outsiders (Okashah, 2008). This is therefore as results of a climate of open crucial communication within a company will modify the members to learn from their mistakes through self-reflective and self-correction. Walter (1999) noted that self-reflective and self-criticisms enable actors to cooperate, resolve conflicts and act responsibly.

Organizational Expectations

The organizational expectations contracted psychological unwritten, which means a set of expectations that defined or expect individual or organization from each other during the work of the individual in the organization, for example, expectations superiors from subordinates, and subordinates from superiors, and colleagues from other colleagues and represented the appreciation and mutual respect, and to provide an organizational environment and regulatory climate helps and supports the needs of the individual psychological and economic factor (Alamiyan, 2002; and Rousseau, 2004).

Encourage Development

Employers benefit once employees develop new skills and technique for improving performance at work. A method to enhance productivity is to teach employees new strategies for handling job responsibilities. Employers that encourage worker development are far more probable to benefit from that training and growth (Alamiyan, 2002). Employees can benefit from employers that supply job training programs. Employers should provide clear ways for advancement that outline what training and knowledge are needed to advance. Employees who have a clear understanding of the necessities are a lot of doubtless to take on the quest to get the training and development needed (Okashah, 2008).

Behavioral Styles

As learning and development professionals have long recognized, totally people are different in terms of diversity in ways of learning. For example, some individuals learn additional effectively by observation videos or visual displays, whereas others answer an additional active tactical approach. However, what concerns variations in behavioral style? Even as individual method info in numerous ways in which, they additionally behave and answer the behavior of others in distinctive however predictable ways (Okashah, 2008). Unlike learning designs — how individuals cognitively process information — behavioral style is a lot of simply understood and visual to others as a result of it is a function of people's common behavioral patterns and preferences. Behavioral style can be seen and delineated, whereas it is more difficult to assess people's learning styles — many people themselves are unaware of their own learning preferences. One of the visible indicators of behavioral style is outgoings. Some individuals are naturally friendly and outspoken. Others are quiet and reserved. Another indicator is pace. Some act quickly, whereas others prefer to carefully consider all options.

Policies and Procedures

A set of policies, principles, rules, and tips created or adopted by a company to succeed in its long-term goals and generally printed in a booklet or different kind that is wide accessible. The policies and procedures are designed to influence and confirm all major decisions and actions, and all activities take place within the boundaries set by them. Procedures are the particular strategies utilized to express policies in action in day-after-day operations of the organization. Together, policies and procedures make sure that a point of view held by the governing body of an organization is translated into steps that results in an outcome compatible with that view (Okashah, 2008).

Impact of Culture on Organizational Performance

Lee and Yu (2004) studied on corporate culture and its impact on organizational performance, particularly to validate the cultural construct and its impact on the performance. The study detected that culture influences numerous processes within the organization and its performance. Balthazard et al. (2006) measured behavior norms and expectations and its impact on the performance. The results explain positive influence of deductive culture style, whereas negative impact was illustrated as regard to dysfunction, defensive styles on the people, and organization.

Rashid (2003) suggests that corporate culture and the organization's commitment influence the organization performance. Deshpande and Farely's on corporate and Allens and Meyer's work on commitment were used. Strong but the positive correlation in this variable was found in these studies. Mallak (2003) administrated their research work to measure the relationship among culture, the built environment, and the outcome variable. Findings supported cultural correlation with higher performance and established environment's role, as a moderating variable facilitates in better process and results there from. Specifically, job satisfaction was found to be a positive outcome of the built environment.

Levine (2006) examined the impact of culture and the process as to how it is developed and sustained, and studied the impact on the organization in many industries. It absolutely was found in the study that most of the organizations in their work are responsive to the facts of their processes. However, joint action creates complexity within the organization. It has also been found in the study that people can prevent the conflict prior to it getting in their way. It additionally tested that sustained and acutely aware culture has the potential for enhancing productivity, creativity, and satisfying work experience.

Onken (1999) examined two temporal components of organizational culture, polychronicity and values, and their effect on organizational performance. Polychronicity involves the extent to what activities are organized by scheduling two or additional events at one time. Second, temporal dimension of organizational culture is that the speed with those employees accomplishes tasks and also the values attached to it. Onken found that organizational culture that supported each speeds of decision making and undertaking several tasks directly had a positive impact on firm performance, particularly in hypercompetitive environments.

Ojo (2009) shows from the findings that are an outsized number of respondents 57.7% strongly agree that corporate culture has an effect on employee job performance, and there is 48.7% of the employees additionally agree that corporate culture determines the productivity level of the organization. Kilic and Dursun (2010) investigated the role of organizational culture on client orientation and performance outcomes of client orientation within the promoting context through a theoretical model. The categories of organizational culture were examined throughout this study area, unit market culture, adhocracy culture, clan culture, and hierarchical culture. The performance outcomes that were investigated are relationship development and individual performance. According to the study results, market culture incorporates a positive and vital impact on client orientation. The study results put together discovered that higher level of client orientation leads to higher levels of relationship development and individual performance.

Based on the literature review, we develop and test the following hypothesis:

H: Organizational culture dimensions have positive impact on organizational performance within Jordanian insurance sector.

METHODOLOGY

To find out the impact of organizational culture on organizational performance, research method is divided into two parts. The first part covers literature review, whereas the second part encompasses empirical study. This study evaluates the effects of cultural dimensions on firm performance in the insurance sector.

Survey Instrument

A questionnaire was designed based on an extensive review of the literature on organizational culture and organizational performance (Okashah, 2008; Wendy, 2008). Prior to the survey, the questionnaire was pre-tested with five alliance specialists and managers in small selection from insurance companies. Based on the comments from them, the questionnaire was modified and distributed.

The questionnaire consists of 7 sections as shown in Table 1. Section A of the questionnaire is to gather information about the profile of respondents such as gender, age, and organizational tenure. Section B is measuring items of Organizational performance while sections C, D, E, F and G south to measure items that are related to organizational culture dimensions. Two scales were used in this study. First is a nominal scale utilized in section A, Likert scale used for sections B, C, D, E, F and G. Likert scale is employed once responses to numerous items that measure a variable can be tapped on 5 points scale that can thereafter be summated across the items. Respondents are given five point Likert scale with 1 = Strongly Disagree, 2 = Disagree, 3 = Uncertain, 4 = Agree and 5 = Strongly Agree.

Sample

This study is based on a sample of managers and employees from 18 insurance companies that were selected randomly. The data were collected by the researcher himself by end of 2012, and the sample was obtained from all the management levels by randomly selected to participate in this survey. Out of 300 questionnaires distributed, 240 questionnaires were used for the final analysis.

Method of Analysis

Frequency analysis was used to examine the demographic characteristics of the respondents. This was followed by the reliability analysis that thinks about with the steadiness and consistency measurement to access the goodness of a measure further, it answers the queries on a selected conception. In addition, a computation of suggests that and the variance was in deep trouble all variables utilized in this

Table 1. Measurement of the variable

Section	Variables	Number of items	Scale
A	Respondent Background • Age • Gender • Organizational tenure • Education level	5	Multiple choices
B, C, D, E, F, and G	(B) Organizational performance (C) Organizational Expectations (D) Encourage Development (E) Behavioral Styles (F) Policies and procedures (G) Stability & Communication	18 4 4 3 4 4	Five-point Likert scale (1-Strongly disagree, 2-Disagree, 3-Uncertain, 4-Agree and 5-Strongly agree)

study. Finally, multiple correlation analysis was conducted to look at that among the five dimensions of the experimental variable is that the most significant to clarify organizational performance. All method of analysis will be conducted by using Statistical Package for Social Sciences (SPSS) version 20.

RESULTS

Demographic of the respondents are summarized in Table 2. Table 2 indicates that from the total of 240 respondents, 54.8% of the respondents are males while the remaining are females. In terms of age, only a minority of the respondents aged between 20-29 years (32.1%), whereas almost half of the respondents were within the age of 30-39 years (42.5%), 19.0% were within the age of 40-49 years. 6.3% of the respondents were over 50 years. Table 2 also suggests that, as far as education level are concerned, the majority of the respondents had bachelor degree 73.0%, followed by diploma degree (14.3%) and (11.5%) for master degree. With regard to organizational tenure, the majority of respondents have been working for 1-5 years with 52.8%; followed by 25% respondents have been working for 6-10 years. 15.5% of respondents have been working for more than 11 years and only 6.7% of respondents are operating for less than one year.

Table 3 shows the Cronbach Alpha values and the number of items for both dependent and independent variables. Sekaran (2003) said reliabilities with less than 0.60 are deemed poor whereas those in the range of 0.70 ranges, is acceptable and those above 0.80 is considered good. On the overall, the reliability of all the

Table 2. Respondents demographic

Demographic	Categories	%
1. Gender	Male	54.8
	Female	45.2
2. Age Group	20-29	32.1
	30-39	42.5
	40-49	19.0
	Above 50	6.3
3. Education level	Diploma	14.3
	Bachelor	73.0
	Master	11.5
	Other	1.2
4. Organizational tenure	Below 1yr	6.7
	1-5 yrs	52.8
	6-10 yrs	25.0
	More than 11 yrs	15.5

measures was comfortably above 0.60, ranging from 0.90 to 0.63. Therefore the scale is satisfactory reliable for subsequent analysis.

Descriptive analysis which includes the mean and standard deviation for the independent and dependent variables are attained and recorded in Table 4. The results show that the mean for organizational performance was 3.58 with SD = 0.45. While the mean for stability and communication was 3.69 with SD = 0.89. As illustrated in Table 4, the mean for organizational expectations was 3.64 with SD = 0.87, encourage development (M = 3.61; SD = 0.79), behavioral styles (M = 3.95; SD = 0.71), and the mean and standard deviation for policies and procedures was 3.93 and 0.99 respectively. Organizational expectations and encourage development indicates that most of the respondents were neither agree nor disagree.

Next, Multiple Regression analysis was conducted to identify which among the independent variables that explained most the dependent variable. Table 5 shows that only 44.9% of the R square in organizational performance is significantly explained by the five dimensions of organizational culture: stability and communication, organizational expectations, encourage development, behavioral styles, and policies and procedures. Based on table 5, only one variable in the regression equation emerged as significant predictors of organizational performance. The predictor is stability and communication. It is evident that the most influential independent

Table 3. Reliability analysis

Variables	No. of Items	Reliability Alpha
Organizational performance	18	0.901
Stability and communication	4	0.632
Organizational Expectations	4	0.723
Encourage Development	4	0.685
Behavioral Styles	3	0.732
Policies and procedures	4	0.646

Table 4. Descriptive statistics of the dependent and independent variables

Variable	Mean	Standard Deviation
Organizational performance	3.58	0.45
Stability and communication	3.69	0.89
Organizational Expectations	3.64	0.87
Encourage Development	3.61	0.79
Behavioral Styles	3.95	0.71
Policies and procedures	3.93	0.99

variable on organizational performance is the culture of Policies and procedures β= 0.122. The other cultural dimensions are not significant to predict organizational performance in this study. However, the result from regression analysis indicated that five dimensions in organizational culture (stability and communication, organizational expectations, encourage development, behavioral styles, and policies and procedures) are significant. The Policies and procedures is the strongest influence to organizational performance.

Table 5. Regression result for independent variables and job satisfaction

Variables	Beta	Sig
Stability and communication	.106	.182
Organizational Expectations	.048	.491
Encourage Development	.049	.474
Behavioral Styles	.071	.303
Policies and Procedures	.122	.076
F Value	21.87	
R^2	.449	
Adjusted R^2	.428	

DISCUSSION

Having measured the five dimensions of organizational culture had allowed us to understand the link between the dimensions of organizational culture as a whole and organizational performance. Several empirical studies have supported the positive link between culture and performance (Kotter & Heskett, 1992; Anne & Maaja, 2007; Kilic & Dursun, 2010). Moreover, studies done by Murugan (2010), Reddy (2011) and Kotter and Heskett (1992), have contributed significantly to the field of culture and performance studies whereby culture has been treated as variable for a specific research purpose.

With respect to the results of this study unconcealed that employees' perceptions of organizational culture were important and absolutely associated with performance, with those perceiving a bigger degree of awareness of organizational culture dimensions exhibiting a lot of positive reactions towards organization performance. The results additionally indicate that wherever Policies and procedures were perceived as dominant culture characteristics, there was a robust association with organizational performance. The results show that the employees who participated in this study have the educational qualifications, more than 70% hold a Bachelor's degree and more than 10% hold a master's degree, and this gives an advantage to companies on the ability of employees to have to learn and adapt with everything that is new. On the other hand, the results give us that more than 50% of employees who participated in the study does not exceed the period of their service in the insurance companies more than five years, while the owners of long experience does not exceed the percentage of 16%. Some considered a positive advantage for companies to inject blood and new experiences, while others see it as a negative advantage to afford companies the extra cost in terms of training the new employees, and the instability which affects the competitiveness in the market.

From the other side there is match between this study and Okashah (2008) study, that there is a clear commitment by staff with policies and procedures adopted in the company, the researcher believes that the commitment of personnel policies and procedures contributes to the completion of job tasks required, and helps to achieve administrative efficiency within the companies, and contributes to the climate regulatory environment that works to achieve the objectives of the company.

As previously discussed there is a positive relationship between organizational culture dimensions and organization performance, which organizational culture dimensions directly impact on organizational performance. The discussion that was provided earlier led to accept the hypothesis of the study.

CONCLUSION

Each one or employee in the organization has own completely different values and beliefs that he/she works with them. Whenever be part of any organization he/she allowed himself to assign initial with the organization's culture to grasp whether he come up with them or not. Indeed the study of organizational culture and performance is advantageous since the cultural values are observable and measurable. Thus, in studying of cultural effect on performance, it is vital that both financial and non financial measures are used to get more comprehensive results. Overall, there is a strong view in the literature that organizational culture leads to increased organizational performance. However, studies on this relationship often differ as to the extent a practice is likely to be positively or negatively related to performance.

Culture is being investigated to variety impact of organizational process. Organizational culture has a deep impact on the employee's performance that may cause to boost in the productivity and enhance the organizational performance. This study is an exploratory investigation of the relationship between organizational culture dimensions that adopted by Okashah (2008) and organizational performance among Jordanian insurance sector. This study endeavors contribute most to the practicing managers and it conjointly contains many implications for additional research. The findings enhance our understanding of the organizational culture and the effect on organizational performance among the insurance companies. The findings of this study shed light upon the importance of organizational cultural dimensions to predict organizational performance and improve the competitive advantages of the companies since this study might provide assistance to Jordanian government by providing a strong backbone in order to achieve its plans to improve the Jordanian insurance sector. Furthermore, this study added the knowledge to the body of business and organizational development as the first empirical study, which was carried out on insurance sectors in Jordan context.

This study provided us with a more robust understanding of the relationship between these factors in order to enhance managerial effectiveness and organizational success. The findings stress the necessity to observe the organizational culture and to evolve better management practices so that organizational performance is maintained at a high level. Hence, employees will be likely to perform better and feel a high level of performance, and in turn will develop more commitment towards their organizations. Finally, the higher levels of employees' performance in such organizations may give an advantage over other organizations in attracting and retaining employees in a competitive environment.

REFERENCES

Ahmad, S. M. (2012). Impact of Organizational Culture on Performance Management Practices in Pakistan. *Business Intelligence Journal*, *5*(1), 50–55.

Alamiyan, M. (2002). *Organizational Behavior in Business Organizations* (in Arabic). Amman: Dar Wael for publication.

Anne, A., & Maaja, V. (2007). Relationships between Organizational Culture and Performance in Estonian Schools with Regard to Their Size and Location. *Baltic Journal of Economics*, *7*(1), 3–17. doi:10.1080/1406099X.2007.10840438

Balthazard, A. P., Cooke, A. R., & Potter, E. R. (2006). Dysfunctional Culture, Dysfunctional Organization: Capturing the Behavioral Norms that Form Organizational Culture and Drive Performance. *Journal of Managerial Psychology*, *21*(8), 709–732. doi:10.1108/02683940610713253

Barney, J. B. (1991). Firm Resources and Sustained Competitive Advantage. *Journal of Management*, *1*(17), 99–120. doi:10.1177/014920639101700108

Cascio, W. F. (2006). *Managing Human Resources: Productivity, Quality of Life, Profits*. McGraw-Hill Irwin.

Child, J. (1974). What Determines Organization Performance? The universals vs. the it-all-depends. *Organizational Dynamics*, *3*(1), 2–18. doi:10.1016/0090-2616(74)90001-1

Chow, C. W., Harrison, G. L., Mckinnon, J. L., & Wu, A. (2001). *Organizational culture: Association with affective commitment, job satisfaction, propensity to remain and information sharing in a Chinese cultural context.* San Diego University CIBER Working Paper Series.

Daft, R. L. (2000). *Organization Theory and Design* (7th ed.). South-Western College Publishing. Thomson Learning.

Denison, D. R., & Mishra, A. K. (1995). Toward a Theory of Organizational Culture and Effectiveness. *Organization Science*, *6*(2), 204–223. doi:10.1287/orsc.6.2.204

DeWaal, A. (2002). The Power of world-class performance management: Use It! *Measuring Business Excellence*, *6*(3), 9–19. doi:10.1108/13683040210441940

Farashahi, M., Hafso, T., & Molz, R. (2005). Institutionalized norms of conducting research and social realities: A research synthesis of empirical works from 1983 to 2002. *International Journal of Management Reviews*, *7*(1), 1–24. doi:10.1111/j.1468-2370.2005.00104.x

Flapper, S. D., Fortuin, L., & Stoop, P. M. (1996). Towards consistent performance management systems. *International Journal of Operations & Production Management*, *6*(7), 27–36. doi:10.1108/01443579610119144

Forehand, G. A., & Von Haller, G. (1964). Environmental variation in studies of organizational behavior. *Psychological Bulletin*, *62*(6), 361–382. doi:10.1037/h0045960 PMID:14242590

Gary, C. (2009). *Performance Management - Integrating Strategy Execution, Methodologies, Risk, and Analytics*. John Wiley & Sons, Inc.

Ginevicius, R., & Vaitkunaite, V. (2006). Analysis of organizational culture dimensions impacting performance. *Journal of Business Economics and Management*, *7*(4), 201–211.

Handy, C. B. (1993). *Understanding Organizations*. New York: Oxford University Press.

Heffernan, M. M., & Flood, P.C. (2000). An Exploration of the Relationship between Managerial Competencies Organizational, Characteristic and Performance in an Irish organization. *Journal of European Industrial Training*, 128-136.

Horwitz, F. M., Anderssen, K., Bezuidenhout, A., Cohen, S., Kirsten, F., Mosoeunyane, K., & Van Heerden, A. (2002). Due diligence neglected: Managing human resources and organizational culture in mergers and acquisitions. *ŃSA. Journal of Business and Management*, (33): 1–10.

Kandula, S. R. (2006). *Performance Management*. New Delhi: Prentice Hall of India private limited.

Kangis, P., Williams, D., & Gordon, S. (2000). Organizational climate and corporate performance: An empirical investigation. *Management Decision*, *38*(8), 531–540. doi:10.1108/EUM0000000005371

Khan, A. (2005). *Matching people with organizational culture*. Business Management Group.

Kilic, C., & Dursun, T. (2010). The Effect of Organizational Culture on Customer Orientation. *The Journal of American Academy of Business, Cambridge*, *15*(2), 1–7.

Kotter, J. P. (2008). *Corporate culture and performance*. Simon and Schuster.

Kotter, P., & Heskett, L. (1992). *Corporate Culture and Performance*. New York: Free Press.

Larsson, R., & Finkelstein, S. (1999). Integrating Strategic, Organizational, and Human Resource Perspectives on Mergers and Acquisitions: A Case Survey of Synergy Realization. *Organization Science: A Journal of the Institute of Management Sciences*, *10*(1), 1–26.

Lee, S. K. J., & Yu, K. (2004). Corporate culture and organizational performance. *Journal of Managerial Psychology*, *19*(4), 340–359. doi:10.1108/02683940410537927

Levine, S. (2006). High Performance Organizations: Creating a Culture of Agreement. Handbook of Business Strategy, 7(1), 375–380.

Mallak, A. L., Lyth, M. D., Olson, D. S., Ulshafer, M. S., & Sardone, J. F. (2003). Culture, the Built Environment and Healthcare Organizational Performance. *Managing Service Quality*, *13*(1), 27–38. doi:10.1108/09604520310456690

Marcoulides, G. A., & Heck, R. H. (1993). Organizational Culture and Performance: Proposing and Testing a Model. *Organization Science*, *4*(2), 209–225. doi:10.1287/orsc.4.2.209

Muafi, M. (2009). The effects of alignment competitive strategy, culture, and role behavior on organizational performance in service firms. *International Journal of Organizational Innovation*, *2*(1), 106–134.

Murugan, M. (2010). A Study on Organizational Culture and Its Impact on the Performance of IT Employees in Chennai. The ICFAI Journal of Management Research, 8(5), 7-16.

Muya, J. N., & Wesonga, J. N. (2012). The Impact of Organizational Culture on Performance of Educational Institutions. *International Journal of Business and Social Science*, *3*(8), 211–217.

Nahavandi, A., & Malekzadeh, A. R. (1993). *Organizational Culture in the Management of Mergers*. Westport, CT: Quorum Books.

Nahm, A. Y., Vonderembse, M., & Koufteros, X. A. (2004). The Impact of Organizational Culture on Time-Based Manufacturing and Performance. *Decision Sciences*, *35*(4), 579–607. doi:10.1111/j.1540-5915.2004.02660.x

Okashah, A. (2008). *The impact of organizational culture at the level of functionality: an empirical study on telecommunication company "paltel" in palestine* (Thesis). The Islamic University-Ghaza. (in Arabic)

Olie, R. (1994). Shades of Culture and Institutions in International Mergers, (Periodical style). *Organization Studies*, *15*(3), 381–405. doi:10.1177/017084069401500304

Onken, M. (1999). Temporal elements of organizational culture and impact on firm performance. *Journal of Managerial Psychology*, *14*(3/4), 231–243. doi:10.1108/02683949910263756

Patterson, M. G., West, M. A., Lawthom, R., & Nickell, S. (1997). *Impact of people management practices on business performance*. London: Institute of Personnel and Development. The Cromwell Press.

Pettigrew, A. (1979). Studying organizational culture. *Administrative Science Quarterly*, *24*(4), 570–581. doi:10.2307/2392363

Rashid, A. Z., Sambasivan, M., & Johari, J. (2003). The Influence of Corporate Culture and Organizational Commitment on Performance. *Journal of Management Development*, *22*(8), 708–728. doi:10.1108/02621710310487873

Reddy, A. (2011). Cultural Dimensions and Impact on Performance Management. *International Journal of Multidisciplinary Research*, *1*(6), 300–311.

Ricardo, R. (2001). *Corporate Performance Management: How to Build a Better Organization through Measurement Driven Strategies Alignment*. Butterworth Heinemann.

Robbins, S. P., & Sanghi, S. (2007). *Organizational Behavior*. New Delhi: Pearson Education.

Rousseau, D. (1991). Quantitative assessment of organizational culture. *Group & Organization Studies*, *15*(4), 448–460. doi:10.1177/105960119001500408

Rousseau, D. (2004). Psychological Contracts in the Workplace: Understanding the Ties that Motivate. *The Academy of Management Executive*, *18*(1), 120–127. doi:10.5465/AME.2004.12689213

Sarros, J., Gray, J., Densten, I., & Cooper, B. (2005). The Organizational Culture Profile Revisited and Revised: An Australian Perspective. *Australian Journal of Management*, *30*(1), 159–182. doi:10.1177/031289620503000109

Schein, E. H. (1990). Organizational culture. *The American Psychologist*, *43*(2), 109–119. doi:10.1037/0003-066X.45.2.109

Schein, E. H. (1992). *Organizational culture and leadership* (2nd ed.). San Francisco: Jossey-Bass Publishers.

Sekaran, U. (2003). *Research methods for business* (4th ed.). Hoboken, NJ: John Wiley & Sons.

Spivak, V. A. (2001). *Organizational culture: theory and practice*. Sankt-Peterburg: Piter. (in Russian)

Stannack, P. (1996). Perspective on Employees Performance. *Management Research News*, *19*(4/5), 38–40. doi:10.1108/eb028456

Stephen, P. R. (2009). *Organizational Behavior* (8th ed.). Pearson Education South Africa.

Tichy, N. M. (1982). Managing Change Strategically: The Technical, Political, and Cultural Keys. *Organizational Dynamics*, *11*(2), 59–80. doi:10.1016/0090-2616(82)90005-5 PMID:10298937

Titiev, M. (1959). *Introduction to Cultural Anthropology*. New York: Henry Holt & Company.

Van der Post, W. Z., de Coning, T. J., & Smit, E. M. (1997). An instrument to measure organizational culture. *South African Journal of Business Management*, *28*(4), 147–168.

Vayanos, P., Hamilton, B. A., & Hammoud, M. (2007). *Promoting the growth and competitiveness of the insurance sector in the Arab world*. Academic Press.

Veiga, J., Lubatkin, M., Calori, R., & Very, P. (2000). Measuring organizational culture clashes: A two-nation post-hoc analysis of a cultural compatibility index. *Human Relations*, *53*(4), 539–557. doi:10.1177/0018726700534004

Walter, A. (1999). Relationship promoters: Driving forces for successful customer relationships. *Industrial Marketing Management*, *28*(5), 537–551. doi:10.1016/S0019-8501(99)00081-4

Williams, R. S. (2002). *Managing Employee Performance: Design and implementation in organizations*. London: Thomson Learning.

Chapter 4
A Knowledge Creation Model for Graduate Research

Yoshiteru Nakamori
Japan Advanced Institute of Science and Technology, Japan

Jing Sun
Japan Advanced Institute of Science and Technology, Japan

Jianguo Wu
Wuhan University of Technology, China

Jing Tian
Wuhan University of Technology, China

Van-Nam Huynh
Japan Advanced Institute of Science and Technology, Japan

ABSTRACT

This chapter proposes a model of knowledge creation and knowledge validation for the research of postgraduate students, and then introduces an evaluation method on research ability and research environment based on this model. The chapter reports the results of statistical analysis, using the data obtained by a questionnaire survey at a graduate school of computer science and technology. Among several discoveries, the great differences in the evaluation of research environment between teachers and students, and in the self-evaluation of research ability between female students and male students are noteworthy. It was confirmed that the proposed model and the evaluation method are effective for supporting research of graduate students.

DOI: 10.4018/978-1-5225-5655-8.ch004

INTRODUCTION

Evaluation of higher education institutions is usually done by objective and quantifiable indicators, such as the number of research papers, the amount of research funds, the securing of excellent students, and the employment situation of students, etc. The reason why the objectivity of evaluation must be ensured is to use it for public decision making, such as determining the amount of subsidies from the government.

On the other hand, however, it is difficult to evaluate the improvement of students' abilities and the effect of research environment on ability improvement. Most of the literatures on the evaluation of educational institutions are related to the evaluation of courses and teaching (Dunegan & Hrivnak, 2003; Rahman, 2006), and the knowledge survey of the increase in knowledge amount of students (Feldman, 1998; Fink, 2003). However, the importance of graduate education is not only the increase in the amount of knowledge but also the increase in abilities of knowledge management and creation. This is exactly the motivation to promote this research.

Knowledge management is a management method to systematically manage the process of discovering, accumulating, exchanging, sharing, creating, and utilizing knowledge as an asset of an organization such as a company (see, for instance, Davenport et al., 1998; Alavi & Leidner, 2001). Recent trends in knowledge management are described in detail in Dwivedi et al. (2011) and Iqbal & Mahmood (2012).

A number of studies on implementing knowledge management in school environment have been reported (Zhao, 2010; Leung, 2010; Chu, Wang & Yuen, 2011; Cheng, 2012). Studies that examined the SECI (Socialization → Externalization → Combination → Internalization) spiral, which is an organizational knowledge creation model proposed by Nonaka and Takeuchi (1995), in the educational sites are found in Joia (2002) and Wu, Lee & Shu (2013). These studies analyzed the strength and weakness of the SECI spiral model in training teachers and in transferring knowledge.

As knowledge management models for individual graduate students, although not as famous as the SECI model, some academic knowledge creation models were proposed in Wierzbicki and Nakamori (2006). Evaluations of these models were reported in Tian et al. (2009) and Sun et al. (2016). Whereas the SECI spiral is a knowledge creation model by a group, these are models of individual knowledge creation supported by a group.

In this research, basing on the above-mentioned models, we develop a new model of knowledge management in graduate research, and try to quantify the abilities of students and the efforts of teachers, which are hard to evaluate quantitatively. Therefore, this paper first proposes a knowledge management model to promote knowledge creation and validation in postgraduate research. Then, the paper reports the analyzed results of a questionnaire survey on the research ability and research environment using the evaluation table derived from the model. This questionnaire survey was conducted from October to December 2016 at the School of Computer Science and Technology, Wuhan University of Technology in China.

One of the purposes of the survey is, in addition to the validation of the proposed model, to grasp the special situation of the respondents and support the promotion of better graduate school education by the self-evaluation of student's growth and the evaluation of research environment provided by faculty. This paper reports two major discoveries in this survey. One is that the teachers were proud of the research environment they provided, but the students did not appreciate it. The other is that female students gave considerably lower self-assessment scores than male students.

This paper is organized as follows. Section 2 explains the background that led to the proposal of a new model, which is a process to learn from existing academic knowledge creation models and organizational knowledge creation models to construct a graduate school research model. Section 3 proposes a model of knowledge creation and validation for graduate research, called the three-stage EDIS spiral, and introduces a list of questions based on this model. Section 4 reports the survey results using this list, which shows the usefulness of the three-stage EDIS spiral model. Section 5 discusses the findings by this research, and finally, Section 6 gives some conclusions and possible future research.

STUDY OF EXISTING KNOWLEDGE CREATION MODELS

Academic Knowledge Creation Models

The first and most famous organizational knowledge creation model is the SECI model proposed by Nonaka and Takeuchi (1995). This model suggests that knowledge is created in an organization through four processes: Socialization, Externalization, Combination, and Internalization. New knowledge is created while explicit knowledge and tacit knowledge are alternatively converted and exchanged between individuals and a group. For the purpose of this study, the SECI model cannot be used directly as this research deals with knowledge creation by individuals supported by a group.

Some studies on knowledge creation practice among school leaders and educators have been reported under the influence of the SECI model (e.g., Bereiter 2002; Harris 2008). As studies related to the theme of this paper, knowledge creation in research (e.g., Knorr-Cetina 1999) and knowledge creation in education (e.g., Scardamalia & Bereiter 1999, 2006) have been argued in parallel. This paper aims to develop a knowledge creation model of the SECI model type in the integrated domain of these two studies, since graduate school researches inevitably include educational perspective as well as research perspective.

In this context, as knowledge creation models similar to the SECI model, only the following three models are published by Wierzbicki & Nakamori (2006), which are helpful for developing a knowledge creation model for graduate students. These three models are: the inter-subjective EDIS spiral, the experimental EEIS spiral, and the hermeneutic EAIR spiral. All of these three spirals begin with having an idea, called "Enlightenment" (illumination, aha, or eureka), as shown in Fig.1.

The EDIS (Enlightenment – Debate – Immersion – Selection) spiral emphasizes "Debate" (discussions in a group about research underway). Unlike the SECI model, this creative model is aimed not at organizational knowledge creation, but just at the individual creation of scientific and technological knowledge supported by a group. No knowledge is lost during all these processes and each process can add new perspectives, ideas, or insights, contributing to "Enlightenment" on the next spiral repetition. Thus, this model guarantees knowledge creation, in smaller or bigger steps, depending on the situation.

It is necessary to extend the EDIS spiral to represent the situation when the verification of a new idea occurs not through "Debate" but through "Experiment." The EEIS (Enlightenment–Experiment–Interpretation–Selection) spiral emphasizes "Experiment" (test of ideas and hypotheses by experimental research). In fact, the process "Experiment" simply represents experimental research – the verification of an idea, not necessarily of an already formed theory. However, every researcher with experimental experience knows that raw experimental data do not provide much new knowledge. Their "Interpretation" is necessary in the sense of the immersion of experimental data based on their experience.

The hermeneutic circle describes the relation of a researcher to the object of his/her study represented by historical or literary texts, objects of art, etc. In this relation, the researcher must immerse himself/herself in the time and culture represented by the objects of the study and use empathic reflection based on the traditions of his/her discipline. The EAIR (Enlightenment – Analysis – Hermeneutic Immersion – Reflection) spiral emphasizes "Hermeneutic Immersion," which is related to

Figure 1. Three academic knowledge creation models (Wierzbicki & Nakamori 2006)

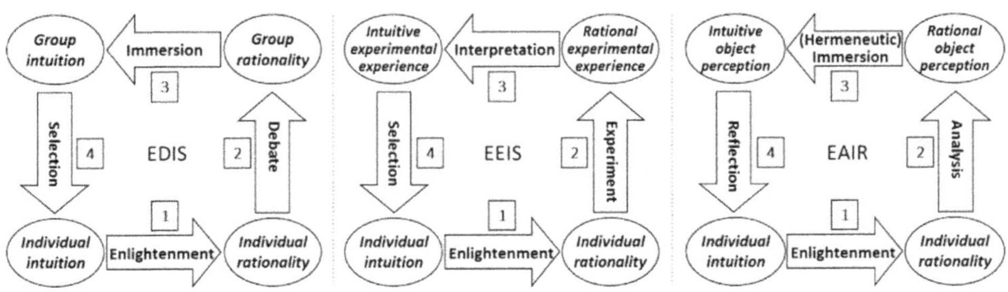

collecting scientific information and knowledge from the literature and other sources, and interpreting and reflecting on these materials.

Tian et al. (2009) examined these models in a graduate university (Japan Advanced Institute of Science and Technology, called JAIST) and found that young researchers of information science and materials science have universal difficulties in doing research, including acquiring necessary knowledge and forming an idea, conducting experiments, as well as participating in further debate and discussion.

Sun et al. (2016) examined three models (EDIS, EEIS, EAIR spirals) by interviewing graduate students and their supervisors in the School of Knowledge Science at JAIST, and found similar difficulties reported in Tian et al. (2009). This school consists of three basic fields, which are management science, information science, and systems science. Sun et al. (2016) assumed that the models in Figure 1 could represent the research processes in these three areas, respectively.

From this survey, they discovered the following:

- Students of all fields are very poor at performing the process "Enlightenment". The processes "Immersion" and "Interpretation" are not effectively used in all fields. These processes are treated differently depending on the respective fields.
- There is a knowledge gap and a strategic gap between professors and students. Knowledge gap refers to the gap between what you must know and what you know, and strategic gap refers to the gap between what you must do and what you can do (Zack 2002). The factors of the gaps are lack of previous experience-based knowledge, knowledge accumulation sensibility, presentation skills, and awareness of the research support environment.

Based on the opinions of graduate students and their supervisors in the research of Sun et al. (2016), this paper tries to develop an integrated knowledge creation model that can be utilized in any field of research, which adopts the following ideas:

- It is better to use the same four nodes used in the SECI model, which indicate four types of knowledge: individual tacit knowledge, individual explicit knowledge, group tacit knowledge, and group explicit knowledge.
- Most importantly, it is necessary to deal differently with the different stages of research: the planning stage, the execution stage, and the achievement presentation stage.

From these considerations, this paper proposes a new model based on the EDIS spiral. But, before that, let us compare the EDIS spiral with the famous SECI model.

Comparison With the SECI Model

A famous organizational knowledge creation model called the SECI spiral (Nonaka and Takeuchi, 1995) suggests that knowledge increases by changing its epistemological and ontological (or social) forms on the two-dimensional plane. One is the epistemological dimension that is related to the explicit degree of knowledge. If one goes to one side on this dimension, the explicit aspect of knowledge will become stronger; but if one proceeds to the other side, the tacit aspect of knowledge will become stronger. The other is the ontological (or social) dimension that is related to affiliation of knowledge. When advancing to one side on this dimension, the personal aspect of knowledge increases, whereas when going to the other side, the social aspect of knowledge becomes stronger.

For example, let us imagine a scene of creating ideas for a new product within a group. Individual tacit knowledge based on individual experience is converted to group tacit knowledge by the process of socialization, and group tacit knowledge is converted to group explicit knowledge by the process of externalization. In this way, a number of ideas emerge in the organization, and through the process of combination a new product will be created. Then, knowledge received as explicit knowledge individually is internalized as individual tacit knowledge, and thus knowledge increases personally as well.

In summary, knowledge will gradually increase while changing the epistemological and ontological forms through four processes as follows:

- Knowledge: Individual tacit
  - (Process: Socialization) → Knowledge: Group tacit
  - (Process: Externalization) → Knowledge: Group explicit
  - (Process: Combination) → Knowledge: Individual explicit
  - (Process: Internalization) → Knowledge: Individual tacit

On the other hand, Wierzbicki and Nakamori (2006) proposed several academic knowledge creation models as shown above. Among them the EDIS (Enlightenment → Debate → Immersion → Selection) spiral uses similar knowledge forms as SECI spiral. In other words, knowledge is converted on the plane spanned by the two dimensions that are similar to those of the SECI spiral. However, whereas the SECI spiral is an organizational knowledge creation model for a company pursuing profit, the EDIS spiral models the process of creating and validating knowledge for research and development by individuals while being supported by a group.

In most cases a new research starts with inspiration based on personal experiences. Research progress or results are presented to other researchers at seminars or academic meetings, some revised plans are discussed within the group, and appropriately selected proposals are used to revise the research or advance to new research. The EDIS spiral is a model of such processes and can be described as follows:

- Knowledge: Individual intuition
  - (Process: Enlightenment) → Knowledge: Individual rationality
  - (Process: Debate) → Knowledge: Group rationality
  - (Process: Immersion) → Knowledge: Group intuition
  - (Process: Selection) → Knowledge: Individual intuition

Figure 2 shows these two models: the SECI spiral model in the left, and the EDIS spiral model in the right. Here, since it is possible to regard intuition as tacit knowledge and regard rationality as explicit knowledge, it can be said that the EDIS spiral is a knowledge conversion model similar to the SECI spiral. However, it should be noted that the order of conversion is exactly opposite.

Since this research deals with personal knowledge creation, we focus on the EDIS spiral. But it is a little abstract as a knowledge creation model for graduate students. For students, it is necessary to promote the above four processes just to create research plans. In addition, even while actually conducting research, the above four processes must be repeatedly executed. Moreover, it is necessary for students to carry out the above four processes further when summarizing the research results and submitting a paper to a journal or presenting the results at a conference. Based on the above consideration, this paper proposes the three-stage EDIS spiral model in the following.

Figure 2. The SECI spiral model (left) and the EDIS spiral model (right)

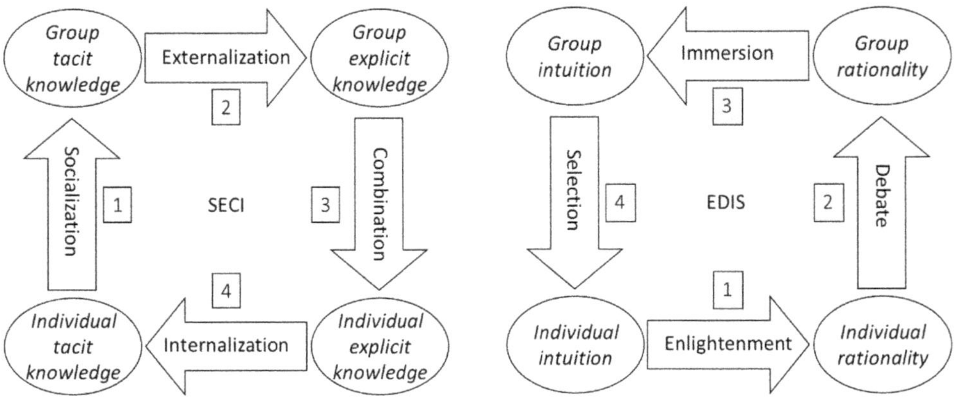

PROPOSAL OF A KNOWLEDGE CREATION MODEL FOR GRADUATE RESEARCH

Three-Stage EDIS Spiral Model

Three-stage EDIS spiral shown in Figure 3 is a knowledge creation and validation model, which is the main proposal in this paper. The roles of three stages are explained below:

Stage 1: (The first round of the spiral) Explains the process of making research proposal. Here, students are required the massive literature review and the deep discussion with supervisors in order to decide the research goal and contents with confidence. The second process of Stage 1: Discussion on research plan with the supervisors and colleagues in the lab is a quite important process in Stage 1.

Stage 2: (The second round of the spiral) Explains the process of doing the main body of research. Students are required to select suitable methodologies to achieve the purpose of the research before performing the second process of Stage 2: Development, the execution of research, which also includes presentation of research results in the lab or in the school to make research better.

Stage 3: (The third round of the spiral) Explains the process of publishing research results. Students are required to write and present papers logically to make members of the academic society understand and accept their research. Here, to enhance the skill of the second process of Stage 3, Dissemination of research results, is quite important.

Through these three stages, new ideas will be created, developed and published. And both the knowledge creation processes (the processes 1 and 2) and the knowledge validation processes (the processes 3 and 4) are included in the model. The details of processes of each stage are given below:

(1-E) Stage 1, Process 1 (Enlightenment): Based on the experience-based knowledge, together with new information, a student must find the direction of new research and write a research plan. This process is quite difficult for a student who is a beginner of research. An appropriate guidance is necessary for him/her to write a research plan.

(1-D) Stage 1, Process 2 (Discussion): Preparing presentation materials, the student must explain his/her research proposal to his/her supervisors, seniors and colleagues at seminars in the lab.

Figure 3. Three-stage EDIS spiral model for knowledge creation and validation

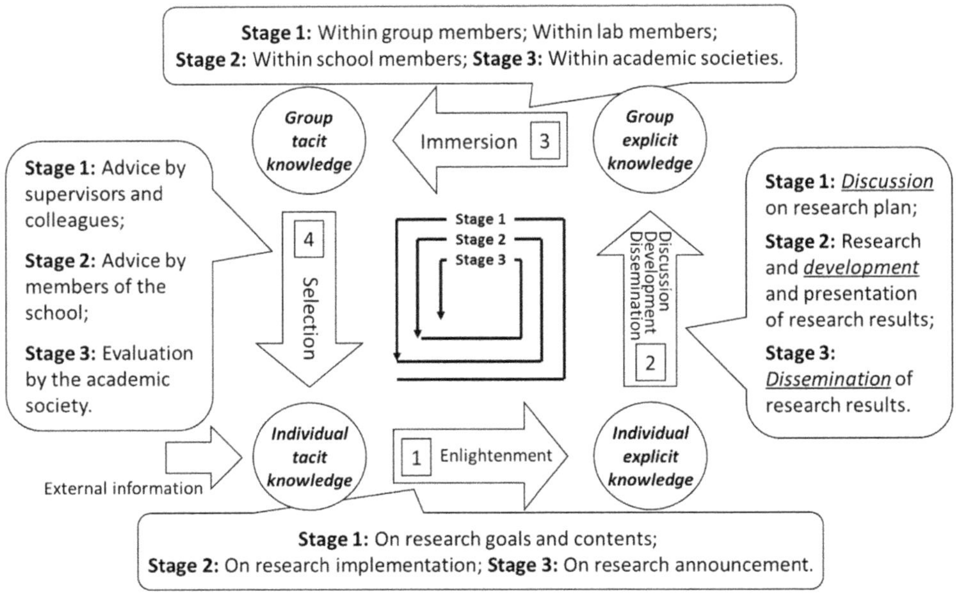

(1-I) Stage 1, Process 3 (Immersion): In an ideal situation, supervisors, seniors and colleagues consider the presented research plan as if it is their own research plan and express their opinions based on their deep understanding of the presented research plan. But actually, the success of this process is not easy because people usually do not have incentive to consider plans of other persons deeply.

(1-S) Stage 1, Process 4 (Selection): The student must examine opinions given by the supervisors or seniors in order to modify the research plan or proceed to the main body of research.

(2-E) Stage 2, Process 1 (Enlightenment): Accepting the suggestions of the supervisors, the student must write an improved, concrete research plan to perform the main body of research. The student has to study some research methodologies before doing research, such as literature review, computer simulation, data mining, questionnaire survey, chemical experiment, etc., depending on the research area.

(2-D) Stage 2, Process 2 (Development): Here the student must perform his/her research carefully. Actually, this is the main process of the research. But, usually, the student does not stay in this process until the end of the research. Instead, he/she will proceed to the next process with the partial progress of the research and come back to this process with a better idea to perform the research.

(2-I) Stage 2, Process 3 (Immersion): When the research reaches a certain conclusion, the student must present the result in the lab or the school. He/she must show it passionately, but also logically, so that the supervisors are willing to give him/her a lot of useful suggestions to improve the research.

(2-S) Stage 2, Process 4 (Selection): In order to understand and select some advice, the student must know the principles of research evaluation. He/she must reflect whether the research is novel, useful, or reproducible. He/she must understand that the evaluation is a part of the research. He/she must study the methods of evaluating research, such as falsification test, intersubjective discussion, etc.

(3-E) Stage 3, Process 1 (Enlightenment): It is the obligation of the graduate student to publish the research as a paper if the research reaches a certain level. In particular, a student seeking to obtain a doctoral degree is required to submit papers to scientific journals and international conferences. The third stage starts with the planning to write an academic paper; here, the student must learn how to write an academic paper.

(3-D) Stage 3, Process 2 (Dissemination): The student has to disseminate the research results to the public by submitting a paper to a journal or presenting a paper at a conference. Writing academic papers directly leads to acquiring the ability to explain research logically. For graduate students, this training is extremely important, and therefore polite instruction by the supervisors is required.

(3-I) Stage 3, Process 3 (Immersion): The student must answer, with full effort, questions of the reviewers of his/her paper, or questions at a conference. This is extremely important to make the research better. Serious communication with reviewers is a very important challenge for a graduate student to grow as a researcher. This model insists on including this process in the graduate education program.

(3-S) Stage 3, Process 4 (Selection): In this final process the student must deeply reflect on the current research and try to discover the next new challenge. It is indispensable for the future research life to acquire the ability to evaluate research results. This model requires students to recognize the importance of this process which has not been discussed sufficiently so far.

Model-Based Questionnaire

This section introduces a list of questions to evaluate the research abilities of graduate students and the research environment around them. Although it is possible to design different questionnaires based on the proposed model, this paper just focuses on the research ability of students and the research environment including guidance by the supervisors.

The proposed model has 3 stages, and each stage consists of 4 processes, therefore there are 12 processes in total. Four items of questions are introduced to evaluate the ability and the environment of each process, which are:

- A(A): Ability of doing research: the diagnosis of the student's ability to carry out each process;
- A(B): Importance of ability: the importance of developing such research ability;
- E(A): Environment of doing research: the evaluation of the provided research environment of the university; and
- E(B): Necessity of environment: the necessity of the environment to carry out each process.

Therefore, the total number of questions is 48. Table 1 shows the labels of 48 questions.

The meanings of labels are, for instance, as follows:

- 1-E-A(A): Stage **1**, **E**nlightenment, research **A**bility, **(A)** satisfaction level;
- 1-E-A(B): Stage **1**, **E**nlightenment, research **A**bility, **(B)** importance level;
- 1-E-E(A): Stage **1**, **E**nlightenment, research **E**nvironment, **(A)** sufficiency level; and
- 1-E-E(B): Stage **1**, **E**nlightenment, research **E**nvironment, **(B)** necessity level.

The above four questions are shown in Figure 4 and Figure 5, in which the questions on the left side are for students, while those on the right side are for teachers. The meanings of the levels 1 to 5 in Figure 4 and Figure 5 are, for example, as follows: 1=strongly dissatisfied; 2=dissatisfied; 3=neither; 4=satisfied; and 5=strongly satisfied. For reference, item (A) of all questionnaires will be shown in the appendix.

Table 1. The labels of 48 questions

	Stage 1				Stage 2				Stage 3			
	1-E	1-D	1-I	1-S	2-E	2-D	2-I	2-S	3-E	3-D	3-I	3-S
Ability	1-E-A(A)	1-D-A(A)	1-I-A(A)	1-S-A(A)	2-E-A(A)	2-D-A(A)	2-I-A(A)	2-S-A(A)	3-E-A(A)	3-D-A(A)	3-I-A(A)	3-S-A(A)
Importance	1-E-A(B)	1-D-A(B)	1-I-A(B)	1-S-A(B)	2-E-A(B)	2-D-A(B)	2-I-A(B)	2-S-A(B)	3-E-A(B)	3-D-A(B)	3-I-A(B)	3-S-A(B)
Environment	1-E-E(A)	1-D-E(A)	1-I-E(A)	1-S-E(A)	2-E-E(A)	2-D-E(A)	2-I-E(A)	2-S-E(A)	3-E-E(A)	3-D-E(A)	3-I-E(A)	3-S-E(A)
Necessity	1-E-E(B)	1-D-E(B)	1-I-E(B)	1-S-E(B)	2-E-E(B)	2-D-E(B)	2-I-E(B)	2-S-E(B)	3-E-E(B)	3-D-E(B)	3-I-E(B)	3-S-E(B)

SURVEY AND FINDINGS

Survey

We conducted a questionnaire survey at the School of Computer Science and Technology, Wuhan University of Technology, Wuhan, China, from October to December 2016. The respondents were 25 teachers and 65 graduate students. Table 2 shows the breakdown of respondents.

The questionnaire survey has three purposes:

Figure 4. An example of evaluation sheet (Stage 1, Process 1, Ability)

Stage 1: **Ability** of Enlightenment (1-E-A)

Questions for **Students**

A) Are you satisfied with your ability to plan your research? (Do you have a lot of ideas and experience to make your research plan?)

B) Do you think this ability is important for you?

A)	Dissatisfied	1	2	3	4	5	Satisfied
B)	Unimportant	1	2	3	4	5	Important

- If your answer is 'dissatisfied' but 'important', what would you like to do specifically?
- If your answer is 'unimportant', please write the reason.

Questions for **Teachers**

A) Are you satisfied with your students' ability to plan their research? (Do your students have a lot of ideas and experience to make research plans?)

B) Do you think this ability is important for students?

A)	Dissatisfied	1	2	3	4	5	Satisfied
B)	Unimportant	1	2	3	4	5	Important

- If your answer is 'dissatisfied' but 'important', what would you like to ask your students specifically?
- If your answer is 'unimportant', please write the reason.

Figure 5. An example of evaluation sheet (Stage 1, Process 1, Environment)

Stage 1: **Environment** of Enlightenment (1-E-E)

Questions for **Students**

A) When preparing a research plan, do you receive sufficient guidance from your supervisors? (Do you have a good environment in your lab to discuss your research plan?)

B) Do you think this environment is necessary for you?

A)	Insufficient	1	2	3	4	5	Sufficient
B)	Unnecessary	1	2	3	4	5	Necessary

- If your answer is 'insufficient' but 'necessary', what would you like to ask your supervisors or university?
- If your answer is 'unnecessary', please write the reason.

Questions for **Teachers**

A) When your students are preparing research plans, do you give them sufficient guidance? (Do you have a good environment in your lab to help students make research plans?)

B) Do you think this environment is necessary for students?

A)	Insufficient	1	2	3	4	5	Sufficient
B)	Unnecessary	1	2	3	4	5	Necessary

- If your answer is 'insufficient' but 'necessary', what would you like to do specifically?
- If your answer is 'unnecessary', please write the reason.

Table 2. Respondents of the survey

Total number of teachers 25	
Faculty experience: less than 5 years	12
Faculty experience: 5 years or more	13

Total number of students 65			
MSc students (Female)	51 (18)	published a paper	10
		writing a paper	24
		never wrote a paper	17
Doctoral students (Female)	14 (3)	published a paper	10
		writing a paper	3
		never wrote a paper	1

1. The first purpose is to verify whether the proposed model is accepted by respondents. This will be achieved through:
    a. Questions about the importance of ability to perform each research process and the necessity of research environment to support students to carry out research, that is, Question (B) in each sheet; and
    b. Questions to teachers about the importance and necessity of the four processes after the prepared questions of each stage.
2. The second purpose is to extract features of respondents. When all respondents can be divided into several subgroups (for example, students and teachers, female and male students, different research fields, different future expectations, etc.), we can investigate differences in evaluation between subgroups. This will provide useful information for improvement of research guidance and research environment.
3. The third purpose is a social contribution. By answering the prepared questions, students and teachers review the ability and environment of doing research, thereby increasing the motivation for developing research ability and research environment. Therefore, it is our duty to feed back the data analysis results to the respondents.

Data Analysis and Findings

Let us first see how much the second purpose of the survey was attained, which indirectly suggests the achievement level of the first purpose. From the overview of the survey data, it is possible to compare groups as follows:

1. Evaluation gap between young teachers and experienced teachers;
2. Evaluation gap between students and teachers;
3. Evaluation gap between master's students and doctoral students;

4. Evaluation gap by differences in career hope;
5. Evaluation gap by the experience of writing a paper; and
6. Evaluation gap between female students and male students.

Among the above, this section reports two important findings:

- Evaluation gap between students and teachers; and
- Evaluation gap between female students and male students.

Evaluation Gap Between Students and Teachers

It is quite interesting and important to recognize the differences in evaluation between students and teachers. Figure 6 shows the average scores by all students (left: 65 students) and all teachers (right: 25 teachers). Both students and teachers show that there is an extremely large difference in research ability and its importance (i.e. target). This is an important issue to be overcome. The scores of the importance of ability and the necessity of environment are between 4.2 and 4.5, which are so high that they are advantageous facts for model validation, which is the first purpose of the survey.

Here, let us pay particular attention to the following two points.

- (On research ability of students) Students think that they are not good at performing the following three processes:
    - Enlightenment in Stage 1 (1-E): Discovery of new research themes;

Figure 6. Average scores by all students (left) and all teachers (right)

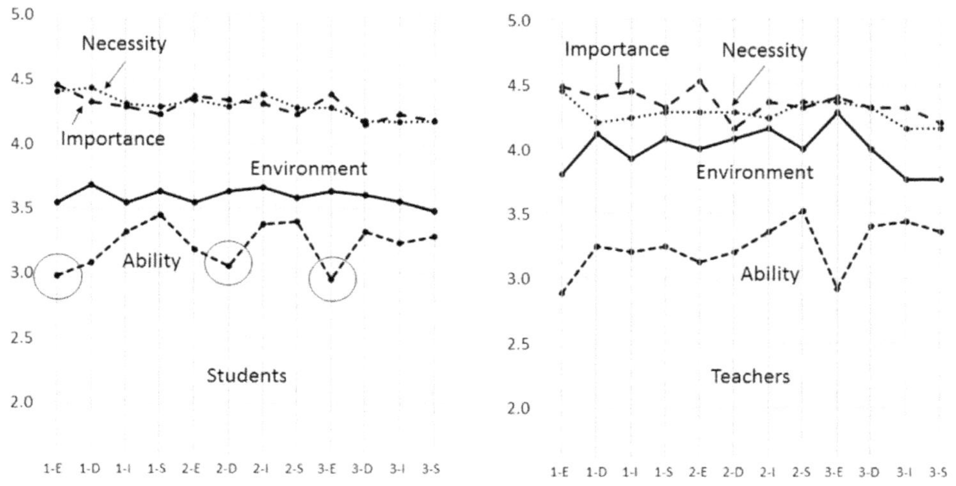

- Development in Stage 2 (2-D): Execution of the main body of research; and
- Enlightenment in Stage 3 (3-E): Preparation of writing and presenting a paper.

- (On research environment) There are fairly big differences in evaluation of research environment between students and teachers. The average values of teachers are 0.21 to 0.65 points higher than the average values of students. We will test to see if there are statistically significant differences below.

Let us examine the following hypothesis:

Null Hypothesis: The true difference in average evaluation scores of research environment between students and teachers is equal to 0.
Alternative Hypothesis: The true difference in average evaluation scores of research environment between students and teachers is not equal to 0.

Table 3 shows the results of statistical hypothesis testing. There are statistically significant differences between evaluation of students and teachers for most processes. It is a quite important finding in this survey that students do not appreciate the research environments to the extent that teachers are proud of them. In other words, students are seeking improvement of research environments. It is also interesting that there are no statistically significant differences in the early and final processes of the research, which are Enlightenment of Stage 1 (1-E: finding the direction of new research), and Immersion and Selection of Stage 3 (3-I and 3-S: learning from reviewers of research papers). For these processes, teachers think that the research environment is insufficient.

Evaluation Gap Between Female Students and Male Students

Let us see the difference in evaluation between female students and male students. Figure 7 shows the average scores by female students (left: 21 students) and male students (right: 44 students).

Apparently, there are big differences in evaluation of research abilities between female students and male students. Actually, the self-evaluations of research abilities by female students are lower than those by male students by 0.35 to 0.97 point. We carried out the following hypothesis test.

Null Hypothesis: The true difference in average evaluation scores of research ability between female and male students is equal to 0.

Table 3. The results of Welch two sample t-test

	Enlightenment	Discussion	Immersion	Selection
Stage 1	t-value = -1.345 p-value = 0.183	t-value = -2.505 p-value = **0.0144**	t-value = -1.898 p-value = 0.0625	t-value = -2.509 p-value = **0.0145**
Stage 2	t-value = -2.577 p-value = **0.0122**	t-value = -2.846 p-value = **0.00558**	t-value = -2.942 p-value = **0.00454**	t-value = -2.433 p-value = **0.0174**
Stage 3	t-value = -3.858 p-value = **0.000263**	t-value = -2.101 p-value = **0.0396**	t-value = -1.127 p-value = 0.264	t-value = -1.420 p-value = 0.160

Discussion → (Stage 2) Development; (Stage 3) Dissemination

Figure 7. Differences of evaluation by female students (left) and male students (right)

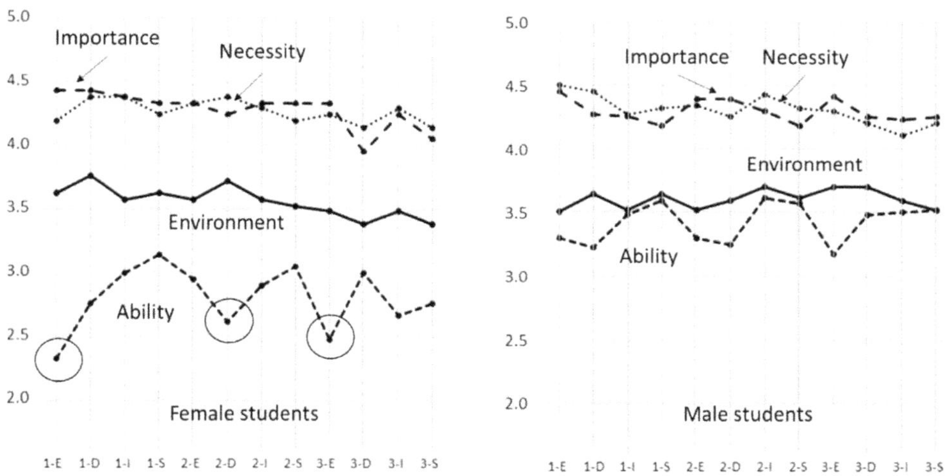

Alternative Hypothesis: The true difference in average evaluation scores of research ability between female and male students is not equal to 0.

Table 4 shows the results of hypothesis testing. In most processes, the average scores by female students are lower than those by male students statistically. Especially, the processes with large differences in self-evaluation are the early and final processes of research (1-E, 3-I, and 3-S). Teachers should pay attention to the support for female students, especially in the implementation of these processes.

Other Findings

Figure 8 shows the average scores by teachers with experience of less than 5 years (left: 12 teachers) and over 5 years (right: 13 teachers). At first glance, it can be seen

Table 4. The results of Welch two sample t-test

	Enlightenment	Discussion	Immersion	Selection
Stage 1	t-value = -3.968 p-value = **0.000249**	t-value = -1.711 p-value = 0.0954	t-value = -2.016 p-value = **0.0496**	t-value = -1.800 p-value = 0.079
Stage 2	t-value = -1.496 p-value = 0.141	t-value = -2.541 p-value = **0.0149**	t-value = -2.687 p-value = **0.0105**	t-value = -2.297 p-value = **0.0262**
Stage 3	t-value = -2.538 p-value = **0.015**	t-value = -1.587 p-value = 0.122	t-value = -3.660 p-value = **0.00063**	t-value = -3.057 p-value = **0.00404**

Discussion → (Stage 2) Development; (Stage 3) Dissemination

Figure 8. Average scores by young teachers (left) and experienced teachers (right)

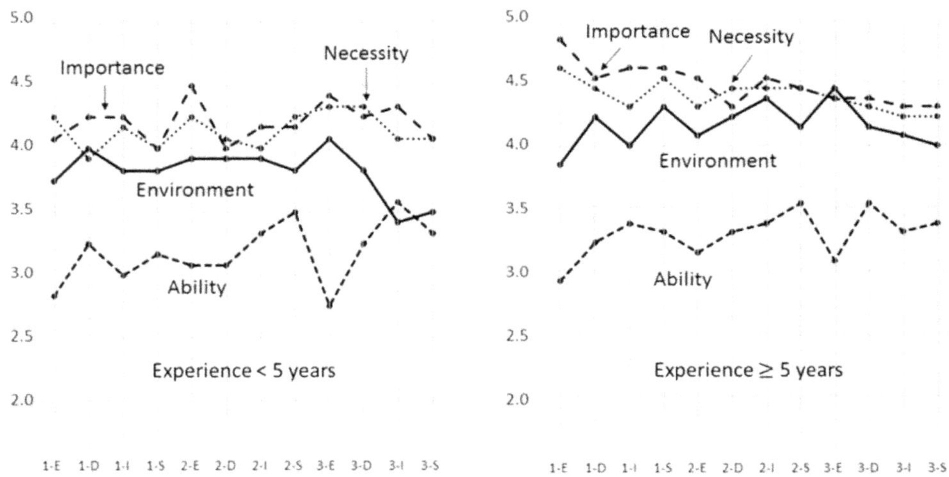

that the evaluation scores of experienced teachers are generally higher than those of young teachers. In particular, it seems that experienced teachers are convinced that they spent time developing the research environment.

Figure 9 shows the average scores by masters' students (left: 51 students) and doctoral students (right: 14 students). Graduate students seem to feel that they are short in their abilities to perform the process of Enlightenment in all stages (1-E, 2-E, and 3-E). In addition, they want to develop ability of executing the main body of research, that is, the process of 'Development' in Stage 2 (2-D).

Figure 10 shows the average scores by students who wish to work at universities or research institutes in the future (left: 9 students), and who wish to work at companies, etc. (right: 56 students). Although we cannot conclude definitely because not many students want to be researchers, students who are planning to work at companies gave

Figure 9. Average scores by masters' students (left) and doctoral students (right)

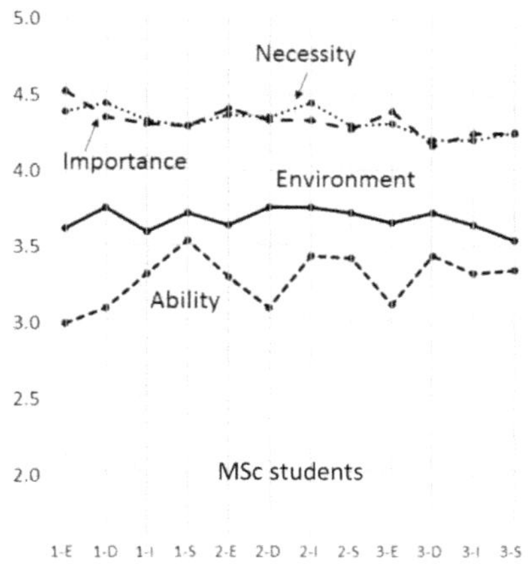

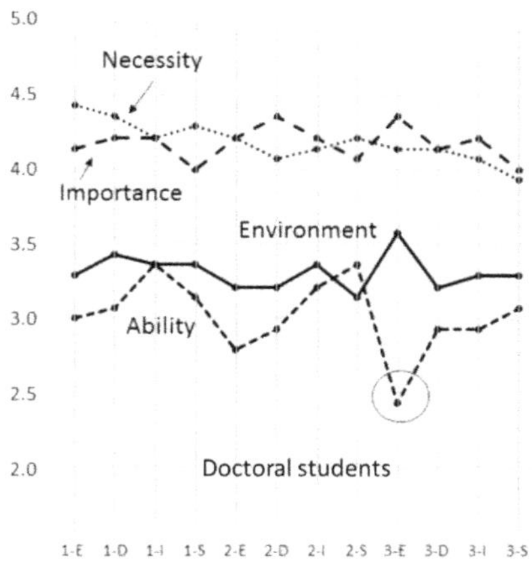

higher scores in most processes. Conversely speaking, students who wish to become researchers tend to estimate their own research ability and research environment low.

Figure 11 shows the average scores by students who never wrote a paper (left: 18 students) and who published a paper (right: 20 students). Twenty-seven students responded that they were writing papers, but this data is not reflected in Figure 11. Students who have never written a paper are keenly aware of the low ability to create research plans (1-E: Stage 1, Process 1: Enlightenment). Students who have written a paper have relatively low scores on their ability to prepare paper plans (3-E: Stage 3, Process 1: Enlightenment).

Figure 10. Average scores by students who wish to be researchers (left) and others (right)

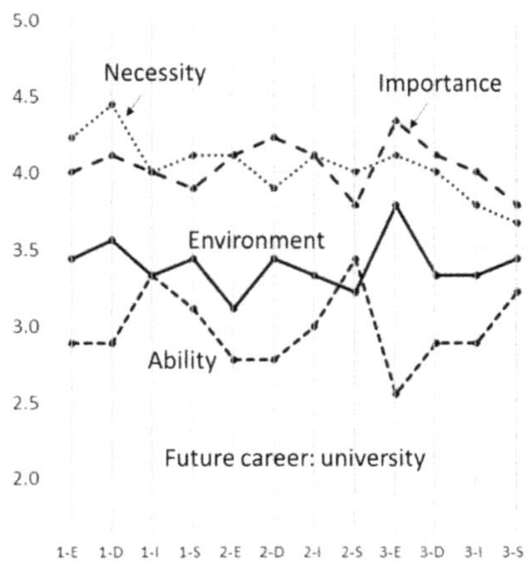

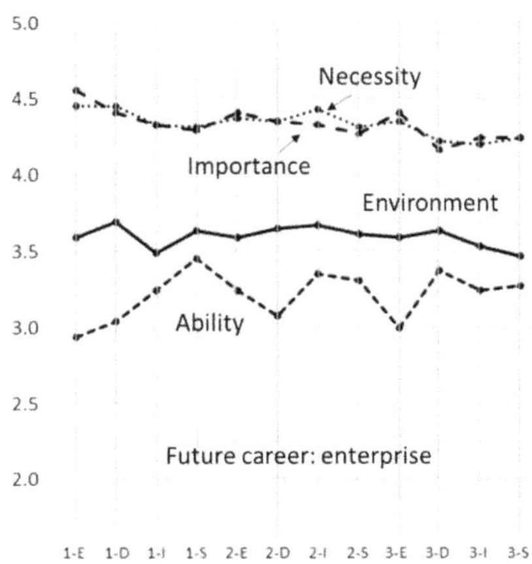

Figure 11. Average scores by students who have never written a paper (left) and who published a paper (right)

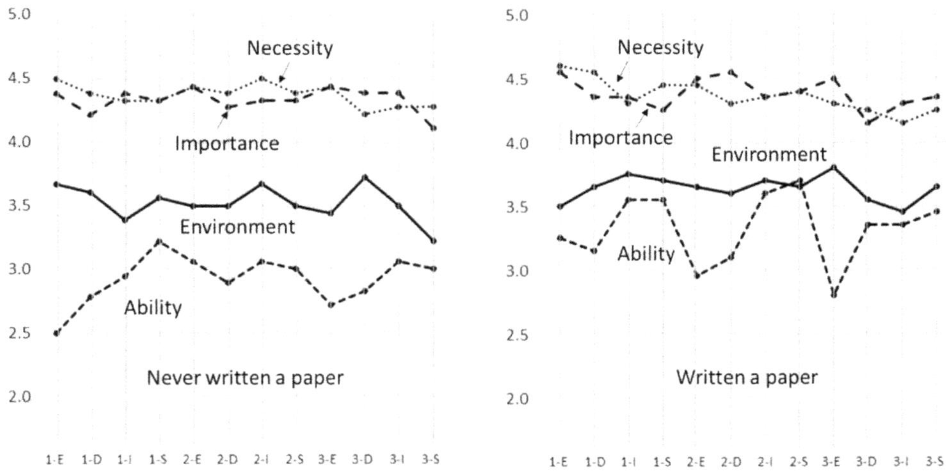

DISCUSSION

The second purpose of the survey was discussed in the previous section. But, it is hard to say that the effectiveness of the proposed model, which is the first purpose, was completely justified, but we would like to report the following fact:

- From Figure 6 to Figure 11, we recognize that most students as well as teachers gave scores greater than 4 about the importance of research ability and the necessity of research environment.

In the survey, we asked teachers the following questions after each stage:

1. Do you agree with the importance and necessity of four processes for writing a research proposal? If not, please suggest modification.
2. Do you agree with the importance and necessity of four processes for performing the main part of research? If not, please suggest modification.
3. Do you agree with the importance and necessity of four processes for disseminating research results? If not, please suggest modification.

The number of teachers who agreed to each question was 11, 8, and 7, respectively, out of 25 teachers. These figures are not high enough to justify the model, but there is no suggestion for modification in the survey this time.

In the following we will discuss the two discoveries in the previous section.

- Difference in evaluation of research environment between students and teachers.
- Inferiority complex of female students studying computer science and technology.

Difference in Evaluation of Research Environment Between Students and Teachers

In *Evaluation Gap Between Students and Teachers*, we confirmed that students evaluated the research environment considerably lower than teachers. In order to explore this reason, let us look at another data collected in the questionnaire survey. In the survey, after a set of questions of each stage, the following questions were asked to students and teachers, respectively:

Questions to students:

1. What kinds of research environment do you need to make your research plan?
2. What kinds of research environment do you need to perform your research?
3. What kinds of research environment do you need to publish your research results?

Questions to teachers:

1. What kinds of research environment do you prepare for students to make their research plans?
2. What kinds of research environment do you prepare for students to perform their research?
3. What kinds of research environment do you prepare for students to publish their research results?

Representative answers to these questions are listed below:

Answers from students:

- Students are requesting independent places for research, detailed research guidance, open discussion environments, appropriate hardware facilities, sufficient raw data, guidance on paper writing, and very good research teams.

Answers from teachers:

- Teachers are providing students with the necessary guidance, relevant literature, regular meetings, hardware, software, and financial support.
- Teachers are asking students to obediently accept guidance, learn individually, compete with each other, write research papers, and participate in academic conferences.

Although most students require the independent and quiet place to do research, there is no big difference in what the students and teachers talk about the research environment. How can you explain the big difference in evaluation between teachers and students? In this paper, we set the following:

Hypothesis A: In general, students do not feel the value as much as teachers think about the research environment services provided by the faculty and university.

The recognition gap between service providers and service recipients exists everywhere in our society. Needless to say, for students, being able to convert the prepared research environment into their own value is the way to success. However, in reality it is difficult, and teachers are probably aware that efforts to improve the research environment are not rewarded.

It is an important task to develop a co-creation method of value between students and teachers based on the recent development of service management.

The concept of a service system has recently attracted attention in the service economy (see for instance, Demirkan et al., 2011; Taylor & Tofts, 2012). The term service management was proposed firstly in computer science, and now it is used in many fields such as service science (e.g., Maglio et al., 2010), service marketing (e.g., Lusch & Vargo, 2006; Lovelok & Wirtz, 2010), and service innovation (e.g., Kosaka & Shirahada, 2013). The main idea in service management is that both service providers and service recipients co-create values cooperatively.

From such consideration, we naturally consider the following:

Future Challenge A: What is important for the future task is to develop a method of value co-creation by students and teachers on the service of research environment.

Inferiority Complex of Female Students Studying Computer Science and Technology

In *Evaluation Gap Between Female Students and Male Students,* we confirmed that female students estimated their research abilities considerably lower than male students. In order to explore this reason, let us look at another data collected in the questionnaire survey. After each question in the survey, the students had chance to consider how to improve their own ability to perform research. A summary of opinions by female and male students is given below.

Opinions of female students:

- I must prepare everything well in advance and develop my own self-confidence.
- I must convince my research plan and explain it to my classmates and teachers.
- I must improve my conversation skills in order to have self-confidence.
- I must learn from others, understand different research methods, and find my own research method.
- I must spend more time reading papers, studying the results of predecessors, and exploring new challenges.

Opinions of male students:

- I must pay attention to training my ability to express my idea.
- I must communicate with others, carefully elaborating my views of research.
- I must read more papers and learn from others to generate own ideas.

From the above, there is no big difference between female and male students about their will and desire, except that female students seem to be less confident about research activities than male students. But, how can you explain the big difference in evaluation of research abilities by female and male students? In this paper, we make the following:

Hypothesis B: In the field of computer science and technology, female students generally tend to make their self-evaluation of research ability lower than male students.

Here, there is a report to support this hypothesis. Quoting the OECD's 2009 report, Kay & Shipman (2014) reported that female students in most countries have lower self-efficacy than male students despite high scores on mathematics. In other words, male students have positive opinions on their abilities compared to female students, but conversely female students feel greater anxiety, helplessness, and stress than male students in the mathematics class.

Is this discovery a general trend? Or is it a special trend of this survey group? Anyway, we will feed back the survey results to the faculty of the school, and decide to set

Future Challenge B: Examine hypothesis B, and if this is correct, we will develop a method to adjust the differences of self-assessment between female and male students in future surveys.

CONCLUSION

This paper proposed a model of knowledge creation and validation for graduate students to carry out research with confidence, and a questionnaire to both students and teachers based on the model. The questionnaire consists of the evaluation of student's research ability, the importance of research ability, the evaluation of research environment, and the necessity of research environment. We carried out a questionnaire survey at a graduate school of computer science and technology, and collected answers from 65 graduate students and 25 teachers.

We made two big discoveries by data analysis. One is the difference in evaluation of research environment between students and teachers. The students did not appreciate the research environment as much as the teachers thought. The other finding is the difference in the self-evaluation of research ability between female students and male students. Female students evaluated their research abilities considerably low compared to male students. Considering the importance of these two facts, we set the following as future important themes: one is to develop a method of co-creation of values by students and teachers about the service of research environment, and the other is to develop a method to adjust the differences of self-assessment between female and male students in future surveys.

It is necessary to mention the effectiveness of the proposed model. Although we have not obtained direct evaluation of the proposed model itself, both the students and teachers have confirmed the importance of research ability and the need for research environment in all processes of the model. This suggests that the proposed model is justified to some extent, but its effectiveness must be verified by further application. In addition to this purpose, we need to carry out questionnaire surveys in many other research fields to find out facts other than the discoveries in this paper.

By doing so, we can proceed with the validation and correction of the proposed model, and highlight the difference between ability evaluation and environmental evaluation due to differences in fields and experience. Furthermore, it is also a challenge to pursue the possibility of the second-round questionnaire survey for the same respondents in order to follow up the growth of the students and the improvement of the research environment. It is also interesting to create a question list for supervisors and to see the difference in the consciousness between students and faculty. Through such activities, we would like to contribute to the improvement of the quality of education and research at graduate schools as knowledge creation systems.

ACKNOWLEDGMENT

We greatly appreciate the graduate students and faculty members of School of Computer Science and Technology, Wuhan University of Technology, Wuhan, China, who cooperated in the questionnaire survey and gave valuable comments.

REFERENCES

Alavi, M., & Leidner, D. E. (2001). Knowledge management and knowledge management systems: Conceptual foundations and research issues. *Management Information Systems Quarterly*, *25*(1), 107–113. doi:10.2307/3250961

Bereiter, C. (2002). *Education and Mind in the Knowledge Age*. Mahwah, NJ: Lawrence Erlbaum Associates.

Cheng, E. C. K. (2012). Knowledge strategies for enhancing school learning capacity. *International Journal of Educational Management*, *26*(6), 557–592. doi:10.1108/09513541211251406

Chu, K. W., Wang, M., & Yuen, A. J. K. (2011). Implementing knowledge management in school environment: Teachers' perception. *Knowledge Management & E-Learning: An International Journal*, *3*(2), 139–152.

Davenport, T. H., De Long, D. W., & Beers, M. C. (1998). Successful knowledge management projects. *Sloan Management Review*, *39*(2), 43–57.

Demirkan, H., Spohrer, J. C., & Krishna, V. (Eds.). (2011). *The Science of Service Systems*. New York: Springer. doi:10.1007/978-1-4419-8270-4

Dunegan, K. J., & Hrivnak, M. W. (2003). Characteristics of mindless teaching evaluations and the moderating effects of image compatibility. *Journal of Management Education*, *27*(3), 280–303. doi:10.1177/1052562903027003002

Dwivedi, Y. K., Venkitachalam, K., Sharif, A. M., Al-Karaghouli, W., & Weerakkody, V. (2011). Research Trends in Knowledge Management: Analyzing the Past and Predicting the Future. *Information Systems Management*, *28*(1), 43–56. doi:10.1080/10580530.2011.536112

Feldman, K. A. (1998). Identifying exemplary teachers and teaching: Evidence from student ratings. In K. A. Feldman & M. B. Paulsen (Eds.), *Teaching and Learning in the College Classroom* (2nd ed.; pp. 391–414). Needham Heights, MA: Simon and Schuster.

Fink, L. D. (2003). *Creating Significant Learning Experiences: An Integrated Approach to Designing College Courses*. San Francisco: Jossey-Bass.

Harris, A. (2008). Leading innovation and change: Knowledge creation by schools for schools. *European Journal of Education*, *43*(2), 219–228. doi:10.1111/j.1465-3435.2008.00343.x

Iqbal, J., & Mahmood, Y. (2012). Reviewing knowledge management literature. *Interdisciplinary Journal of Contemporary Research in Business*, *4*(6), 1005–1026.

Joia, L. A. (2002). Assessing unqualified in-service teacher training in Brazil using knowledge management theory: A case study. *Journal of Knowledge Management*, *6*(1), 74–86. doi:10.1108/13673270210417718

Kay, K., & Shipman, C. (2014). *The Confidence Code: The Science and Art of Self-Assurance – What Women Should Know*. New York: HarperCollins.

Knorr-Cetina, K. (1999). *Epistemic cultures: How the sciences make knowledge*. Cambridge, MA: Harvard University Press.

Kosaka, M., & Shirahada, K. (Eds.). (2013). *Progressive Trends in Knowledge and System-Based Science for Service Innovation*. Hershey, PA: IGI Global.

Leung, C. H. (2010). Critical factors of implementing knowledge management in school environment: A qualitative study in Hong Kong. *Research Journal of Information Technology*, *2*(2), 66–80. doi:10.3923/rjit.2010.66.80

Lovelock, C. H., & Wirtz, J. (2010). *Service Marketing: People, Technology, Strategy* (7th ed.). Upper Saddle River, NJ: Prentice Hall.

Lusch, R. F., & Vargo, S. L. (Eds.). (2006). *The Service Dominant Logic of Marketing: Dialog, Debate, and Directions*. New York: Taylor & Francis.

Maglio, P. P., Kieliszewski, C. A., & Spohrer, J. C. (Eds.). (2010). *Handbook of Service Science*. New York: Springer. doi:10.1007/978-1-4419-1628-0

Nonaka, I., & Takeuchi, H. (1995). *The Knowledge-Creating Company: How Japanese Companies Create the Dynamics of Innovation*. Oxford, UK: Oxford University Press.

Rahman, K. (2006). Learning from your business lectures: Using stepwise regression to understand course evaluation data. *The Journal of American Academy of Business, Cambridge*, *19*(2), 272–279.

Scardamalia, M., & Bereiter, C. (1999). Schools as knowledge building organizations. In D. Keating & C. Hertzman (Eds.), *Today's children, tomorrow's society: The developmental health and wealth of nations* (pp. 274–289). New York: Guilford.

Scardamalia, M., & Bereiter, C. (2006). Knowledge building: Theory, pedagogy, and technology. In R. K. Sawyer (Ed.), *The Cambridge handbook of the learning sciences* (pp. 97–118). Cambridge, UK: Cambridge University Press.

Sun, J., Nakamori, Y., Tian, J., & Xiang, J. W. (2016). Exploring academic knowledge creation models for graduate researches. *Proceedings of the 2016 IEEE International Conference on Software Quality, Reliability and Security*, 202-209. doi:10.1109/QRS-C.2016.31

Taylor, R., & Tofts, C. (2012). *Managing Complex Service Systems.* New York: Springer.

Tian, J., Nakamori, Y., & Wierzbicki, A. P. (2009). Knowledge management and knowledge creation in academia: A study based on surveys in a Japanese research university. *Journal of Knowledge Management, 13*(2), 76–92. doi:10.1108/13673270910942718

Wierzbicki, A. P., & Nakamori, Y. (2006). *Creative Space: Models of Creative Processes for the Knowledge Civilization Age.* Berlin: Springer-Verlag. doi:10.1007/b137889

Wu, W. L., Lee, Y. C., & Shu, H. S. (2013). Knowledge management in education organization: A perspective of knowledge spiral. *The International Journal of Organizational Innovation, 5*(4), 7–13.

Zack, M. H. (2002). Developing a knowledge strategy. *California Management Review, 41*(3), 125–145. doi:10.2307/41166000

Zhao, J. (2010). School knowledge management framework and strategies: The new perspective on teacher professional development. *Computers in Human Behavior, 26*(2), 168–175. doi:10.1016/j.chb.2009.10.009

APPENDIX

This appendix lists the main parts of the survey table.

Table 5. Questions for students about research ability

1-E-A	Are you satisfied with your ability to plan your research? (Do you have a lot of ideas and experience to make your research plan?)
1-D-A	Are you satisfied with your ability to explain your research plan? (Do you have a lot of ideas and experience to explain your research plan?)
1-I-A	Are you satisfied with your ability to make your group understand your research plan deeply? (Do you have a lot of ideas and experience to make your research plan deeply in the group?)
1-S-A	Are you satisfied with your ability to select good suggestions on your research plan? (Do you have a lot of ideas and experience to select good suggestions on your research plan?)
2-E-A	Are you satisfied with your ability to plan your actual research? (Do you know methods and ways to perform your actual research?)
2-D-A	Are you satisfied with your ability to perform your research? (Do you have a lot of ideas and experience to perform your research?)
2-I-A	Are you satisfied with your ability to make your group understand your research results deeply? (Do you have a lot of ideas and experience to make your group understand your research results deeply in the group?)
2-S-A	Are you satisfied with your ability to select suggestions on your research result? (Do you have a lot of ideas and experience to select suggestions on your research result?)
3-E-A	Are you satisfied with your ability to write research papers or prepare presentations? (Do you know methods and ways to plan research papers or presentations?)
3-D-A	Are you satisfied with your ability to explain your research results in the academic society? (Do you have a lot of ideas and experience to explain your research results?)
3-I-A	Are you satisfied with your ability to make the academic society understand your research results? (Do you have a lot of ideas and experience to make the academic society understand your research results?)
3-S-A	Are you satisfied with your ability to select suggestions on your research from the academic society? (Do you have a lot of ideas and experience to select suggestions on your research from the academic society?)

Table 6. Questions for students about research environment

1-E-E	When explaining your research plan, do you receive sufficient guidance from your supervisors or senior researchers? (Do you have a good environment in your lab (in the university) to discuss your research plan?)
1-D-E	When explaining your research plan, do you receive sufficient guidance from your supervisors or senior researchers? (Do you have a good environment in your lab (in the university) to discuss your research plan?)
1-I-E	When discussing your research plan, do you receive sufficient guidance from your supervisors or senior researchers to elaborate your research plan? (Do you have a good environment in your lab (in the university) to elaborate your research plan?)
1-S-E	When selecting modified ideas about your research plan, do you receive sufficient guidance from your supervisors or senior researchers? (Do you have a good environment in your lab (in the university) to select ideas from your supervisors or seniors?)
2-E-E	When planning your actual research, do you receive sufficient guidance from your supervisors or senior researchers? (Do you have a good environment in your lab (in the university) to plan an actual research?)
2-D-E	When performing your research, do you receive sufficient guidance from your supervisors or senior researchers? (Do you have a good environment in your lab (in the university) to perform your research?)
2-I-E	When discussing your research results, do you receive sufficient guidance from your supervisors or senior researchers? (Do you have a good environment in your lab (in the university) to discuss your research results?)
2-S-E	When selecting modified ideas about your research results, do you receive sufficient guidance from your supervisors or senior researchers? (Do you have a good environment in your lab (in the university) to elaborate your research?)
3-E-E	When writing research papers or preparing presentations, do you receive sufficient guidance from your supervisors or senior researchers? (Do you have a good environment in your lab (in the university) to prepare research papers or presentations?)
3-D-E	When explaining your research, do you receive sufficient guidance from your supervisors or senior researchers? (Do you have a good environment in your lab (in the university) to explain your research results?)
3-I-E	When discussing your research results in order to make them understand your research in an academic society, do you receive sufficient guidance from your supervisors or senior researchers? (Do you have a good supporting environment in your lab (in the university) to discuss your research in an academic society?)
3-S-E	When selecting modified ideas about your research results from the academic society, do you receive sufficient guidance from your supervisors or senior researchers? (Do you have a good environment in your lab (in the university) to elaborate your research based on the suggestions from the academic society?)

Table 7. Questions for teachers about research ability of students

1-E-A	Are you satisfied with your students' ability to plan their research? (Do your students have a lot of ideas and experience to make their research plan?)
1-D-A	Are you satisfied with your students' ability to explain their research plans? (Do your students have a lot of ideas and experience to explain their research plans?)
1-I-A	Are you satisfied with your students' ability to make the group understand the research plans deeply? (Do you students have a lot of ideas and experience to discuss their research plans in order to make the group understand their research deeply?)
1-S-A	Are you satisfied with your students' ability to select suggestions on their research plan? (Do your students have a lot of ideas and experience to select suggestions on their research plan?)
2-E-A	Are you satisfied with your students' ability to plan their actual research? (Do your students know methods and ways to perform their actual research? Can they pick up suitable methodologies for their own research?)
2-D-A	Are you satisfied with your students' ability to perform their research? (Do your students have a lot of ideas and experience to perform their research?)
2-I-A	Are you satisfied with your students' ability to make the group understand their research results deeply? (Do your students have a lot of ideas and experience to make the group understand their research results deeply?)
2-S-A	Are you satisfied with your students' ability to select suggestions on their research result? (Do your students have a lot of ideas and experience to select suggestions on their research result?)
3-E-A	Are you satisfied with your students' ability to plan research papers or presentations? (Do your students know methods and ways to plan research papers or presentations?)
3-D-A	Are you satisfied with your students' ability to explain their research results with papers or presentations in the conferences and the academic society? (Do your students have a lot of ideas and experience to explain their research results in the academic society?)
3-I-A	Are you satisfied with your students' ability to make the academic society understand their research results deeply? (Do your students have a lot of ideas and experience to present and make the academic society understand their research results deeply?)
3-S-A	Are you satisfied with your students' ability to select suggestions on their research from the academic society? (Do your students have a lot of ideas and experience to select suggestions on their research from the academic society?)

Table 8. Questions for teachers about research environment

1-E-E	When your students are preparing research plans, do you give them sufficient guidance? (Do you have a good environment in your lab (in the university) to help students make research plans?)
1-D-E	When your students are explaining their research plans, do you give them sufficient guidance? (Do you have a good environment in your lab (in the university) to discuss your students' research plans?)
1-I-E	When discussing students' research plans, do you give them sufficient guidance to elaborate their research plans? (Do you have a good environment in your lab (in the university) to elaborate students' research plan?)
1-S-E	When students are selecting modified ideas about their research plan, do you give them sufficient guidance? (Do you have a good environment in your lab (in the university) to help students select ideas from supervisors or seniors?)
2-E-E	When your students are planning their actual research, do you give them sufficient guidance? (Do you have a good environment in your lab (in the university) to help students plan actual researches?)
2-D-E	When your students are performing their research, do you give them sufficient guidance? (Do you have a good environment in your lab (in the university) to help students perform their researches?)
2-I-E	When discussing your students' research results, do you give them sufficient guidance? (Do you have a good environment in your lab (in the university) to help students discuss their research results?)
2-S-E	When students are selecting modified ideas about their research results, do you give them sufficient guidance? (Do you have a good environment in your lab (in the university) to help students elaborate their researches?)
3-E-E	When students are preparing research papers or presentations, do you give them sufficient guidance? (Do you have a good environment in your lab (in the university) to help students prepare research papers or presentations?)
3-D-E	When students are explaining their research in the academic society, do you give them sufficient guidance? (Do you have a good environment in your lab (in the university) to help students explain their research results?)
3-I-E	When discussing your students' research results in an academic society, in order to let other members understand their research, do you give them sufficient guidance? (Do you have a good environment in your lab (in the university) to help students discuss their research in an academic society?)
3-S-E	When students are selecting modified ideas about their research results from the academic society, do you give them sufficient guidance? (Do you have a good environment in your lab (in the university) to help students elaborate their research based on the suggestions from the academic society?)

Chapter 5
Knowledge Retention in Smaller Firms

Susanne Durst
University of Skövde, Sweden

Guido Bruns
Incipere, Germany

Ingi Runar Edvardsson
University of Iceland, Iceland

ABSTRACT

In environments that are characterized by skill shortage, knowledge retention should be a key activity. Our understanding of this knowledge practice, however, is underdeveloped. The purpose of this chapter is to contribute to the further development of this field. The authors are interested in studying how smaller firms approach the retention of critical knowledge. Semi-structured interviews with organization members of five smaller Austrian firms operating in the building and construction industry were conducted. The findings presented in this chapter advance the limited understanding of knowledge retention in general and regarding smaller businesses. They also contribute to the further development of the study of knowledge management in the building and construction sector. Based on the study's findings, suitable measures to better manage the process of knowledge retention in smaller firms are derived. These measures may be useful for smaller firms operating in similar industries as well.

DOI: 10.4018/978-1-5225-5655-8.ch005

INTRODUCTION

In industries that rely to a large extent on complex, semi-structured projects such as the building and construction industry the existence of different types of knowledge is vital. In such environments, the retention of relevant knowledge is very important for the functioning of organisations and thus their sustainability. Knowledge retention has been defined as "maintaining, not losing, knowledge that exists in the minds of people (tacit, not easily documented) and knowing (experiential action manifesting in behaviour) that is vital to the organization´s overall functioning" (Martins and Meyer, 2012, p. 80). Even though in general organizations acknowledge the benefits of knowledge management for their business operations (McAdam, 2000; Chatzoudes et al., 2015), it appears that the practice of knowledge retention is still rather neglected. The same seems to be true research-wise. Compared to other knowledge management (KM) practices, studies of knowledge retention are rare (March and Stock, 2006; Hislop, 2013); a situation that refers to all organizations, regardless of size. If one addresses the study of knowledge retention in small and medium-sized enterprises (SMEs), however, there is a shortage of research, which refers to both conceptual and empirical papers. For example, Durst and Edvardsson (2012) who conducted a literature review on KM in SMEs identified only one paper on the topic of knowledge retention. Against the prevalence of smaller business to the majority of economies, this situation is unsatisfactory. Therefore, a better understanding of knowledge retention is important both in practice and in theory.

The purpose of this chapter is to develop our understanding of how smaller businesses address the issue of knowledge retention. The chapter deals with the following issues: How is knowledge treated in the company? How is critical knowledge retained? What methods are applied to retain relevant knowledge? How is knowledge shared in the company? The study is conducted in the building and construction sector, which on one side is often dominated by smaller companies. On the other side, it represents a sector in which the study of knowledge management has started only recently (Maqsood and Finegan, 2009).

The remaining chapter is organised as follows. In the next section, the related literature is outlined. Then the method employed is outlined. This is followed by a presentation of the findings. The chapter terminates with the conclusion and implications of the study.

THEORETICAL BACKGROUND

Knowledge Retention

Knowledge retention "deals with cases where expert knowledge workers leave organizations after long periods of time" (Levy, 2011, p. 582). Taking a broader view Martins and Meyer (2012) define knowledge retention as "maintaining, not losing, knowledge that exists in the minds of people (tacit, not easily documented) and knowing (experiential action manifesting in behaviour) that is vital to the organization´s overall functioning" (p. 80). In contrast to knowledge management solutions, knowledge retention takes place in a limited period of time and addresses the challenge of transforming an "expert´s most valuable knowledge" to an organizational asset (Levy, 2011, p. 583). This situation creates a huge challenge as the reduced time available complicates knowledge retention even more (Lahaie, 2005). This also shows the need for setting clear priorities (Jackson, 2010).

Although voluntary and involuntary turnover of staff is a normal part of any business and also knowledge is in a constant state of change e.g. knowledge that has been relevant in the past may become obsolete over time or it has simply been forgotten because of time elapses (Tan et al., 2006), there is always some knowledge that should remain in the organization. Consequently, organizations need to implement systematic approaches to knowledge retention "because knowledge not embodied in the existing product or technology may decay due to the nature of the knowledge itself, the available means of storage, or lack of maintenance or use" (Marsh and Stock, 2006, p. 425). Knowledge that is successfully retained makes a company less dependent on external knowledge, a situation some companies are striving for.

Not all knowledge is critical to organizations and thus should be retained. Moreover, it is rather usual that knowledge that has been relevant in the past may become obsolete over time or it has simply been forgotten because of time elapses (Tan et al., 2006). Therefore, knowledge is in a constant state of change and should continuously be updated. Consequently, critical knowledge is that knowledge which is most at risk of being lost (Frigo, 2006). According to DeLong (2004), lost knowledge can occur at individual, group, or organizational levels, have either anticipated or unanticipated effects, have tangible or intangible impacts and create immediate or delayed costs. Daghfous et al. (2013) identified employee turnover, rushed reallocation of the workforce, rushed handovers, heavy workloads, outsourcing, resistance to learning, and reluctance to share knowledge as the main drivers of knowledge loss.

Previous research has identified various consequences of lost knowledge, which are summarized in Table 1.

When it comes to the issue of knowledge retention and hence knowledge loss, most firms do not have a systematic approach and the approaches used are few, ad

Table 1. Consequences of lost knowledge

Author(s)	Consequences
DeLong (2004)	Lost knowledge undermines organizational strategies in the form of reduced capacity to innovate; ability to pursue growth strategies threatened; reduced efficiency undermines low-cost strategies; losing knowledge can give competitors an advantage, and losing specific knowledge at the wrong time increases vulnerability
Treleaven and Sykes (2005)	Loss of tacit and heuristic knowledge, i.e. loss of staff´s ability to continue social interaction processes and reflectivity
Geisler (2007)	Time and costs related to replacement activities
Boone et al. (2008)	…"ongoing erosion in the knowledge stock" (p. 1232)
Chalkiti and Sigala (2010)	Reduced service quality levels, increased costs and time related to staff recruiting and training
Daghfous et al. (2013)	Cost increase, loss of productivity, reduction in customer satisfaction, increased supply chain risks, reduction in core competencies

hoc and reactive (Daghfous et al., 2013). In addition, despite the apparent meaning of employees (including managers) to organizations, it is still observable that workforce is considered as easily replaceable (Geisler, 2007). Such an approach is rather short-sighted as it prevents the organization from developing its organizational knowledge (Tsoukas and Vladimirou, 2001). Moreover, by considering the demographic challenges ahead this proceeding can be assessed as rather dangerous. Finally, one should not forget the costs involved with people replacement.

To address knowledge retention different strategies and measures have been suggested (Table 2). According to Chalkiti and Sigala (2010), knowledge retention strategies are intended to benefit organizations even though an organization member leaves the organization.

Delong (2004) considers storytelling, mentoring and coaching as effective ways for transferring both implicit and tacit knowledge. As mentoring is concerned Aiman-Smith et al. (2006), however, point out that this approach often fails because of over-structuring and lacking accountability regarding the individuals involved. About storytelling, Whyte and Classen' (2012) study highlighted some challenges with its application: the participants' ability to tell a story in an appropriate manner (some are unable to keep it short, some are poor storytellers) and time (capturing many years of work experience is a time-consuming process which could take months or even years, so storytelling is not a short-term solution).

Notwithstanding the number of different retention approaches and methods, Peet (2012) accentuates that less is known about the efficacy or impact of these methods (in her case methods relating to tacit knowledge identification and sharing). In order to address this gap, Peet tested a specific approach to tacit knowledge sharing (i.e.

Table 2. Overview of knowledge retention strategies

Author(s)	Retention Strategies/Measures
De Long and Davenport (2003)	Interviews/Videotaping, Mentoring, Storytelling, Communities of Practices, Training and Education
Rappaport et al. (2003)	1) Conduct a situation analysis and workforce planning; 2) create new jobs, career paths, and employment arrangements; 3) apply creative recruitment strategies and reconfigure training and development. Also, implement a KM strategy and align the culture. Concrete measures: pairing experienced workers with less experienced workers, creation of mentoring roles, involvement of experienced workers in design of training programs, rehirement of retirees, use of intranet and application of flexible work arrangements
Rothwell (2004)	Critical incident interviews and questionnaires
Liebowitz (2004)	Phased retirement, retiree job bank, emeritus program, part-time retired annuitant, mentoring program, knowledge sharing forums, rehearsal retirement, job sharing, facilitator of an online community of pratice and knowledge capture/retention program
Gammelgaard and Ritter (2005)	Individual memory, databases, social capital, and virtual communities of practice
Lahaie (2005)	Pre-retirement planning, casual employment, term/specified period employment, facilitated monitoring, networking and succession planning
Frigo (2006)	Apply tools for determining the scope of the effect of workforce changes, e.g. retirement profiles, tenure profiles, critical knowledge at risk profiles
March and Stock (2006)	Formal audits, memos, and presentations
Aiman-Smith et al. (2006)	Hire retired individuals from other organizations, phased retirement, global skills resource management, employee rotation, sharing of lessons learned, communities of practices, storytelling, training and educating programs, improved recruiting relationships, and existence of knowledge management department/knowledge manager
Geisler (2007)	Socialization, tutoring, mentoring and continuous reporting; also use of incentives
Blankenship and Brueck (2008)	Use of social network analysis, document repository, incorporation of retirees, mentoring, and organizational learning and training
Arif et al. (2009)	1) socialization, 2) codification, 3) knowledge construction, and 4) knowledge retrieval
Mishra and Bhaskar (2011)	Exit interview and informally by interactions with immediate managers
Levy (2011)	1) scope definition, 2) knowledge transfer, and 3) knowledge integration
Peet (2012)	Knowledge interviewing

generative knowledge interviewing) and explored its short and longer-term impact on knowledge creation and organizational functioning. The study succeeded in showing the usefulness of knowledge interviewing. The participating organization not only was able to reduce the costs relating to talent management and leadership transition but also able to organize its human resource and talent management efforts more efficiently. Another drawback highlighted in the literature is that some retention approaches do not focus on the network of relationships which are critical to getting work done but are limited to individual knowledge independent of the relevance of the former (Parise et al., 2006). Jackson (2010) stresses that static capture of knowledge is insufficient in addressing the dynamic nature of knowledge.

Consequently, one must not underestimate the barriers to knowledge retention. Some more barriers are displayed in Table 3.

However, even though knowledge has been captured and stored it may not be found or interpreted in the right way (Parise et al., 2006). Indeed, very often there is a lack of contextual knowledge preventing the individuals concerned from interpreting extant documents in a proper way (Markus, 2001). Finally, when one considers knowledge retention activities, one should look not only for internal solutions but for external ones as well, as the former might be too costly for an organization (Miller et al., 2011).

Building and Construction Industry and Knowledge Retention

The building and construction industry used to be regarded as somewhat conservative (Maqsood and Finegan, 2009). Growing challenges in the business environment due to fierce (price) competition and/or increased requirements on the part of the customers, however, have caused many firms to reconsider their past practice. These developments generally call for a better management of knowledge within the construction industry (Hari et al., 2005).

As the industry is a project-based one (Maqsood and Finegan, 2009), projects are often unique which in turn require a new set of knowledge and/or skills. For the industry reputation and established relationships are of considerable importance, because every product is adjusted to the individual customer (Rathswohl and Franz,

Table 3. Barriers to knowledge retention

Delong (2004)	Lost knowledge costs are usually hidden; leaders don´t know where the organization is vulnerable; no one owns the problem of lost knowledge; there is no slack for knowledge sharing activities; management must do more than just capture knowledge
Gammelgaard and Ritter (2005)	Fragmentation, overload and de-contextualization
Blankenship and Brueck (2008)	Time, other competing priorities, missing understanding of how to retain knowledge
Fong and Lee (2009)	Missing support by top management as key factor, besides a lack of time
Bessick and Naicker (2013)	Lacking job satisfaction and talent management

2013). Bigliardi et al. (2010) highlight the following factors, which underline the importance of KM in project-based organizations: "the turbulence of the construction industry, where the demand is generally characterized by low predictability; the temporal and economic relevance of each project; the uncertainty that characterize the realization phase, mainly related to the heterogeneity of the technological processes involved and to the site location; the low standardization of the construction product and process; and the management and organizational firm's complexity" (p. 20). Taking advantage of developed knowledge from prior projects can help the firms to improve the execution of following projects (Maqsood and Finegan, 2009). Additionally, given the growing complexity of projects, a number of different actors are involved; therefore, knowledge sharing and knowledge transfer are important (Yun et al., 2011). The involvement of different actors also forms the potential for new knowledge sources and innovation (Du Plessis, 2007). Nowadays, not only the planner, architect, and contractors but also consulting firms, energy technicians, suppliers and building contractors are involved.

The study of KM in the building and construction industry, in general, is still new (Maqsood and Finegan, 2009). Yet, it is expected that it could "enhance individual, group, and organizational learning, improve information circulation, and even support innovation" (Ribeiro and Ferreira, 2010, p. 362). The dynamic structure of the industry complicates an organization´s ability to transfer knowledge between individuals and organization. Likewise, companies have difficulties in implementing formal and systematic approaches to knowledge sharing. Indeed, research suggests that interactive approaches such as networking, brainstorming or recaps by projects are more common (Chen and Mohammed, 2010). Not surprisingly, documentation is neglected (Matt, Rauch and Franzellin, 2012; Ihlau, Duscha and Gödecke, 2013). Ribeiro and Ferreira (2010) who studied ways by which construction projects are prepared showed that the informants involved did not make use of experiences for the preparation and execution of new projects. Lessons learned, errors/mistakes etc. are not documented, and engineers working on construction sites often lack time or motivation to write down detailed accounts of problem solving solutions. Construction knowledge is shared and discussed personally among the different actors involved, yet is not documented for possible future use.

Knowledge retention is of significant importance for the industry, because of the many uncertainties involved. Compared to manufacturing companies, where all production stages take place one after another, standardization in the sector is much more difficult to reach and implement, as the phases which run in parallel are flexible in the course of the construction planning. Therefore, project planning can change quickly, which must be responded to in a flexible manner (i.e., by experience and knowledge generated from previous projects). Consequently, newly generated knowledge should be shared as soon as possible to avoid unnecessary disruptions, financial losses or damage to the company´s reputation (Rathswohl and Franz, 2013). Finally, firms operating in the building and construction industry are said to be people-reliant (Bishop et al., 2008), which represents another reason why these firms should put an emphasis on knowledge retention. In a rare study of knowledge retention in the construction industry, Arif et al. (2012) tested the framework developed by Arif et al. (2009) on three case studies of large construction consultancies in the UAE. The findings demonstrate the importance of motivation, leadership support and the availability of suitable platforms/infrastructure for knowledge retention.

In many countries (e.g., the German speaking countries) the building and construction industry is dominated by small firms, thus their approaches to knowledge retention might be completely different to those found in larger companies. Besides these companies' contribution to the countries and their prosperity, the issue of KM and knowledge retention, in particular, is underdeveloped. Indeed, we know little about their approaches to knowledge retention. The aim of this study was to address this situation.

METHODOLOGY

Given the study's aim, an exploratory (qualitative) research approach appeared appropriate. A qualitative approach allows researchers to get closer to the participants and their way of thinking to scrutinize the entire research problem in depth (Maykut and Morehouse, 1994).

The companies selected for the study included small firms from the province of Vorarlberg in Austria. Semi-structured interviews were conducted with the managing directors and other staff members of the firms; thus, a multi-level analysis approach was pursued. In total, five firms with eight persons were involved in the study. An interview guide supported the interview process. The guide was tested prior to the interviews and as a result, the wording of some questions was amended. The final interview guide focused on the following points: general facts concerning the organization, the understanding of knowledge retention, its application in the firm, as well as the measures used to support knowledge retention. The interviews were conducted in March 2014 and took place in the interviewees' work environment. The interviews lasted anywhere from 30 minutes to 1 hour, were tape recorded and later transcribed. Note-taking during and after the interviews took place as a means to bring forward thinking and to write down seemingly important aspects related to the phenomenon under investigation. Data analysis involved reading the transcripts several times to become familiar with the data. The overall data analysis approach followed the data display and analysis approach as proposed by Miles and Huberman (1994). The authors divide the analysis process into three sub-processes. First, the data are reduced, meaning it is summarized. Second, data display is conducted by organizing the data into charts, diagrams or other visual forms. After all of the interviewees' data were displayed, an overview over the most relevant collected data was present. This helped in comparing the data, find similarities or differences and draw conclusions. The data analysis process involved all researchers, which allowed for multiple perspectives for each interview and for cross-case analysis (Eisenhardt, 1989).

PRESENTATION OF FINDINGS

In this section, the findings of the study will be presented. For the identification of interviewees and companies, the following abbreviations were used: MD = managing director, SM = staff member, and C = company. The themes of interest were grouped under headings corresponding to the research questions.

In the surveyed companies, the turnover rate of personnel is rather low, thus the companies have many long-term employees. A departure of one of these employees inevitably leads to a loss of knowledge. Particularly negative is the consequence for the company if the retiring people have accumulated critical knowledge.

Hence, the interviewees were asked to specify what relevant knowledge means for the company and in which business areas they see the company's relevant knowledge. This information could identify areas of specific interest regarding knowledge retention.

Most interviewees (SM3, MD3, MD4, and MD5) stated that relevant knowledge can be found in all areas of the company:

... if I have the good technicians here and in the production people who can apply it... all that plays together - then this is actually our strength. It actually belongs to the whole company (MD3).

If different types of knowledge are considered, employees see firm-specific knowledge more in the practical, manual work, whereas managers focus more on the knowledge that contributes to business success.

For SM1, for example, firm-specific knowledge is the one that can be used productively, i.e., on construction sites. SM2 remarked that firm-specific knowledge enables the firm to be successfully in the market, i.e., to deliver a good job for the customer. Whereas MD5 stated, "firm-specific knowledge is for me the knowledge which has an impact on the success of the company".

Furthermore, the interest was in obtaining ideas of what makes the building and construction sector special with regard to knowledge management. On that matter,

MD3 said, "Knowledge is publicly available on the sites and thus it can easily be copied". SM2 stated: "... production plants work with automated machines which require no special knowledge, while in the construction sector experts are working on the machines." In the same vein/a similar manner, SM1 declared that "manufacturing firms in other industries have relatively few factors that affect operations. In our industry, there are many different factors, why it is not possible to write down everything. There is mass production, whereas we have individual production or special production". The statements confirm prior research regarding the limited opportunities for standardization compared with other industries (e.g. Bigliardi et al., 2010).

Regarding knowledge retention, the majority of interviewees (MD2, SM2, MD3, SM3, SM4, and MD4) had not explicitly dealt with this issue. SM1 and MD5 stated that they deal partially with knowledge retention. For instance, MD5 said: "Not really. Well, from a sales point of view, I am thinking about knowledge retention."

The question of what the interviewees meant by knowledge retention resulted in two different perspectives. SM4 and MD3 claimed that knowledge retention helps in keeping knowledge within the company and thus not disclosing it to third parties, whereas MD2, MD4, MD5, and SM3 had similar understandings of the term. For example, MD5 described knowledge retention as the "storage and easy retrieval of experiences and knowledge that are relevant to the company's success." MD2 stressed that "it is the knowledge that is stored so that it can be found again. "

As internal factors that may influence knowledge retention, SM1 and SM2 named fluctuation and the retention of old ways of doing things. SM2 specified "an internal thing is for sure keeping the old, which has always been good." External factors, on the other hand, are fast-paced developments in the industry (MD4) or more elaborate and complex documentation, which is required by authorities (SM2). The labor market´s development was named as a major factor by SM1, SM2, and SM3. MD2 and MD5 highlighted the lack of control in the industry. Manufacturing firms in other industries have more control. Consequently, errors are easier to fix since no spatial separations are present. Additionally, in manufacturing plants, there are fixed processes while in the building construction industry it is more complicated to plan operations because every project needs a special solution.

It was further investigated whether knowledge in the surveyed companies is collected and organized in a systematic way. The interviewees SM1, MD5, and MD2 mentioned that a systematic capture is partially given, i.e., knowledge is handled based on a given process. MD3 and SM4 confirmed similar proceedings.

This takes place verbally, of course also by databases, but more often verbally (SM4).

That what we develop in the company is on the computers, then it is filed on CDs and hard disks. Otherwise, there are talks every day or it happens just according to demand (MD3).

The interviewees MD4 and SM2 stated that they would systematically store their knowledge. SM2 referred to databases to which every organization member has access. They were described as important tools which are used every single day.

There is a quality matrix... a design matrix for standard situations must be performed. This is documented and then it is developed constantly, experiences are added and updated. All this is stored in the database and is then constantly updated and discussed in the context of employee training ...especially in the area of production (MD4).

Regarding the responsibility for knowledge retention, the interviewees listed a number of different persons. SM1 and SM2 named the managing directors. The production managers were mentioned by SM2, MD4, and SM4. Both members of company C2 said it is the secretary because she already performs the backup and she has free access to all files. MD5 stated that this task is currently taken by various department heads. In the course of developing a new IT-system that responsibility is delegated to the Head of IT. This person would then be responsible for the complete retention process. The interviewees described the tasks to be conducted like: "documenting, organizing and designing these meetings"(MD4). It also covers database maintenance. MD2 described the knowledge retention task in more detail:

The one who does this must be able to assess what is important and what is not and what is current and what is not. So there must be someone who has the necessary expertise in a particular area. I see that person at the management level. However, this does not mean that there are not very important and worth preserving information at the employee level. (MD4)

No respondent stated that all members of the organization are responsible for knowledge retention. Additionally, the findings suggest that knowledge retention is mainly equated with data storage.

The interviewees were also asked about any consequences of knowledge retention that has not been carried out. Most of them stated that no financial losses were incurred. SM2 mentioned negative experiences because of the death of an employee. A loss of knowledge and know-how was the consequence. The interviewee also claimed that the company did not suffer any financial loss. This was explained with the situation that no jobs and customer relationships were lost. The respondent however also mentioned the massive effort needed to close the knowledge gap that had been created by the former employee.

MD3 mentioned a crash of the entire computer system a few years ago and spoke of the financial impact of 12.000-15.000 EUR, which also resulted in a partial loss of knowledge. The managing director of company C5 described a situation in which a longtime employee had left the company, which required a lot of time and efforts to regenerate this knowledge. MD2 reported a previously unrecognized but positive effect of knowledge loss:

Even if one leaves the company without a malicious intent, such as a retirement, there is, of course, a certain amount of experience and knowledge loss. This does not mean that this is necessarily a bad thing. If an employee is longer there, many rely on him and for them, it is then not necessary to think for themselves and to acquire knowledge. However, after he is gone, the remaining ones have to acquire knowledge on their own. Now, when a new and young employee is assuming a thing - especially today, where a lot of electronic occurs - then it is a different approach to a job compared to an employee who is close to retirement. And these are things that are bad for a company on the one hand if such a person leaves the company; on the other hand, there are new opportunities and challenges. (MD2)

This statement is in line with previous research underlining that turnover per se is not negative but can provide the basis for new ideas and avoid standstill in the company (e.g., Chalkiti & Sigala, 2010).

Knowledge sharing is an essential part of knowledge retention; thus, this issue was addressed as well.

Not surprisingly, sharing of new knowledge happens differently in small companies (C1, C2, and C3) and medium-sized ones (C4 and C5)). The findings showed that the smaller ones share their knowledge mostly verbally. All companies, however, seem to emphasis knowledge sharing in the companies.

We are so small that knowledge more or less shares itself, simply by talking. For entering new knowledge in databases, we are too small. But usually, we read something, talk about it and then pass it on. This is quite appropriate for our company size (MD2).

The larger SMEs use more sophisticated systems. Within the company, C4 written handouts are distributed electronically, for example. "As a rule, after each meeting, written protocols are filed under a given data path. Then everyone can access the log under that path"(MD5).

The interviewees MD2, SM2, MD3, SM3, MD4, and SM4 referred to the lack of time as an obstacle to knowledge storage. SM4 declared "it's just simply much faster to explain something to someone than to write it down somewhere," which is in line with findings from Arif et al. (2012) who also stressed the link between the time needed to document knowledge and knowledge retention activities.

If asked for effective methods for knowledge sharing most respondents indicated that mutual communication by team members' results in the best results. SM1 pointed to the "Lesson Learned" method: "There are so many factors which you can never write down or which are not applicable one-to-one. This is why teamwork on the construction site is the most effective way." SM4 explained the effectiveness of conversations as follows: "Effectiveness is for me if you can explain and give an example at the same time. So, in short, by the speed."

Two statements from the interviewees MD3 and SM3 emphasized the meaning of a people-centered approach in in the given context (Bishop et al., 2008). Moreover, MD3 specified his distrust towards electronic data storage. He declared "emails are too impersonal to me - it's better if I can look someone in the eyes." SM3 stated that "it's just more pleasant to talk to someone, as to save the knowledge somewhere." MD4 stated that data is not stored systematically, instead regular meetings with the staff are held to promote an active exchange. SM2 mentioned the use of smartphones. They are used in addition to the conversations "to serve as a contact point between the sites and the office. This is usually done via a fast data exchange of documents or pictures for documentation."

Company C5 uses a CRM system on tablet computers for their sales activities "to make the necessary data available with the least effort. And this applies to all employees concerned" (MD5). The accessibility of stored data for the organization members is also considered. "We also have older employees in sales and therefore it must be handled easily, without much training. It must also be intuitive. The simpler it is, the more often it will also be applied" (MD5). Companies C2 and C5 make use of another method of knowledge sharing. They employ retired former employees on a part time basis to have their knowledge available at a certain time (Rappaport et al., 2003). These employees visit the companies at specific dates in order to answer questions of the organization members.

The interviewees reported a positive experience with measures for knowledge sharing and knowledge retention. Negative experience was not disclosed. Among the advantages improved speed, the use of old and proven techniques, systematic order and reduced effort for knowledge extraction as a consequence of staff were named. The findings clarified that the willingness to invest time in knowledge retention activities depends on the "perceived" utility.

Finally, the question of how to handle the timeliness of knowledge was of interest. All interviewees reported that documents are continuously updated or if needed. The interviewees from the company C4 and SM2 indicated that it is particularly important to keep the documents in the production area up to date. On the one hand, because it is the core of the company, on the other hand for holding the products on the cutting edge of technology. Another reason was: "because here are the biggest sources of errors, and therefore the most money can be lost. Here are the biggest friction losses if the communication does not work"(SM2*)*. Furthermore, he said:

If you use older databases, you must always ask yourself: 'Is that even still relevant? So, you should only store those things which are known for not changing quickly. It's dangerous if you trust too much on your own data files, then you are missing some changes. And changes happening fast in today's world. Important documents should always be up to date (MD2).

This statement points at the link between knowledge retention activities and the economic incentives behind.

CONCLUSION

This chapter provides insights into how smaller companies from Austria operating in the building and construction industry handle the practice of knowledge retention. Against the role of knowledge for firms' sustainability on one hand and the lack of knowledge retention studies on the other hand, our understanding of this knowledge practice needs to be developed to make possible improved knowledge management. The present study's intention was to contribute to the knowledge retention literature regarding both small businesses and the building and construction industry.

The findings illustrate that data storage, personal communication and Lesson Learned approaches are the most common measures used for knowledge retention. Through them the companies can capture different types of knowledge (explicit and tacit). In addition, the findings indicate that people-oriented approaches are likely to be used more often for explicit knowledge. This calls for some actions as explicit knowledge is easier to retain; thus, firms should consider the usefulness of IT-oriented approaches too.

About possible negative consequences of missing knowledge retention activities, the findings imply that the participants are not willing to admit or unaware of the indirect costs, i.e., time and resources needed to reduce the knowledge gap, involved in the danger of losing relevant knowledge. Even though the interviewees mentioned that the firms did not suffer from financial costs, they were mainly addressing direct costs. Regarding the issue of responsibility, the firms may consider embedding knowledge retention activities in their day to day operations. As critical knowledge can reside everywhere in the company and so the dangers of knowledge attrition or in the worst-case knowledge loss could be handled in a better way, i.e., responsibility would be shared among all organization members and not restricted to one person.

The findings are in line with prior research demonstrating that the building and construction industry is highly project-based and consequently the activities are more difficult to standardize (e.g., Maqsood and Finegan, 2009; Bigliardi et al., 2010). Thus, the challenge for any KM activities will be to find a proper trade-off between losing and retaining knowledge. This also refers to the situation that projects often do not follow a linear (step by step) process but rather one that is characterized by overlapping stages, which add another challenge to KM activities.

From a theoretical point of view, the findings provide some fresh insights into how smaller firms deal with the practice of knowledge retention. These insights are important as continuous knowledge retention activities are considered as relevant to survival and competitiveness and support KM in organizations. Additionally, the findings provide a better understanding of knowledge retention activities in smaller businesses operating in the building and construction industry.

From a practical point of view, this study stresses the benefits for firms from engaging in activities related to knowledge retention to ensure the firm's well-being and improved KM. These activities are ideally built around a knowledge management strategy that aims at the efficient use and reuse of knowledge in the long term. The existence of collaboration and communication found in the companies surveyed presents a sound basis for any next and more concrete steps. During daily meetings, the managing directors could highlight the benefits of knowledge retention for every single employee and the firm. Moreover, given the fact that SMEs are said to be good at networking (O´Donnell, 2014), the managing directors should also consider the benefits of external knowledge retention, i.e., retaining and maintaining knowledge externally, for the company (Miller et al., 2011). In fact, it could be an interesting option for companies with fewer resources on their own.

NOTE

The authors of this chapter are aware that the presented study has several limitations. Firstly, the results were gained from a small number of firms; therefore, the reliability of our findings is limited. As outlined, (empirical) studies on the topic are rather rare, that is why this research is explorative in nature. Nevertheless, future studies should focus on the incorporation of a larger number of firms. Secondly, researchers should also consider alternative research approaches and research techniques to further enhance our understanding of knowledge retention. As the consequences of knowledge loss/attrition are only known in the long-term, there is a clear need for longitudinal studies.

REFERENCES

Arif, M., Egbu, C., Alom, O., & Khalfan, M. M. A. (2009). Measuring knowledge retention: A case study of a construction consultancy in the UAE. *Engineering, Construction, and Architectural Management*, *16*(1), 92–108. doi:10.1108/09699980910927912

Arif, M., Khalfan, M., Barnard, J. H., & Heller, N. A. (2012). Assessing Knowledge Retention in Construction Organisations: Cases from the UAE. *The Australasian Journal of Construction Economics and Building*, *12*(2), 55–71. doi:10.5130/ajceb.v12i2.2461

Bessick, J., & Naicker, V. (2013). Barriers to tacit knowledge retention: An understanding of the perceptions of the knowledge management of people inside and outside the organization. SA Journal of Information Management, 15(2).

Bigliardi, B., Dormio, A. I., & Galati, F. (2010). ICTs and knowledge management: An Italian case study of a construction company. *Measuring Business Excellence*, *14*(3), 16–29. doi:10.1108/13683041011074182

Blankenship, L., & Brueck, T. (2008). Planning for knowledge retention now saves valuable organizational resources later. *Journal - American Water Works Association*, *100*(8), 57–61.

Boone, T., Ganeshan, R., & Hicks, R. L. (2008). Learning and Knowledge Depreciation in Professional Services. *Management Science*, *54*(7), 1231–1236. doi:10.1287/mnsc.1070.0849

Chalkiti, K., & Sigala, M. (2010). Staff turnover in the Greek tourism industry: A comparison between insular and peninsular regions. *International Journal of Contemporary Hospitality Management*, *22*(3), 335–359. doi:10.1108/09596111011035945

Chatzoudes, D., Chatzoglou, P., & Vraimaki, E. (2015). The central role of knowledge management in business operations: Developing a new conceptual framework. *Business Process Management Journal*, *21*(5), 1117–1139. doi:10.1108/BPMJ-10-2014-0099

Chen, L., & Mohamed, S. (2010). The strategic importance of tacit knowledge management activities in construction. *Construction Innovation*, *10*(2), 138–163. doi:10.1108/14714171011037165

Delong, D. W. (2004). *Lost Knowledge - Confronting the Threat of an Aging Workforce*. New York: Oxford University Press, Inc. doi:10.1093/acprof:oso/9780195170979.001.0001

Delong, D. W., & Davenport, T. (2003). Better practices for retaining organizational knowledge: Lessons from the leading edge. *Employment Relations Today*, *30*(3), 51–63. doi:10.1002/ert.10098

Du Plessis, M. (2007). The role of knowledge management in innovation. *Journal of Knowledge Management*, *11*(4), 20–29. doi:10.1108/13673270710762684

Durst, S., & Edvardsson, I. R. (2012). Knowledge Management in SMEs: A Literature Review. *Journal of Knowledge Management*, *16*(6), 637–649. doi:10.1108/13673271211246194

Eisenhardt, K. M. (1989). Building Theories from Case Study Research. *Academy of Management Review*, *14*(4), 532–550.

Fong, P. S. W., & Lee, H. F. (2009). Acquisition, reuse and sharing of knowledge in property management firms. *Facilities*, *27*(7), 291–314. doi:10.1108/02632770910956148

Frigo, M. (2006). Knowledge retention: A guide for utilities. *Journal - American Water Works Association*, *98*(9), 81–84.

Gammelgaard, J., & Ritter, T. (2005). The knowledge retrieval matrix: Codification and personification as separate strategies. *Journal of Knowledge Management*, *9*(4), 133–143. doi:10.1108/13673270510610387

Geisler, E. (2007). The metrics of knowledge: Mechanisms for preserving the value of managerial knowledge. *Business Horizons*, *50*(6), 467–477. doi:10.1016/j.bushor.2007.07.001

Hislop, D. (2013). Knowledge Management in Organizations: A Critical Introduction (3rd ed.). Oxford University Press.

Ihlau, S., Duscha, H., & Gödecke, S. (2013). *Besonderheiten bei der Bewertung von KMU: Planungsplausibilisierung, Steuern, Kapitalisierung*. Wiesbaden: Springer Fachmedien. doi:10.1007/978-3-658-00945-8

Jackson, P. (2010). Capturing, structuring and maintaining knowledge: A social software approach. *Industrial Management & Data Systems*, *110*(6), 908–929. doi:10.1108/02635571011055117

Lahaie, D. (2005). The impact of corporate memory loss. What happens when a senior executive leaves? *Leadership in Health Services, 18*(3), xxxv–xlvii. doi:10.1108/13660750510611198 PMID:16167654

Levy, M. (2011). Knowledge retention: Minimizing organizational business loss. *Journal of Knowledge Management, 15*(4), 582–600. doi:10.1108/13673271111151974

Liebowitz, J. (2004). Bridging the knowledge and skills gap: Tapping federal retirees. *Public Personnel Management, 33*(4), 421–448. doi:10.1177/009102600403300406

Maqsood, T., & Finegan, A. F. (2009). A knowledge management approach to innovation and learning in the construction industry. *International Journal of Managing Projects in Business, 2*(2), 297–307. doi:10.1108/17538370910949310

Markus, M. L. (2001). Toward a theory of knowledge reuse: Types of knowledge reuse situations and factors in reuse success. *Journal of Management Information Systems, 18*(1), 57–93. doi:10.1080/07421222.2001.11045671

Marsh, S. J., & Stock, G. N. (2006). Creating Dynamic Capability: The Role of Intertemporal Integration, Knowledge Retention, and Interpretation. *Journal of Product Innovation Management, 23*(5), 422–436. doi:10.1111/j.1540-5885.2006.00214.x

Martins, E. C., & Meyer, H. W. J. (2012). Organizational and behavioral factors that influence knowledge retention. *Journal of Knowledge Management, 16*(1), 77–96. doi:10.1108/13673271211198954

Matt, D. T., Rauch, E., & Franzellin, V. (2012). Wissensarbeit in Kleinunternehmen am Beispiel des Baugewerbes. *Industrial Management (Des Plaines), 12*(28), 21–24.

Maykut, P., & Morehouse, P. S. (1994). *Beginning Qualitative Research. A Philosophic and Practical Guide*. London: Falmer Press.

McAdam, R. (2000). Knowledge management as a catalyst for innovation within organizations: A qualitative study. *Knowledge and Process Management, 7*(4), 233–241. doi:10.1002/1099-1441(200010/12)7:4<233::AID-KPM94>3.0.CO;2-F

Miles, M. B., & Huberman, A. M. (1994). *An Expanded Sourcebook. Qualitative Data Analysis* (2nd ed.). Thousand Oaks, CA: Sage.

Miller, K., McAdam, R., Moffett, S., & Brennan, M. (2011). An exploratory study of retaining and maintaining knowledge in university technology transfer processes. *International Journal of Entrepreneurial Behaviour & Research, 17*(6), 663–684. doi:10.1108/13552551111174729

Mishra, B., & Bhaskar, A. U. (2011). Knowledge management process in two learning organisations. *Journal of Knowledge Management, 15*(2), 344–359. doi:10.1108/13673271111119736

O'Donnell, A. (2014). The Contribution of Networking to Small Firm Marketing. *Journal of Small Business Management, 52*(1), 164–187. doi:10.1111/jsbm.12038

Parise, S., Cross, R., & Davenport, T. H. (2006). Strategies for Preventing a Knowledge-Loss Crisis. *MIT Sloan Management Review, 47*(4), 31–38.

Peet, M. (2012). Leadership transitions, tacit knowledge sharing and organizational generativity. *Journal of Knowledge Management, 16*(1), 45–60. doi:10.1108/13673271211198936

Rappaport, A., Bancroft, E., & Okum, L. (2003, Winter). The Aging Workforce Raises New Talent Management Issues for Employers. *Journal of Organizational Excellence*, 55-66.

Rathswohl, S., & Franz, V. (2013). Auf Wissen bauen: Besonderheiten im Baugewerbe. *Wissensmanagement - Das Magazin für Führungskräfte, 1*, 12-13.

Ribeiro, F. L., & Ferreira, V. L. T. (2010). Using knowledge to improve preparation of construction projects. *Business Process Management Journal, 16*(3), 361–376. doi:10.1108/14637151011049403

Rothwell, W. J. (2004, February). Capturing the Lessons of Experience: Knowledge Transfer: 12 Strategies for Succession Management. *IPMA-HR News*, 10-11.

Tan, H. C., Carrillo, P., Anumba, C., Kamara, J. M., Bouchlaghem, D., & Udeaja, C. (2006). Live capture and reuse of project knowledge in construction organisations. *Knowledge Management Research and Practice, 4*(2), 149–161. doi:10.1057/palgrave.kmrp.8500097

Treleaven, L., & Sykes, C. (2005). Loss of organizational knowledge. From supporting clients to serving head office. *Journal of Organizational Change Management, 18*(4), 353–368. doi:10.1108/09534810510607056

Tsoukas, H., & Vladimirou, E. (2001). What is organizational knowledge. *Journal of Management Studies, 38*(7), 973–993. doi:10.1111/1467-6486.00268

Whyte, G., & Classen, S. (2012). Using storytelling to elicit tacit knowledge from SMEs. *Journal of Knowledge Management, 16*(6), 950–962. doi:10.1108/13673271211276218

Yun, G., Shin, D., Kim, H., & Lee, S. (2011). Knowledge-mapping model for construction project organizations. *Journal of Knowledge Management*, *15*(3), 528–548. doi:10.1108/13673271111137475

Chapter 6
Knowledge-Scientific Evaluation of Social Service Systems

Yoshiteru Nakamori
Japan Advanced Institute of Science and Technology, Japan

Fei Meng
Japan Advanced Institute of Science and Technology, Japan

Van-Nam Huynh
Japan Advanced Institute of Science and Technology, Japan

ABSTRACT

This chapter proposes a new evaluation method for social service systems in order to support participants to create new knowledge and value. The main proposal is a rating scale method that can evaluate the current level (as-is) and the future desire (to-be) of participants, and with which we can analyze achievements of individuals. By using a concrete example of an education program, which is a worth living discovery seminar for retired men, the chapter reports the results of hypothetical tests on the relationships between items of evaluation and an interesting finding related to the reason of differences in attitude of participants.

INTRODUCTION

The term "service management" was proposed firstly in computer science, and now it is used in many fields such as service science (e.g., Maglio et al., 2010), service marketing (e.g., Lusch & Vargo, 2006; Lovelok & Wirtz, 2010), and service

DOI: 10.4018/978-1-5225-5655-8.ch006

innovation (e.g., Kosaka & Shirahada, 2013). The common idea of service systems is that "both service providers and service recipients co-create values cooperatively".

When we talk about a service system, we always have the image of services that are primarily associated with the products in the company. In contrast, this paper focuses on the services which have the purposes of activating people in the community. We call them "social service systems". Even if economic effect is not expected, a social service system helps the involved people act lively with a definite aim in life.

Such systems give a variety of information to the participants, who, however, have to convert the information into their own knowledge, and create values together with the service providers. Thus, the perspective of value co-creation has become a common concept in service science in recent years.

One problem which is closely related to the aging population is the decline of the local communities. Japan's population is concentrated in some of the metropolitan areas. Especially that rural young people flow out to urban areas is a major problem to lead to rural decline. The government has implemented a number of policies along with the slogans of "regional revitalization" or "regional creation".

These policies, however, have not raised the remarkable achievements. Therefore, it seems to be a good idea to use the power of retired men for regional activation. To that end, social service systems that can help retired men create concrete ideas to activate local communities as well as themselves are necessary. One of such efforts is the "*worth living discovery seminar for retired men*" which will be introduced in this paper. The seminar was conducted for three years with the support of the Ministry of Health, Labor and Welfare of Japan. It accepted more than 50 people as students (in the following we will used the term "participants" instead of students) (Fujimori, 2012, 2014).

This paper regards this seminar as a social service system, and proposes an evaluation framework for such a system. Since its central role is to create ideas, the evaluation framework to be proposed is based on a theory of knowledge construction systems (Nakamori et al., 2011). As further new attempts, the paper will introduce the viewpoints of how the willingness of the participants has changed, and what kinds of value have been created as a service system.

The rest of this paper is organized as follows. The next section introduces a "worth living discovery seminar for retired men", including new concepts to implement action plans and traditional evaluation methods for such activities. In Section 3, we review existing evaluation techniques for education systems, for quality life, and for service systems in order to show the need for a new evaluation technique. Section 4 proposes an evaluation framework based on the theory of knowledge construction systems (Nakamori et al, 2011). Section 5 explains the questionnaire survey using the proposed evaluation framework, with some hypotheses which will be examined by the survey. Then, Section 6 will show the statistical analysis of the survey data for the verification of hypotheses. Finally some conclusions will be given in Section 7.

A SOCIAL SERVICE SYSTEM

This section introduces a social service system which is an education program through which old men can find their reason for living after the retirement. Since Japanese society is heading rapidly to an aging society, we have to realize a society in which the elderly can have a life worth living.

The tendency of declining birthrate and aging population is a big problem in many countries of the world. Especially its progress is remarkable in Japan. According to the statistics of the Cabinet Office, Japan's population of October 1, 2012 is 127.52 million people. And the population of people who are over 65 years old is 30.79 million, which is equivalent to 24.1% of the total population (Cabinet Office, Government of Japan, 2015).

According to "Japan's Future Estimated Population" published by the National Institute of Population and Social Security Research in January 2012, the total population of Japan will decrease continuously and become less than 90 million in 2060 (See Figure 1). Among them 40% of the population is 65 years old or older and 25% of the population is 75 years old or older (National Institute of Population and Social Security Research, 2012) (See Figure 2). Thus, Japan is about to enter an aging society that no country in the world has ever experienced.

The Japanese government began to implement a variety of initiatives from around 1990. Actually, the government has launched measures against the declining

Figure 1. Long-term trends of Japan's population (adapted by authors; National Institute of Population and Social Security Research, 2012)

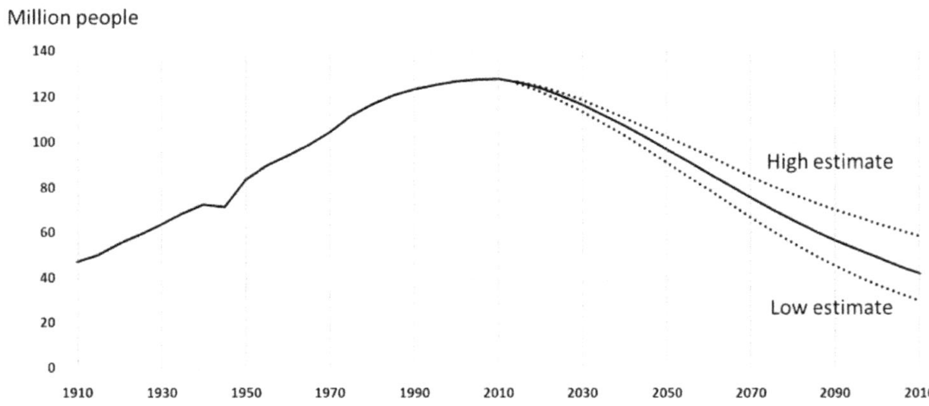

Figure 2. Medium-term trend of Japan's population structure (adapted by authors; National Institute of Population and Social Security Research, 2012)

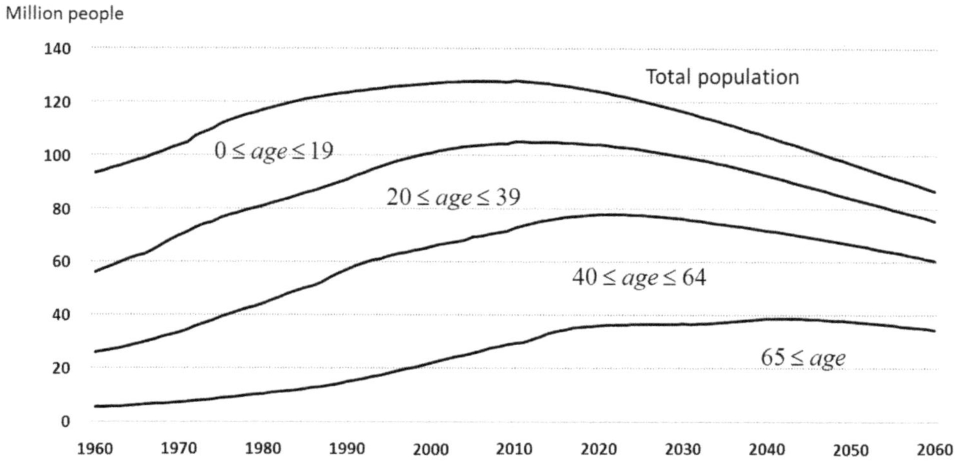

birthrate one after another. However, these measures do not seem to have significant effects up to the present. The government has also been working to promote "social participation of the elderly" for aging society measures. Nevertheless, the suicide rate among elderly men in Japan is the top level in the world (OECD, 2013).

Under the circumstances of such Japanese society, the first major challenge is to help every retired man find a new purpose of life and thereby contribute to the society. This is an extremely important initiative because it contributes to not only the healthy and bright second life of elderly people but also the burden reduction of young people.

One of the important activities is "the worth living discovery seminar for retirement men" (Fujimori, 2012, 2014). This paper regards this as one of the social activation systems, which is a kind of service system in a wide sense. Although the concept of service system is not new, it has attracted attention mainly in the service economy in recent years (see for instance, Demirkan et al., 2011; Taylor & Tofts, 2012).

Fujimori (2012) defined the decision-making of elders at each stage of their lives until nursing care is needed as "*Care-Will*", which refers to the "Will" of elders in the process of their lives. It is different from "Living Will", which is medical care with the intention of life-prolonging. Fujimori (2012) further defined the series of processes leading to the "Care-Will" as the "*Care-Will Model*". It is a process model that retirees make new relationships with others and thereby participate in communities. To realize this, the model requires retirees to discover ways of overcoming difficulties.

Fujimori (2014) planned a series of lectures to support the creation of "Care-Will" of the male retirees, and carried out the education to more than 50 participants during three years (April 2011-March 2014) with the research fund from the Ministry of Health, Labor and Welfare, Japan. This is actually a service system in which the value must be co-created by the collaboration between experts and participants. The curriculum of the "*Care-Will Course*" was designed by referring to an educationist Havighurst (1969), a psycho analyst Erikson (1986), and a psychologist Bandura (1997).

The course was about 20 hours in total using four Saturdays each year. On the first day and the second day, the course provided information about rich living to the participants. The lectures included the following subjects:

- Daily life after retirement;
- How to stare at mind and body;
- Towards positive aging;
- Home management;
- Concertos of men and women;
- Family and nursing care;
- Community and health;
- Life course and social connectedness;
- Richness and health of modern society; and
- Social systems of the elderly support.

On the third day, the participants were asked to design their own second life carriers ("*Care-Will Planning*") (See Figure 3), and on the last day, the participants were asked to present their plans and mutually evaluate their plans. Because the major concern of this program is to create new ideas by the participants in order to make plans of their second-lives, this program can be described by a knowledge creation spiral as shown in Figure 4, where the interaction between tacit and explicit knowledge and the intensive discussion among participants are emphasized.

The objectives of this education program are firstly the acquisition of "*self-efficacy*" that supports personal intrinsic motivation, and secondly the acquisition of the personal existential "*reason for living*" that supports keeping the health. The third aim is the "*improvement of social quality of life with the emphasis on autonomy*" and "*self-realization in the relationship*" based on the "Care-Will Model".

Based on the interviews with the participants, Tatsuse & Fujimori (2014) investigated how the "Care-Will Course" could contribute to the above objectives. They reported that indicators of the "*social self-efficacy*" (Sherer & Maddux, 1982) and the "*social quality of life*" did not increase. However, the indicator of "*general self–efficacy*" was elevated. Furthermore, although the survey on reason for living

Figure 3. The second-life planning sheet developed by Fujimori (2014)

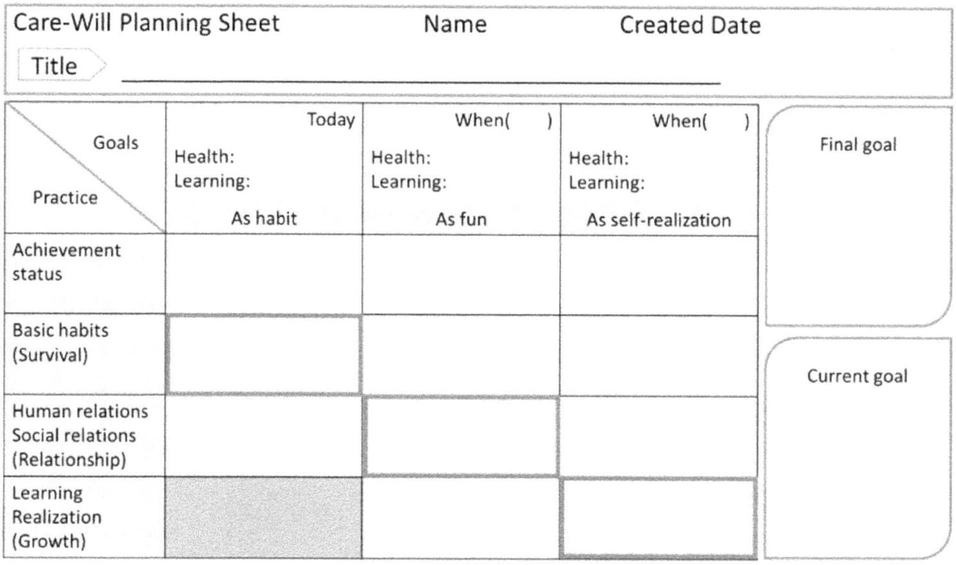

has been carried out with respect to the "*life positive feelings*", "*future active level*" and "*self-existence meaning*", only the score of the last one was elevated.

Based on the analysis, Tatsuse & Fujimori (2014) concluded that the course was effective in the "*self-activating*", but was not sufficient for the social outreach. Thus, it is possible to examine the change of the subjects' emotion by the psychological evaluation indices. However, it is vague to describe what has improved specifically. It is also difficult for subjects to describe themselves explicitly.

Because the major concern of "Care-Will Course" is to create new ideas by the participants in order to make plans of their second-lives, the traditional psychological approach is not enough to evaluate this course. Therefore, based on the knowledge-scientific methodology, we should explore what kinds of information were converted into knowledge, what kinds of consciousness were changed, and what kinds of value were created.

On the basis of participant observation, Meng et al. (2015) reported a knowledge-scientific approach to evaluate this course from the viewpoints of knowledge creation, consciousness reform, and value co-creation. Meng et al. (2015) found that it was quite difficult for the participants to convert given information into their own knowledge, and the willingness of participants returned to the original level six months after the course.

Meng et al. (2015) also treated the value created in the course qualitatively. However, a quantitative analysis is needed for elaboration of the course and aftercare

Figure 4. The idea of the education program
(Developed by the authors after hearing the idea from Fujimori.)

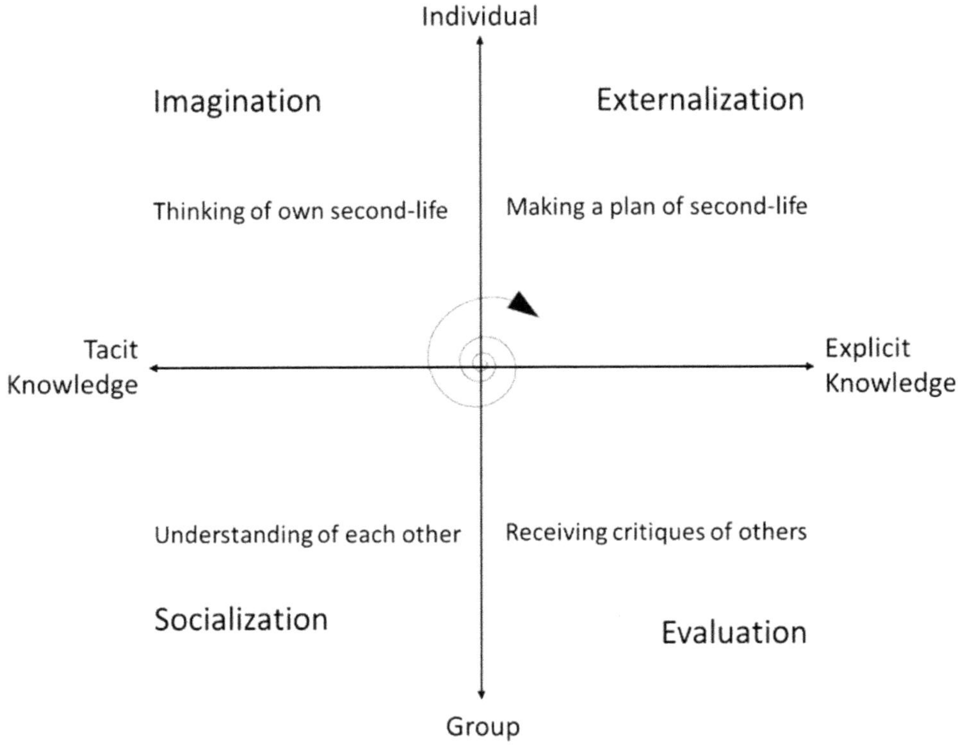

of the participants. Therefore, this paper tries to establish a quantitative evaluation for such service systems. Here, because the major concern of such systems is to create idea, will, and value, this paper proposes an approach from knowledge science and service science. The slogan here is "*from information to knowledge, from knowledge to action, and from action to value*".

The project continued three years from April 2011 to March 2014. Since April 2014, the project leader, Ms. Fujimori, has been organizing the so-called "*Care-Will Study*" which is the seminar where some of the graduates gather together once a month to discuss how they are executing their "Care-Will Plans". The purpose of this paper includes to find the differences of knowledge, willingness, and value of those who are participating in the "Care-Will Study" and those who are not participating in it.

REVIEW OF EVALUATION TECHNIQUES

Because evaluation is a fundamental activity of human beings to obtain insight into existing initiatives and to consider future change, these exist a huge amount of literatures; for instance, see House (1978), Stufflebeam & Webster (1980), Ross, et al. (2004), or Reeve & Paperboy (2007). Evaluation methods may be qualitative or quantitative, and include case studies, survey research, statistical analysis, model building, etc.

This section reviews evaluation techniques from three fields: education systems, quality of life, and service systems, since the subject of this study is a project, which can be regarded as a social service system to create the value of life, or an education system for the elderly to improve the quality of life.

Evaluation Techniques for Education Systems

Evaluation of education systems includes two main purposes: one is evaluation of students and the other is evaluation of courses. Since this study is related to the latter, we should refer to the "knowledge survey" that is a method of evaluating a course based on the knowledge which the students acquired after the instruction.

Knowledge survey is effective in helping students learn, instructors improve their delivery, and department explore new curricula and pedagogies (Feldman, 1998). Examples of questions in knowledge survey are (Fink, 2003):

I know the topic quite well, I know at least 50% of the topic partially, and I know where I can find more information about it. Within 20 minutes, I am confident I can find the complete answer", and "I am not confident I can answer the question.

In this way, the evaluation of education systems usually focuses on how students increased knowledge, but does not focus on *how students created new knowledge*. It is important to develop an evaluation method for the latter purpose. This research proposes a new evaluation method that focuses on whether data or information is converted into knowledge.

Evaluation Techniques for Quality of Life

Quality life means the quality of the contents of life or the quality of social life. It measures how a human feels happy in his/her life. Here happiness is measured from various viewpoints such as physical and mental health, good relationships, rewarding work, comfortable living environment, adequate education, recreational activities, etc.

Standard indicators of the quality of life include not only wealth and employment but also the built environment, physical and mental health, education, recreation and leisure time, and social belonging (Gregory et al., 2009; Nussbaum & Sen, 1993).

The most important keyword used in the project introduced above is "self-efficacy". It is the recognition of possibility to carry out the necessary action in a certain situation (Bandura, 1977, 1997). Or, it is the extent or strength of one's belief in one's own ability to complete tasks and reach goals (Ormrod, 2006). Self-efficacy affects every area of human endeavor, especially health (Luszczynska & Schwarzer, 2005).

There exist many indices to measure quality of life. Among them, "CASP19" and "Ikigai9" are often used recently for elderly worth living survey. The CASP-19 is a quality of life measure, consisting of 19 items, comprising four domains: control, autonomy, pleasure and self-realization, developed initially in a population aged 65-75 years (Hyde et al., 2003; Sim et al., 2011).

"Ikigai9" is a measure of worth living consciousness of the elderly developed in Japan (Imai et al., 2009, 2012). It consists of 9 items of evaluation: "I often feel I am happy", "I think I will learn and want to start something new", "I think I am useful for others and society", "I think there is room in my heart", "I am interesting in many things", "I think I am a need for something or someone", "My life is fulfilling rich", "I want to stretch out my potential", and "I think I have an effect on someone".

The aim of this research is not to measure general quality of life, but to measure increased desires of participants in the course for knowledge and value creation. That is, the existing indices to measure the quality of life are not sufficient for our purpose. Therefore, this paper adopts the method of measuring the actor's action level in three-valued logic at each node of the knowledge construction model (Wierzbicki & Nakamori, 2006; Nakamori, 2011).

Evaluation Techniques for Service Systems

A service system is a configuration of resources (including people, information, and technology) connected to other systems by value propositions, and service science is the study of service systems and of the co-creation of value within complex configurations of resources (Vargo et al., 2008; Maqlio et al., 2010).

In a business to business context, service encounter quality is directly related to customer satisfaction and service quality perceptions, and indirectly to perceived value and loyalty (Jayawardhena, 2010). But generally, evaluation of the quality of service is determined from the difference between the expectations and actual results, and "service" itself is evaluated by the "customer value" (Sakao & Lindahl, 2012).

This paper treats service systems, the value of which cannot be created without the great efforts of the customer. Typical examples are hospitals and schools in which the user's self-effort is required to create value.

A "social service system" consists of service providers and service recipients as elements, and creates knowledge and value as the result of emergence by the interaction between elements of the system. A concrete example treated in this paper is an education providing system for retirees to search the purpose of their second lives.

We will analyze the value creation by using the three-valued logic: personal value, group value, and social value. We will perform questionnaire survey to participants in the course by asking "Do you wish to create your own value, or family/community value, or the value for the society?"

A NEW EVALUATION FRAMEWORK

Similar to the work of Meng et al. (2015), this paper is also based on the theory of knowledge construction systems (Nakamori, Wierzbicki & Zhu, 2011). The main model in this theory is the knowledge construction model (Nakamori, 2000, 2003), which is a procedural, but virtually systemic approach to knowledge creation.

The five ontological elements (nodes or subsystems) of this knowledge construction model are *Intervention* (the will to solve problems), *Intelligence* (existing scientific knowledge), *Involvement* (social motivation), *Imagination* (an important aspect of creativity), and *Integration* (systemic knowledge).

As shown in Figure 5, if you face a problematic situation, you have to identify problems at *Intervention*, which includes requirements and perspectives, as well as ideas on how to collect knowledge, how to create knowledge, how to integrate knowledge, and how to justify knowledge. At *Intelligence* you collect objective knowledge, and at *Involvement* and *Imagination* you collect experience-, wisdom-, or insight-based knowledge, and finally at *Integration* you synthesize knowledge and justify the created knowledge. Here, in addition to solutions to the problem, you usually find new problems by enlightenment.

Nakamori and Zhu (2004) gave an interpretation of the above model from the point of view of the sociology. They regarded the scientific, social, and creative dimensions as the *scientific-actual*, *social-relational*, and *cognitive-mental fronts*, respectively. See Figure 6. These fronts correspond to the structures of our society, which are the systemic, collective contexts and their underlying principles which constrain and enable human action.

As shown in Figure 6, in the scientific-actual front we consider established theories, concepts, influential design, etc. In the social-relational front we consider social norms, values, expectations, obligations, and power relationships, etc. Then, in the cognitive-mental front we consider habit behavior, assumptions, and beliefs, etc. In such an interpretation, the three nodes: *Intelligence*, *Involvement*, and *Imagination* in Figure 5 and Figure 6 are regarded as agencies (abilities) of actors.

Agency is the ability with which actors, who are socio-technologically embedded, reproduce and transform the world. The *Intelligence* agency includes logic, experience, technical skill, rationality, etc. The *Involvement* agency includes interest, emotion, intent, faith, trust, empathy, social capital, social skills, political skills, etc. Then, the *Imagination* agency includes intuition, innocence, the skill of enlightenment, etc.

The remaining two nodes in Figure 5, *Intervention* and *Integration*, are a pair that cannot be separated. A famous Chinese philosopher Wang Yangming (1472-1529) said that knowledge and action are but one, implying that knowledge is not knowledge if you do not use it. Here we interpret this from a constructivist viewpoint. That is, if you want to find good knowledge, you have to set up a good problem; but if you wish to set up a good problem, you have to have some knowledge already.

There exist several applications of this methodology in the areas of education system (Nakamori, 2006; Ma et al., 2007), technology development (Yamashita et al., 2009), regional vitalization (Nakamori, 2013), etc. These applications have gained high praise from participants, but the methodology has not been verified by

Figure 5. The knowledge construction model (Nakamori, 2000, 2003)

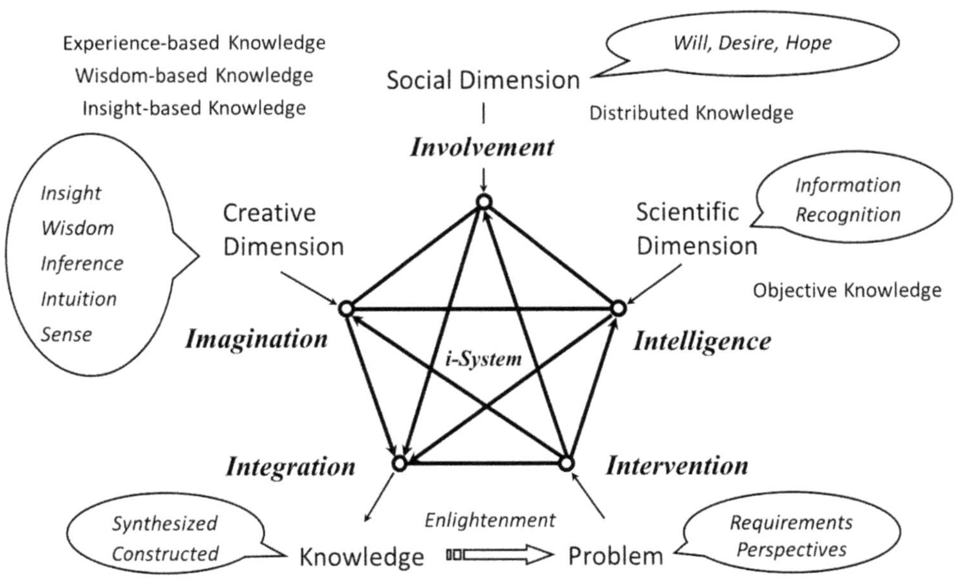

Figure 6. A sociological interpretation of the i-System

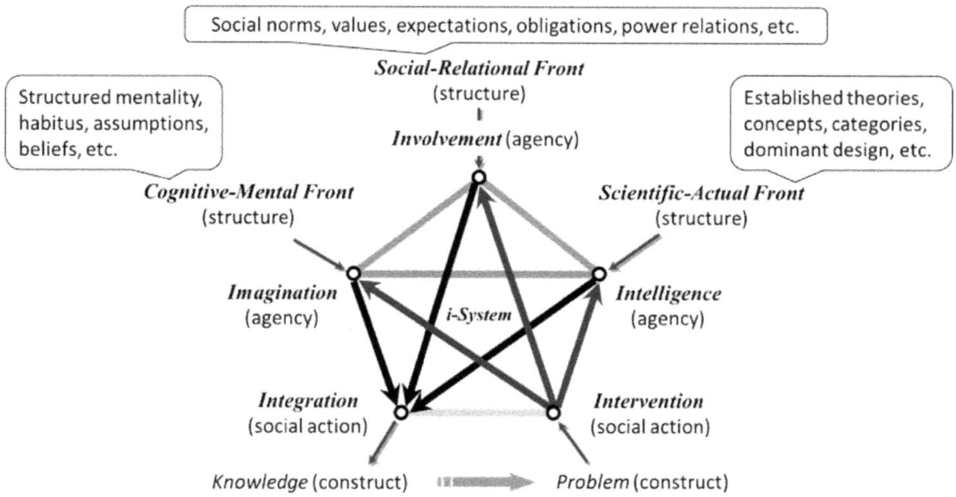

the scientific meaning. This is because the knowledge construction model is not an explanatory model but rather a normative model. The latter can only be justified through applying it to many cases. In this sense, this model is still in the process of justification.

In respective nodes of the knowledge construction model (*Intervention*, *Intelligence*, *Involvement*, *Imagination*, *Integration*), we consider three aspects of evaluation (*knowledge creation*, *willingness creation*, *value creation*) as shown in Table 1.

We consider 5 questions for knowledge, willingness, and value creation, respectively.

- In the column of "*Knowledge Creation*", we ask the subject how he is dealing with collected data, summarized information, and created knowledge.
- In the column of "*Willingness Creation*", we consider the three levels of activity at respective nodes (Nakamori, 2011):
  - Intervention: Drive → Determination → Dedication
  - Intelligence: Emotional → Intuitive → Rational
  - Involvement: Individual → Group → Society
  - Imagination: Routine → Diversity → Fantasy
  - Integration: Specialized → Interdisciplinary → Intercultural
- In the column of "*Value Creation*", we ask the subject whether the created value is for himself, for his family/friends, or for the society/local community.

Table 1. Evaluation structure for social service systems

	Knowledge Creation	Willingness Creation	Value Creation
Intervention	(A-1) Mastery of plan creation methods (Data → Information → Knowledge)	(B-1) Strength of enthusiasm (Drive → Determination → Dedication)	(C-1) Value of making a plan (Individual < Group < Society)
Intelligence	(A-2) Collection and understanding of objective information (Data → Information → Knowledge)	(B-2) Accuracy of information (Emotional → Intuitive → Rational)	(C-2) Value of the collected information (Individual < Group < Society)
Involvement	(A-3) Information from the family and friends (Data → Information → Knowledge)	(B-3) Spread of collaboration (Individual → Group → Society)	(C-3) Value of the constructed relationships (Individual < Group < Society)
Imagination	(A-4) Creation of personal ideas (Data → Information → Knowledge)	(B-4) Expansion of imagination (Routine → Diversity → Fantasy)	(C-4) Value of the created ideas (Individual < Group < Society)
Integration	(A-5) Plan creation and evaluation (Data → Information → Knowledge)	(B-5) Complexity of integration (Specialized → Interdisciplinary → Intercultural)	(C-5) Value of the plan execution (Individual < Group < Society)

Here, let us introduce the most important proposal in this paper, which is an as-is/to-be rating scale method. Three examples are shown in Figure 7, Figure 8 and Figure 9, in which the subject is asked for an answer in five-level evaluation of current situation as well as goal in the future. The five-level evaluation allows a vague answer between data and information, information and knowledge. All of the questionnaires: (A-1) to (A-5), (B-1) to (B-5), and (C-1) to (C-5) will be presented in the appendix.

Let us denote the scores of current status [c] of a subject (participant) [s] by

(A-k)[c, s], (B-k)[c, s], (C-k)[c, s],

and the scores of goal [g] of a subject [s] in the future by

(A-k)[g, s], (B-k)[g, s], (C-k)[g, s].

Recall that the symbol (A), (B) and (C) mean "knowledge creation", "willingness creation", and "value creation", respectively. The symbol (k) indicates one of the nodes from "*Intervention*" to "*Integration*" in the knowledge construction model in Figure 5. The symbols [c] and [g] indicate the "current status" and "goal", respectively.

The above six functions from (A-k)[c, s] to (C-k)[g, s] take values 1, 2, 3, 4, or 5. A basic assumption in this paper is that *these figures are obtained from the interval scale*. To this end, we have to strongly suggest the subjects that 1 to 5 in Figure 7, Figure 8, and Figure 9 are equally spaced.

Achievement levels can be defined by

Figure 7. A questionnaire on knowledge creation using the as-is/to-be rating scale

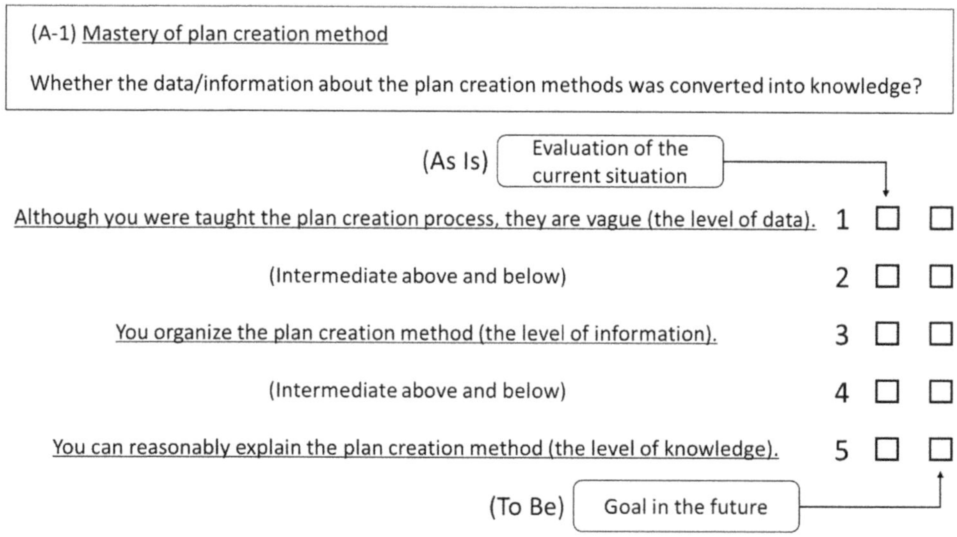

Figure 8. A questionnaire on willingness creation using the as-is/to-be rating scale

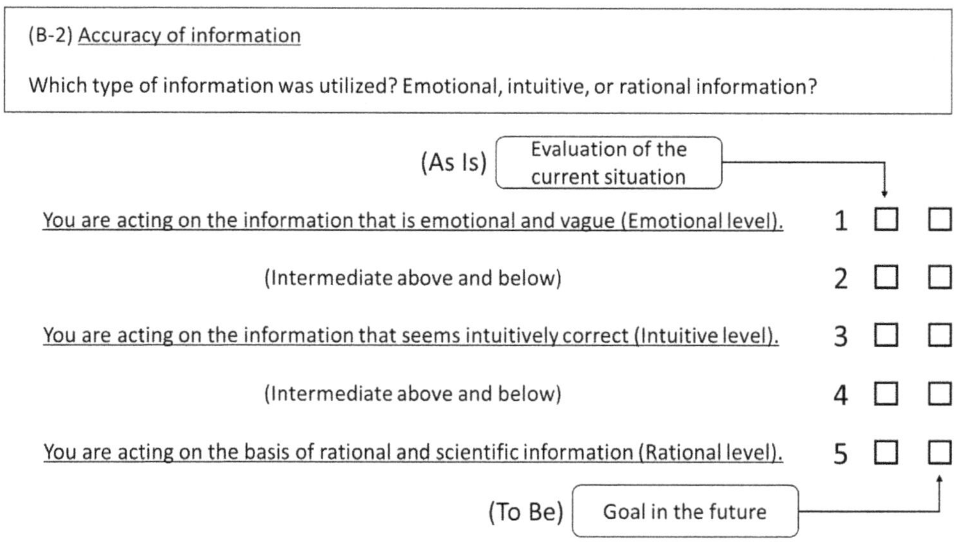

(A-k)[a, s] = (A-k)[c, s] / (A-k)[g, s]

This function takes the value between 0 and 1. If a subject has scored such as the following

Figure 9. A questionnaire on value creation using the as-is/to-be rating scale

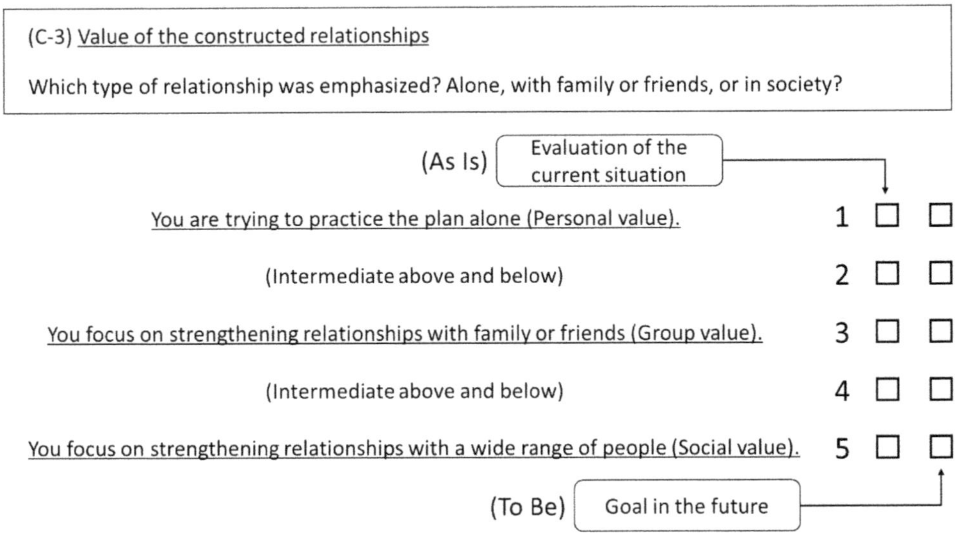

(A-k)[c, s] > (A-k)[g, s],

then let

(A-k)[g, s] = (A-k)[c, s]

from the assumption that this person raised the target.

Fujimori has been organizing the "*Care-Will Study*", a seminar held once a month since April 2014 for aftercare of participants. Let us introduce the notations of two groups:

PS: the group of people who are participating in this seminar. This group consists of about one third of the course participants.
NPS: the group of people who are not participating in this seminar. We do not know precisely whether people in this group are successfully running their plans or not.

When we use the average scores in the analysis below, we use the notations such as

(A-k)[c, PS], (A-k)[g, PS], (A-k)[a, PS];

(A-k)[c, NPS], (A-k)[g, NPS], (A-k)[a, NPS].

We introduce further notations such as

(A)[c, PS]: the average of (A-k)[c, PS] with respect to the nodes (k).

(A)[c, NPS]: the average of (A-k)[c, NPS] with respect to the nodes (k).

We prepared an additional special question, which could be an explained variable in the analysis:

(D) Situation of the plan creation and execution:

1. You have not created any plan.
2. You created a plan, but you are reconsidering it.
3. You created a plan, but have not yet practiced.
4. You created a plan and practiced it, but now you are improving it.
5. You created a plan and now you are steadily practicing it.

Let us denote the current status, the goal, and the achievement level of this extra question scored by the subject [s] by (D)[c, s], and (D)[g, s], and (D)[a, s], respectively. But here we assume:

(D)[g, s] = 5 for all subjects [s].

HYPOTHESES AND QUESTIONNAIRE SURVEY

Prior to sending questionnaires, we set the hypotheses as follows:

H1: Differences between participants and nonparticipants in the "Care-Will Study":
H1a: Participants (PS) are promoting the "Care-Will Plan" more than nonparticipants (NPS).
H1b: Participants (PS) are creating more knowledge than nonparticipants (NPS);
Participants (PS) are creating more willingness than nonparticipants (NPS);
and
Participants (PS) are creating more value than nonparticipants (NPS).
H2: With respect to the causal relationships between the evaluation items (in the current assessment)
H2a: (A-1) has a great influence on (A-2), (A-3) and (A-4); and
(A-2), (A-3) and (A-4) affect (A-5) significantly.
H2b: (B-1) has a great influence on (B-2), (B-3) and (B-4); and

(B-2), (B-3) and (B-4) affect (B-5) significantly.
H2c: (C-1) has a great influence on (C-2), (C-3) and (C-4); and (C-2), (C-3) and (C-4) affect (C-5) significantly.
H3: With respect to the causal relationships between plans and results (in achievement level)
H3a: (A-1) and (A-5) significantly affect (B-1) and (B-5).
H3b: (B-1) and (B-5) significantly affect (C-1) and (C-5).
H3c: (C-1) and (C-5) significantly affect (D).

We carried out a questionnaire survey in July and August 2015 for the participants of the "Care-Will Course". The number of participants during three years is 53 in total. We sent the survey form to all these participants, but we received only 24 answer letters. Among them the number of valid responses is 21, and

$|PS| = 7$, $|NPS| = 14$.

That is, the number of respondents who are the members of the "Care-Will Study" is 7, and who are not the members is 14.

From the survey, we have noticed the following strange match:

(D)[c, PS] = (D)[c, NPS] = 3.714;

(D)[a, PS] = (D)[a, NPS] = 0.743.

Thus, the averages of achievement levels of plan creation and execution are the same between the participants (PS) and nonparticipants (NPS) in the "Care-Will Study", the aftercare seminar. Then, a question arises: why were they divided into participation and nonparticipation in the "Care-Will Study"? Anyway, at this point, the hypothesis H1a (Participants are promoting the "Care-Will Plan" more than nonparticipants) was rejected.

ANALYSIS AND RESULTS

Verification of Hypothesis H1b

We carried out the t-test for verification of the hypothesis H1b. Here we define:

PS = Experimental group (participants in "Care-Will Study");

NPS = Control group (nonparticipants in "Care-Will Study").

Using the t-test, we examined whether there are significant differences between average values in several evaluation items. This paper used R-language (Ihaka and Gentleman, 1996) for statistical analysis. Table 2, 3 and 4 show the results of the t-test. But, before the t-test we ran the equal variance check between every two data sets by the F-test. For the data sets the equal variance cannot be assumed, we wrote (var. equal = F) in the table (Table 4), and have run the appropriate t-test.

Looking at Table 2, 3 and 4, we can summarize the comparison of participants and nonparticipants in "Care-Will Study" as follows:

1. Participants are getting more information than nonparticipants, and might be continuously creating new ideas. However, the goals of participants are higher than those of nonparticipants. Therefore there is no significant difference between two groups in the achievement level.
2. There are no significant differences between two groups in the current status of willingness creation and value creation. However, the level of nonparticipants' goals is lower than that of participants, therefore, the achievement level of nonparticipants is higher than that of participants.

Table 2. Verification by the t-test for knowledge creation

	Average	Statistic	Judgment
Current Status	(A)[c, PS] =3.17 (A)[c, NPS]=2.60	t-value =2.4549 p-value=0.0396	(A)[c, PS] > (A)[c, NPS] Significant difference in the level of 5%
Goal	(A)[g, PS] =3.97 (A)[g, NPS]=3.50	t-value =2.6811 p-value=0.0279	(A)[g, PS] > (A)[g, NPS] Significant difference in the level of 5%
Achievement	(A)[a, PS] =0.808 (A)[a, NPS]=0.752	t-value =1.2572 p-value=0.2441	No significant difference

Table 3. Verification by the t-test for willingness creation

	Average	Statistic	Judgment
Current Status	(B)[c, PS] =2.43 (B)[c, NPS]=2.51	t-value =0.4460 p-value=0.6674	No significant difference
Goal	(B)[g, PS] =3.57 (B)[g, NPS]=3.24	t-value =2.8204 p-value=0.0225	(B)[g, PS] > (B)[g, NPS] Significant difference in the level of 5%
Achievement	(B)[a, PS] =0.687 (B)[a, NPS]=0.788	t-value =2.5052 p-value=0.0366	(B)[a, PS] < (B)[a, NPS] Significant difference in the level of 5%

Table 4. Verification by the t-test for value creation

	Average	Statistic	Judgment
Current Status (var. equal = F)	(C)[c, PS] =2.40 (C)[c, NPS] =2.57	t-value =1.0271 p-value=0.3605	No significant difference
Goal	(C)[g, PS] =3.66 (C)[g, NPS] =3.14	t-value =3.4966 p-value=0.0081	(C)[g, PS] > (C)[g, NPS] Significant difference in the level of 1%
Achievement (var. equal = F)	(C)[a, PS] =0.646 (C)[a, NPS] =0.845	t-value =5.3785 p-value=0.0037	(C)[a, PS] < (C)[a, NPS] Significant difference in the level of 1%

3. Especially, from the viewpoint of value creation as a service system, nonparticipants might consider that they almost reached their targets in the "Care-Will Course" and might judge that they do not need to participate in the aftercare "Care-Will Study".
4. On the other hand, since participants might be those who are not yet satisfied, we should consider the care for them more carefully.

As to the verification of the hypothesis H1b, we can conclude that the participants (PS) are creating more knowledge than nonparticipants (NPS), but there is no difference in willingness creation and value creation between two groups. That is, the hypothesis H1b is partially valid.

We would like to repeat that the goals of nonparticipants in the aftercare seminar are rather low compared with those of participants, therefore, their achievement (satisfaction) levels are high. This is the reason why they are not participating in the aftercare seminar "Care-Will Study". This finding thanks to the introduction of the new evaluation method: the as-is/to-be rating scale method.

Verification of Hypothesis H2

This section explores the relationships between items of evaluation using the data of *current status*. The method used here is the structural equation approach that assumes the effect relationships between measured variables, which is in the class of the covariance structure analysis. Among this approach, we use here the path analysis which tries to explain plural dependent variables by plural independent variables. We use the recursive model in the path analysis, especially, the multivariate regression analysis for determining the correlation between the dependent variables after removing the effect of independent variables.

Figures 10, 11 and 12 are the results of the path analysis (by R language) for knowledge creation, willingness creation, and value creation, using the data of *current status*. In these figures, the coefficient (standardized solution) with an asterisk indicates that the null hypothesis "the coefficient is zero" cannot be rejected in the level of 5%; the coefficient written in bold letter indicates that the null hypothesis can be rejected strongly; and the coefficient written in normal letter indicates that it is in the border to reject the null hypothesis.

From the above path analysis we found some important points that we should improve in the "Care-Will Course" as follows:

Figure 10. The path analysis for knowledge creation

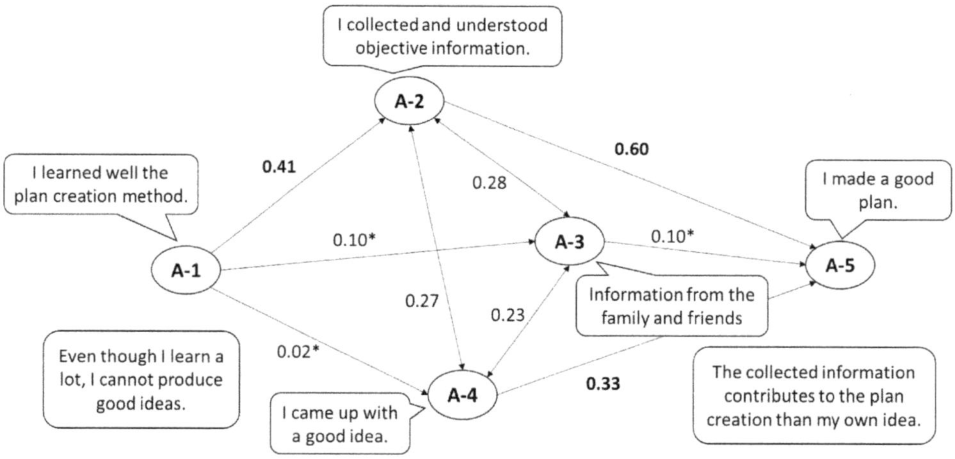

Figure 11. The path analysis for willingness creation

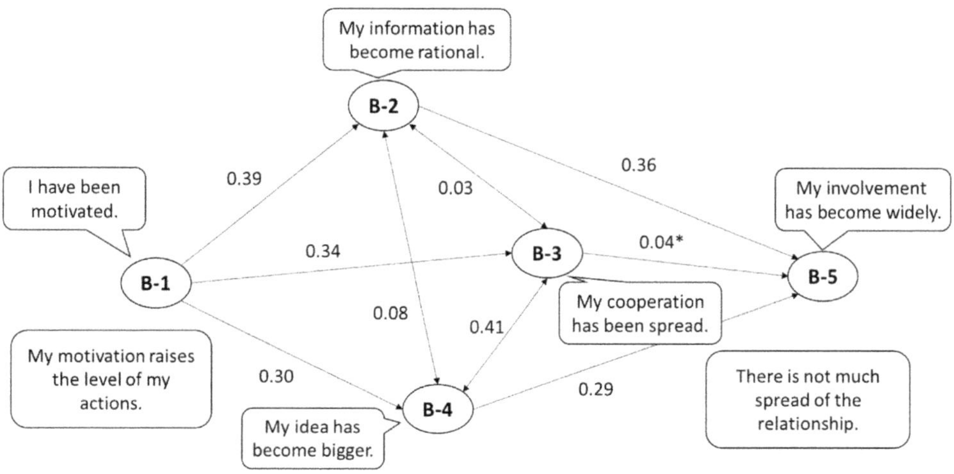

Figure 12. The path analysis for value creation

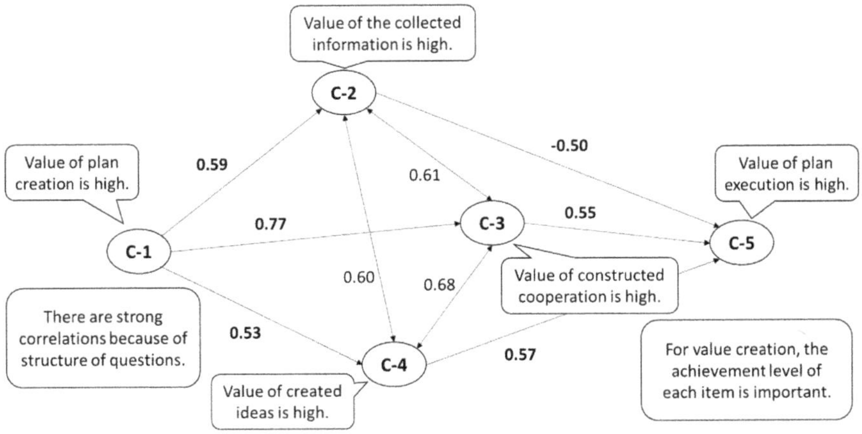

1. For the knowledge creation (Figure 10), a good plan (A-5) can be created by learning the method hard (A-1). The plan creation (A-5) depends on the idea creation (A-4) to some extent. Therefore, we should prepare lectures on knowledge creation by individuals. Since the result (A-5) is highly dependent on the collected information (A-2), the selection of information is very important.
2. In willingness creation (Figure 11), the strength of enthusiasm (B-1) affects weakly other activities of making the plan. We have to run, in parallel, some regional contribution projects and ask the participants to contribute to one of the projects in order to raise the level of complexity of integration (B-5).
3. In value creation (Figure 12), since questions are if the value is worth to you, your group, or the society in all evaluation items, the correlation coefficients between items are rather high. Therefore, the achievement level of each item is important. However, in the current investigation, we could not find the evidence that the participants created big value since the average of (C-5) is about 2.4 in the 5-point scale. This suggests the need of further aftercare.

As to the verification of hypothesis H2, the following relationships were *not* verified:

H2a: (A-1) Mastery of plan creation methods → (A-3) Information from the family and friends
(A-1) Mastery of plan creation methods → (A-4) Creation of personal idea
(A-3) Information from the family and friends →(A-5) Plan creation and evaluation
H2b: (B-3) Spread of collaboration → (B-5) Complexity of integration
H2c: (C-2) Value of the collected information → (C-5) Value of the plan execution

Verification of Hypothesis H3

It is quite important to find relationships between some items in knowledge creation, willingness creation, and value creation in order to elaborate the "Care-Will Course". Here we consider the causal relationships between items (A-1), (A-5), (B-1), (B-5), (C-1), (C-5), and (D), using the *achievement data*. Figure 13 shows the result of the path analysis. Recall that the coefficient (standardized solution) with an asterisk indicates that the null hypothesis "the coefficient is zero" cannot be rejected.

From Figure 13 we can conclude:

1. To learn the plan creation method (A-1) is important to broaden the range of activities (complexity of integration) (B-5).
2. To create the plan actually (A-5) is important to raise the willingness (strength of enthusiasm) (B-1).
3. To raise willingness (B-1) is quite important to create value in making the plan (C-1) and also in executing the plan (C-5).
4. Actual "Care-Will Plan" practice (D) is *not* correlated with value creation (C-1) and (C-5) strongly. This is a quite important finding in this investigation because we found that *there must be personal circumstances which should be overcome to perform the plan actually.*

In the hypothesis H3, we assumed:

H3a: (A-1) and (A-5) significantly affect (B-1) and (B-5).
H3b: (B-1) and (B-5) significantly affect (C-1) and (C-5).

Figure 13. The path analysis between important variables

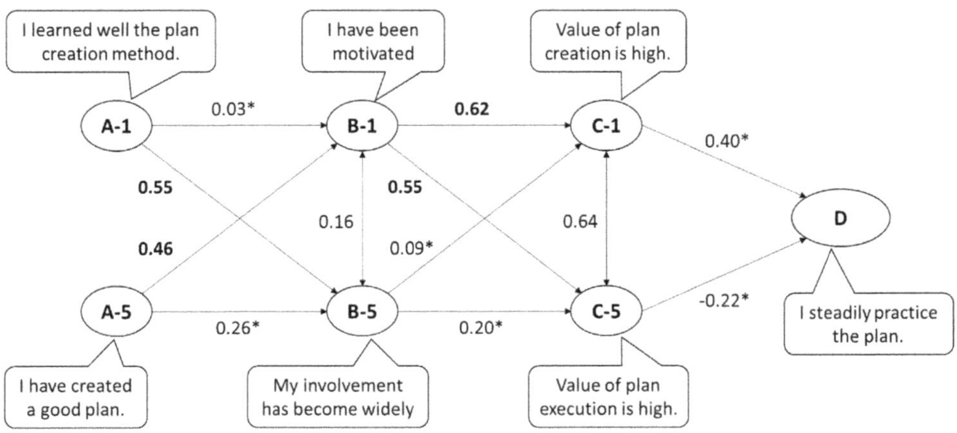

H3c: (C-1) and (C-5) significantly affect (D).

From Figure 13 we can see that H3a and H3b are valid partially, but H3c is *not* valid in this investigation.

CONCLUSION

This paper proposed a new evaluation method for social service systems in order to support participants to create new knowledge and value. The main proposal was a rating scale method that can evaluate the current level (as-is) and the future desire (to-be) of participants, and with which we can analyze achievements of individuals. By using a concrete example of an education program which is a worth living discovery seminar for retired men, this paper reported the results of hypothetical tests on the relationships between items of evaluation, and an interesting finding related to the reason of differences in attitude of participants.

We found that the average values of plan execution were the same between participants and nonparticipants in the aftercare seminar "Care-Will Study". We tried to find the reason why the participants were divided into participants and nonparticipants in the aftercare seminar. This was explained that the goals of nonparticipants were rather low and they might achieve their plans already. Conversely, there is a need to take care of those who have not been satisfied. We also found that there is little correlation between "plan execution" and "plan creation". There might be personal problems to practice the plan actually. There is a need to develop a lecture in order to fill the gap of plan creation and plan execution.

Using the psychological evaluation indices Tatsuse and Fujimori (2014) reported that the course was effective in the self-activating, but was not sufficient for the social outreach. But because the major concern of "Care-Will Course" is to create new ideas by the participants in order to make plans of their second-lives, based on the knowledge-scientific methodology, we explored what kinds of information were converted into knowledge, what kinds of consciousness were changed, and what kinds of value were created. This kind of knowledge-scientific approach does not exist in the past; that is, we do not have existing methods to compare. But, in future, we should try to show the validity of the proposed method by applying it to many similar projects.

This paper introduced the three-valued logic in evaluation of knowledge, willingness, and value creation. From the knowledge-scientific view, it is quite important to distinguish data, information, and knowledge. But because people cannot easily distinguish them clearly, we were struggling in the way of questions. In the evaluation of willingness creation, we followed the three-valued logic in Nakamori

(2011). However, since the respondents seemed a little confused with this logic, we should reconsider the contents of three levels in the future research. In the evaluation of value creation, we used the three levels of individual, group, and society. That is, we asked the respondents that the created value is for yourself, for the family (or the local community), or for the society. Although such distinguish is important to evaluate the created value, we need to consider the levels (or importance) of value in the future research.

REFERENCES

Bandura, A. (1977). Self-efficacy: Toward a unifying theory of behavioral change. *Psychological Review, 84*(2), 191–215. doi:10.1037/0033-295X.84.2.191 PMID:847061

Bandura, A. (1997). *Self-efficiency: The Exercise of Control.* New York: W. H. Freeman Publishers.

Cabinet Office, Government of Japan. (2015). *Measures against the declining birthrate of the past.* Retrieved from http://www8.cao.go.jp/shoushi/shoushika/data/torikumi.html

Demirkan, H., Spohrer, J. C., & Krishna, V. (Eds.). (2011). *The Science of Service Systems.* New York: Springer. doi:10.1007/978-1-4419-8270-4

Erikson, E. H., Erikson, J. M., & Kivnick, H. Q. (1986). *Vital Involvement in Old Age: The Experience of Old Age in Our Time.* Penguin Books Canada Ltd.

Feldman, K. A. (1998). Identifying exemplary teachers and teaching: Evidence from student ratings. In K. A. Feldman & M. B. Paulsen (Eds.), *Teaching and Learning in the College Classroom* (2nd ed.; pp. 391–414). Needham Heights, MA: Simon and Schuster.

Fink, L. D. (2003). *Creating Significant Learning Experiences: An Integrated Approach to Designing College Courses, Revised and Updated.* San Francisco, CA: Jossey-Bass.

Fujimori, J. (2012). Practice and verification of the "Care-Will Model" toward autonomous social support implementation with a focus on male retirees. First Research Report of a Policy Sciences Project, Grant-in-Aid for Scientific Research, Health, Labor and Welfare Ministry, Japan.

Fujimori, J. (2014). Practice and verification of the "Care-Will Model" toward autonomous social support implementation with a focus on male retirees. Final Research Report of a Policy Sciences Project, Grant-in-Aid for Scientific Research, Health, Labor and Welfare Ministry, Japan.

Gregory, D. (Eds.). (2009). *Dictionary of Human Geography* (5th ed.). Oxford, UK: Wiley-Blackwell.

Havighurst, R. J. (1969). *Adjustment to Retirement: A Cross-national Study*. Assen: Van Gorcum.

House, E. R. (1978). Assumptions underlying evaluation models. *Educational Researcher*, *7*(3), 4–12. doi:10.3102/0013189X007003004

Hyde, M., Wiggins, R. D., Higgs, P., & Blane, D. B. (2003). A measure of quality of life in early old age: The theory, development and properties of a needs satisfaction model (CASP-19). *Aging & Mental Health*, *7*(3), 186–194. doi:10.1080/1360786031000101157 PMID:12775399

Ihaka, R., & Gentleman, R. C. (1996). R: a language for data analysis and graphics. *Journal of Computational and Graphical Statistics*, *5*, 299–314.

Imai, T., Osada, H., & Nishimura, Y. (2009). Structure of concepts of purpose in life of retirees of 60 years of age or older: The difference of purpose in life concepts and subjective sense of well-being. [In Japanese]. *Japanese Journal of Gerontology*, *31*(3), 366–377.

Imai, T., Osada, H., & Nishimura, Y. (2012). Verification of the reliability and validity of the measure of purpose in life consciousness (Ikigai-9). [In Japanese]. *Japanese Journal of Public Health*, *59*(7), 433–439. PMID:22991767

Jayawardhena, C. (2010). The impact of service encounter quality in service evaluation: Evidence from a business-to-business context. *Journal of Business and Industrial Marketing*, *25*(5), 338–348. doi:10.1108/08858621011058106

Kosaka, M., & Shirahada, K. (Eds.). (2013). *Progressive Trends in Knowledge and System-Based Science for Service Innovation*. Hershey, PA: IGI Global.

Lovelock, C. H., & Wirtz, J. (2010). *Service Marketing: People, Technology, Strategy* (7th ed.). Upper Saddle River, NJ: Prentice Hall.

Lusch, R. F., & Vargo, S. L. (Eds.). (2006). *The Service Dominant Logic of Marketing: Dialog, Debate, and Directions*. New York: Taylor & Francis.

Luszczynska, A., & Schwarzer, R. (2005). Social cognitive theory. In M. Conner & P. Norman (Eds.), *Predicting Health Behaviour* (2nd ed.). Buckingham, UK: Open University Press.

Ma, T., Wierzbicki, A. P., & Nakamori, Y. (2007). Establish a creative environment for roadmapping in academia - From the perspective of *i*-System methodology. *Journal of Systems Science and Systems Engineering, Springer, 16*(4), 469–488. doi:10.1007/s11518-007-5047-5

Maglio, P. P., Kieliszewski, C. A., & Spohrer, J. C. (Eds.). (2010). *Handbook of Service Science*. New York: Springer. doi:10.1007/978-1-4419-1628-0

Meng, F., Huynh, V. N., Nakamori, Y., Fujimori, J., & Tatsuse, T. (2015). Development and evaluation of a social service system. *The 16th International Conference on Knowledge and Systems Science*, 155-165.

Nakamori, Y. (2000). Knowledge management system toward sustainable society. *International Symposium on Knowledge and Systems Sciences: Challenges to Complexity*, 57-64.

Nakamori, Y. (2003). Systems methodology and mathematical models for knowledge management. *Journal of Systems Science and Systems Engineering*, *12*(1), 49–72. doi:10.1007/s11518-006-0120-z

Nakamori, Y. (2006). Designing, utilizing and evaluating 'technology-creating Ba' in a Japanese scientific research institution. *Systems Research and Behavioral Science*, *23*(1), 3–19. doi:10.1002/sres.705

Nakamori, Y. (Ed.). (2011). *Knowledge Science: Modeling the Knowledge Creation Process*. Boca Raton, FL: CRC Press, Taylor & Francis Group. doi:10.1201/b11054

Nakamori, Y. (2013). Knowledge reconstruction and justification for regional vitalization. *Journal of Systems Science and Systems Engineering*, *22*(4), 457–468. doi:10.1007/s11518-013-5232-7

Nakamori, Y., Wierzbicki, A. P., & Zhu, Z. C. (2011). A theory of knowledge construction systems. *Systems Research and Behavioral Science*, *28*(1), 15–39. doi:10.1002/sres.1046

Nakamori, Y., & Zhu, Z. C. (2004). Exploring a sociologist understanding for the *i*-System. *International Journal of Knowledge and Systems Science*, *1*(1), 1–8. doi:10.4018/jkss.2010010101

National Institute of Population and Social Security Research. (2012). *Population Projections for Japan (January 2012): 2011 to 2060*. Retrieved from http://www.ipss.go.jp/site-ad/index_english/esuikei/ppfj2012.pdf

Nussbaum, M. C., & Sen, A. (Eds.). (1993). *The Quality of Life*. Oxford, UK: Clarendon Press. doi:10.1093/0198287976.001.0001

OECD. (2013). Health at a Glance 2013 Report. OECD 2-13-11-21, Chapter 1.6. doi:10.1787/health_glance-2013-en

Ormrod, J. E. (2006). *Educational psychology: Developing learners* (5th ed.). Upper Saddle River, NJ: Pearson/Merrill Prentice Hall.

Reeve, J., & Paperboy, D. (2007). Evaluating the evaluation: Understanding the utility and limitations of evaluation as a tool for organizational learning. *Health Education Journal*, *66*(2), 120–131. doi:10.1177/0017896907076750

Ross, P. H., Ellipse, M. W., & Freeman, H. E. (2004). *Evaluation: A systematic approach* (7th ed.). Thousand Oaks, CA: Sage.

Sakao, T., & Lindahl, M. (2012). A value based evaluation method for Product/Service System using design information. *CIRP Annals Manufacturing Technology*, *61*(1), 51–54. doi:10.1016/j.cirp.2012.03.108

Sherer, M., Maddux, J. E., Mercandante, B., Prentice-Dunn, S., Jacobs, B., & Rogers, R. W. (1982). The self-efficacy scale: Contribution and validation. *Psychological Reports*, *51*(2), 663–671. doi:10.2466/pr0.1982.51.2.663

Sim, J., Bartlam, B., & Bernard, M. (2011). The CASP-19 as a measure of quality of life in old age: Evaluation of its use in a retirement community. *Quality of Life Research: An International Journal of Quality of Life Aspects of Treatment, Care and Rehabilitation*, *20*(7), 997–1004. doi:10.1007/s11136-010-9835-x PMID:21243529

Stufflebeam, D. L., & Webster, W. J. (1980). An analysis of alternative approaches to evaluation. *Educational Evaluation and Policy Analysis*, *2*(3), 5–19. doi:10.3102/01623737002003005

Tatsuse, T., & Fujimori, J. (2014). Verification of the "Care-Will Model". In J. Fujimori (Ed.), *Research Report on Practice and Verification of the "Care-Will Model" toward Implementation of Autonomous Social Support to Male Retirees*. Research Grants of the Ministry of Health, Labor and Welfare.

Taylor, R., & Tofts, C. (2012). *Managing Complex Service Systems*. New York: Springer.

Vargo, S. L., Maglio, P. P., & Akaka, M. A. (2008). On value and value co-creation: A service systems and service logic perspective. *European Management Journal, 26*(3), 145–152. doi:10.1016/j.emj.2008.04.003

Wierzbicki, A. P., & Nakamori, Y. (2006). *Creative Space: Models for the Knowledge Civilization Age*. Berlin: Springer. doi:10.1007/b137889

Yamashita, Y., Nakamori, Y., & Wierzbicki, A. P. (2009). Knowledge synthesis in technology development. *International Journal of Systems Science and Systems Engineering, 18*(2), 184–202. doi:10.1007/s11518-009-5103-4

APPENDIX

This appendix gives the list of all questionnaires used in this research.

(A-1) Mastery of Plan Creation Methods

1: Although you were taught the plan creation process, they are vague (the level of data).
3: You organize the plan creation method (the level of information).
5: You can reasonably explain the plan creation method (the level of knowledge).

(A-2) Collection and Understanding of Objective Information

1: Although you have a lot of objective information, they are chaotic (the level of data).
3: You are collecting and organizing useful information (the level of information).
5: You are utilizing the collected information to create the plan (the level of knowledge).

(A-3) Information From the Family and Friends

1: You sometimes talk with your family members or friends about the plan (the level of data).
3: You organize your thoughts by hearing the opinions of family members/ acquaintances (the level of information).
5: You utilize diverse opinions for plan creation (the level of knowledge).

(A-4) Creation of Personal Ideas

1: Your idea is still very vague (the level of data).
3: You organize some interesting ideas (the level of information).
5: You are utilizing your ideas for plan creation (the level of knowledge).

(A-5) Plan Creation and Its Evaluation

1: Your idea is not yet organized (the level of data).
3: You announced your plan and received evaluation (the level of information).
5: You are promoting the plan (including in the run-up) (the level of knowledge).

(B-1) Strength of Enthusiasm

1: You are creating a plan in carefree feeling (Drive level).
3: Plan creation is a big decision for you (Determination level).
5: You are dedicated to create and run your plan (Dedication level).

(B-2) Accuracy of Information

1: You are acting on the information that is emotional and vague (Emotional level).
3: You are acting on the information that seems intuitively correct (Intuitive level).
5: You are acting on the basis of rational and scientific information (Rational level).

(B-3) Spread of Collaboration

1: You practice the plan alone (Individual level).
3: You practice the plan with the family, local people, or acquaintances (Group level).
5: Your plan involves a wide range of people (Society level).

(B-4) Expansion of Imagination

1: You have an idea that could be run reliably (Routine level).
3: You have an interesting idea if realized (Diversity level).
5: You have an idea that is difficult to realize as a dream (Fantasy level).

(B-5) Complexity of Integration

1: You are dealing with knowledge in the range of a field (Specialized level).
3: You are dealing with knowledge that spans multiple areas of expertise (Interdisciplinary level).
5: You are dealing with knowledge of a variety of people with different cultures (Intercultural level).

(C-1) Value of Making a Plan

1: You think about the way of your own life in the plan (Personal value).
3: You think happiness of the family or development of the local community (Group value).
5: You are thinking to contribute to the society (Social value).

(C-2) Value of the Collected Information

1: You found the value in your own health and economic information (Personal value).
3: You found the value in the information related to happiness of the family or development of the local community (Group value).
5: You found the value in the general information about the society (Social value).

(C-3) Value of the Constructed Relationships

1: You are trying to practice the plan alone (Personal value).
3: You focus on strengthening relationships with family or friends (Group value).
5: You focus on strengthening relationships with a wide range of people (Social value).

(C-4) Value of Created Ideas

1: You are creating ideas that can be run personally (Personal value).
3: You are creating ideas that can be run by the family, local people, or acquaintances (Group value).
5: You are creating ideas that can be run socially (Social value).

(C-5) Value of Plan Execution

1: Execution of the plan has a significant value to you (Personal value).
3: Execution of the plan has a significant value to the family, local people, or acquaintances (Group value).
5: Execution of the plan has a significant value to the society (Social value).

Chapter 7
Exploring the Determinants Affecting Academics' Knowledge-Sharing Behavior in United Arab Emirates Public Universities

Huda Skaik
Independent Researcher, Malaysia

Roslina Othman
International Islamic University Malaysia, Malaysia

ABSTRACT

This chapter investigates knowledge-sharing practice among academics, examines the relationship between knowledge-sharing behavior and its predictors based on theory of planned behavior, and identifies the motivators and barriers affecting these predictors. Data were collected using a web-based questionnaire. Using SPSS and PLS-SEM, the analysis process involved (1) measuring the extent of knowledge sharing practice, (2) assessing the measurement model, (3) assessing the structural model, (4) testing the hypotheses, (5) validating research model fit. Results showed (1) great extent of knowledge-sharing practice, (2) knowledge-sharing behavior is significantly affected by intention, which is affected by attitude, subjective norms, self-efficacy, but not affected by controllability, (3) attitude is positively affected by trust and reputation as motivators of knowledge sharing, while controllability is negatively affected by lack of time and poor communication as barriers of knowledge sharing. This is the first research on knowledge-sharing behavior in higher education in UAE.

DOI: 10.4018/978-1-5225-5655-8.ch007

INTRODUCTION

Knowledge sharing is a key process of knowledge management (Nonaka & Takeuchi, 1995) and its strategies are keys to organizational and individual development (Alavi & Leidner, 2001; Earl, 2001). Knowledge sharing is defined as the process of exchanging and transferring existing knowledge and ideas among people in order to create new knowledge and ideas (Syed, Zaini, Noormala & Zahairah, 2009). In short, it is known as a process of exchanging knowledge including skills insights, and experience (Durmusoglu Jacobs, Nayir, Khilji & Wang, 2014; Lilleoere & Hansen, 2011). Generally, it occurs when individuals with a common purpose and similar problems come together to exchange ideas and information (McNeil, 2003).

The benefits of knowledge sharing for organizations and individuals are numerous. On the organizational level, the benefits include facilitating knowledge creation (Akhavan, Ghojavand & Abdali, 2012), achieving continuous organizational growth, survival, and development (Durmusoglu et al., 2014), meeting organizational goals and objectives (Wang & Noe, 2010), solving business problems (McDermott & O'Dell, 2001), enhancing performance, maintaining competitiveness and profitability (Hsu, 2008), gaining better understanding of customer needs and identifying new business opportunities (Sandhawalia & Dalcher, 2011), enhancing process efficiency (Chugh, 2012), and improving the knowledge base for decision-making and more balanced policy decisions (Egger, 2013).

On the individual level, the benefits include getting the information easier and faster (Badaracco, 2010), promoting individuals' learning and innovation (Ling, Sandhu & Jain, 2009), transferring knowledge among workers in the same unit or from one unit to another (Burgess, 2005), strengthening capabilities (Egger, 2013), enhancing performance (Xiao & Jin, 2010), improving efficiency (Cummings, 2004), empowering team effectiveness (Pangil & Chan, 2014), developing strategies to encourage organizational knowledge base (Reychav & Weisberg, 2009), reducing loss of individuals' knowledge and expertise (Gurbuz, 2008), and transmitting knowledge and expertise to new generations (Badaracco, 2010).

The growing recognition of the importance of knowledge sharing in several fields including research and development and scientific production (Bousari & Hassanzadeh, 2012; Ensign & Hebert, 2010; Lilleore & Hansen, 2011) is creating a demand for applying it in higher education (Babalhavaeji & Kermani, 2011; Chikoore & Ragsdell, 2013; Dokhtesmati & Bousari, 2013). This is due to the deep-rooted role of academic institutions in higher education as a plant of knowledge creation, scientific inventions, intellectual and cultural production, and innovative researches (Fullwood, Rowley & Delbridge, 2013; Patel & Ragsdell, 2011).

For academics in particular, knowledge sharing is within the core of their work, since they are involved in teaching, writing and publishing scholarly works

(Babalhavaeji & Kermani, 2011), conducting researches, participating in conferences and workshops, providing trainings, and keeping abreast in their fields. They are seen as expert knowledge workers from whom their academic institutions generate value. Bearing in mind that higher education institutions grow and prosper from the knowledge of their academics, it is quite necessary to encourage and promote knowledge sharing among academics considering their role in enhancing education, research, and scholarly work (ibid).

Universities are science centers established to generate and provide knowledge, and to equip people with the best education in order to serve their societies and uplift the well-being of mankind. They grow and prosper from the knowledge of their human capital, particularly the academics (Singer & Hurley, 2005). In the knowledge-based age, universities seek to ensure success and permanence, achieve organizational goals (Sharma, 2010), and have constant performance improvements. In the academic environment, the role of knowledge sharing is becoming quite significant to achieve maximum results for academic institutions (Babalhaveji & Kermani, 2011) due to the important role academics play in providing education, conducting research, and publishing scholarly works. Therefore, universities should promote knowledge sharing among their academics.

Knowledge Sharing in UAE

The United Arab Emirates (UAE) has experienced significant developments in various fields such as business, construction, infrastructure, telecommunications, financial services, public services, transportation, media, health and medicine, information technology, hospitality and tourism, industry, and education. It has been and still leading the way in promoting knowledge, due to its importance in making the positive change that supports the society progress and development (FAHR, 2017). The government has been working consistently and strongly to establish a knowledge-based society (Al-Nahyan, 2012a). It has realized that the adoption of 'knowledge economy' is the only way to achieve economic growth, which contributes to providing distinct products and services. Therefore, the government has allocated a special place for knowledge in 'UAE Vision 2021' entitled 'United in Knowledge' (FAHR, 2017).

Even so, most organizations have not integrated knowledge management initiatives into their operational processes, which could be attributed to the difficulty in locating the knowledge residing within people due to the increasing scale of operations and level of productivity. Thus, there is inability to locate, store, and share knowledge that could help in innovation, development and meeting planned objectives (Ahmad & Daghfous, 2010). Therefore, the government strategy has been recently focusing on human capital (Al-Nahyan, 2012a, 2012b). As a step to achieve this, it has allocated more than 1/3 of its budget to education, research and innovation (Al-Nahyan, 2012a).

UAE is also seeking to become a regional hub for higher education that provides high quality with international standards education in diverse programs and fields (Al-Nahyan, 2012b). As a result, the government launched partnerships with numerous international reputable academic institutions to establish campuses in UAE (Al-Nahyan, 2012a). Moreover, the government is working collaboratively with other partners worldwide, whereby UAE aims to be the hub of knowledge for the Arab region (Abu-Ghazaleh, 2012). This will definitely enable higher education institutions, researchers, and knowledge workers in the Arab world to become part of the world network of research and education (ibid).

The different initiatives launched by the government aim to encourage and motivate organizations in public and private sectors to implement knowledge management processes in order to strengthen UAE competitive position and become increasingly globalized. In addition, it is observed that there is a growing awareness of the benefits and importance of knowledge sharing to create new knowledge, increase capital assets (Al-Adaileh & Al-Atawi, 2011), achieve innovation and success (Ahmad & Daghfous, 2010; Al-Alawi, Al-Marzooqi & Mohammed, 2007), maintain continuous growth, reach objectives, and attain profitability, as well as develop employees' competencies (Seba, Rowley & Lambert, 2012) and enhance individuals' learning, performance, and skills.

Yet apparently there is scarcity or lack of knowledge sharing strategies in UAE due to lack of research on the issue, organizational and local culture, absence of recognition and rewards, low level of willingness to share knowledge or the inclination of people to hoard knowledge as a means of job security, and unavailability of proper systems to capture knowledge of experienced professionals to pass on to the new employees. Such lack is certainly the biggest impediment to progress towards a knowledge-based society.

PROBLEM STATEMENT

As indicated earlier, the government's determination to build a knowledge-based society and promote knowledge in the whole region, in addition to investing in human capital, makes knowledge sharing quite fundamental to generate ideas, create new knowledge, and utilize intellectual capital (Seba et al., 2012). Therefore, it is expected from organizations in UAE to be actively involved in knowledge sharing in order to acquire, develop, and share knowledge to build and refine their knowledge resources and competencies (Siddique, 2012). However, in collective work, where so much depends on intellectual capital, not all people would have the intention to share their knowledge and expertise with seniors, peers, and subordinates. People's willingness and intention to share knowledge can be affected by many factors ranging from organizational, individual, to technological (Seba et al., 2012; Tohidinia & Mosakhani, 2010; Zhang & Ng, 2013).

The fast growth in UAE in various fields and the increasing diversification of operations, seen in the spread of local, regional, and international organizations, have created a situation where organizations are not effectively sharing internal knowledge and experience (Ahmad & Daghfous, 2010; Skok & Tahir, 2010). As a result, researchers have started to shed light on the need of a knowledge sharing strategy in UAE, particularly concentrating on business and management, construction, and police force service (Ahmad & Daghfous, 2010; Rowley, Seba & Delbridge, 2012; Seba et al., 2012; Skok & Tahir, 2010). These researchers either studied the practice of knowledge sharing and its activities, or the effect of some organizational, individual, and technological factors on it.

If UAE is to transfer into a knowledge-based society with a knowledge-based economy, and build the aspired 'Arab Model' (Al Maktoum, 2007) in the region, then the government should encourage and support a culture of knowledge sharing. Moreover, if UAE is to become a regional hub for higher education and play its promising role in creating and spreading knowledge, then promoting a culture of knowledge sharing is a necessity within higher education institutions particularly among academics given their important role in knowledge creation and distribution.

The review of literature on knowledge sharing in UAE indicated that there is lack of its practice in different fields, while in higher education specifically there is nonexistence of any research addressing knowledge sharing in academic institutions. Therefore, this research intends to investigate the practice of knowledge sharing among academics in UAE public universities, and identify the factors influencing their knowledge sharing behaviour based on the theory of planned behaviour.

THEORETICAL FRAMEWORK

The Theory of Planned Behavior (Ajzen, 1985) states that human behavior is guided by three kinds of salient beliefs: behavioral beliefs, normative beliefs, and control beliefs. Behavioral beliefs are about the likely consequences or attributes of the behavior, normative beliefs are about the normative expectations of other people, and control beliefs are about the presence of factors that may facilitate or hinder performing the behavior (Ajzen, 1985, 1991). In their respective aggregates, behavioral beliefs produce a favorable or unfavorable attitude towards the behavior; normative beliefs result in perceived social pressure or subjective norms; and control beliefs give rise to perceived behavioral control, the perceived ease or difficulty of performing the behavior (Ajzen, 1985, 1991). In combination, attitude, subjective norms, and perceived behavioral control lead to the formation of a behavioral intention (Ajzen, 1985, 1991, 2002).

However, due to the conceptual and methodological ambiguities concerning the concept of perceived behavioral control, Ajzen (1991) defined the concept as an individual's confidence that s/he is capable of performing the behavior in concern. He further indicated that such definition is compatible with Bandura's (1997) concept of self-efficacy and so he placed self-efficacy within the framework of the theory. However in 2002, Ajzen stated that perceived behavioral control should be viewed as a unitary, higher-order concept that consists of two interrelated components (Kraft, Rise, Sutton & Roysamb, 2005). In order to resolve these ambiguities, Ajzen (ibid) deconstructed perceived behavioral control into two constructs: self-efficacy and controllability. He defined self-efficacy as the individual's confidence in the ease or difficulty to perform certain behaviour, whereas controllability is referred to as the individual's beliefs, based on the available resources, about the extent to which performing the behaviour is up to him/her.

Behavior is the degree to which an individual actually decides to perform or not perform a specific action and it is determined by the individual's intention to perform it or not (Ajzen, 1991). Intention is an individual's willingness to engage in certain behavior (Ajzen, 1985, 1991, 2002). Attitude is the degree to which an individual has favorable or unfavorable evaluation or appraisal of the behavior (Ajzen, 1991). Attitude towards the behavior is an influential factor to perform that behavior (Ajzen, 1985, 1991, 2002) indirectly by influencing the individual's intention (Ajzen, 1991). Subjective norms are defined as the individual's perceived social pressure to perform or not to perform a given behavior (Ajzen, 1991). Subjective norms refer to the individual's belief that important relevant others, including executive board, senior management, supervisor, and peers expect him/her to engage in the behavior of interest (Chennamaneni, 2006). Self-efficacy is an individual's confidence in the ease or difficulty to perform the behavior in question (Ajzen, 2002) and is considered an important factor influencing an individual's intention to perform the behavior (Ajzen, 1991, 2002). Controllability is an individual's beliefs, based on the available resources, about the extent to which performing a given behavior is up to him/her (Ajzen, 2002) and is seen an important determinant that influences an individual's behavior through intention (Madden, Ellen & Ajzen, 1992).

According to Ajzen (1991, 2006), the more favorable the attitude and subjective norms, and the greater the self-efficacy and controllability, the stronger would be the intention to carry out the behavior. Thus, intention is considered the immediate antecedent of behavior (Ajzen, 1991, 2002, 2006). Having sufficient degree of actual control over the behavior, an individual is expected to perform the behavior when the opportunity arises. Each of the theory elements of intention, attitude, subjective norms, self-efficacy, and controllability is counted as an aspect of the actual behavior (Ajzen, 1991).

The Theory of Planned Behavior, which extends the boundary condition of the individual's volitional control given in the previously founded Theory of Reasoned Action, emerged as one of the most influential and popular theoretical frameworks to study human behavior (Ajzen, 2002). In the domain of knowledge management, knowledge sharing is classified as an organizational behavior and thus knowledge sharing behavior is defined as "an individual's behavior that is not directly or explicitly recognized, and in the aggregate promotes effective functioning of the organization" (Bordia, Irmer, Garden, Phair & Abusah, 2004:130). With this being taken into account, then employing the Theory of Planned Behavior is the most appropriate to study knowledge sharing behavior.

In knowledge sharing context, studies have found that intention significantly affects knowledge sharing behavior (Alajmi, 2011; Babalhavaeji & Kermani, 2011; Chen, Chen & Kinshuk, 2009; Minbaeva & Pedersen, 2010; Tohidinia & Mosakhani, 2010). Moreover, studies have found that attitude, subjective norms, and self-efficacy significantly affect intention (Alajmi, 2011; Babalhavaeji & Kermani, 2011; Chennamaneni, 2006; Elogie & Asemota; 2013; Kuang, Davison & Yao, 2012; Lin & Lee, 2004; Minbaeva & Pedersen, 2010; Ryu, Ho & Han, 2003). Previous studies on knowledge sharing behavior have always adopted the first or second models of the Theory of Planned Behavior (1985, 1991) where perceived behavioral control proved to affect intention (Elogie & Asemota; 2013; Kuang et al., 2012; Tohidinia & Mosakhani, 2010). However, only one study has adopted the amended model of the theory (2002) where controllability is measured as a separate variable. In that study, Alajmi (2011) found that controllability does not affect intention.

Based on all the above, and in terms of the research context on knowledge sharing, it is hypothesized that academics' knowledge sharing behavior is significantly influenced by their intention, which in turn is positively influenced by their attitude, subjective norms, self-efficacy, and controllability.

LITERATURE REVIEW

Motivating Factors

The need to create a knowledge sharing culture in organizations is quite imperative (Tan, Lye, Ng & Lim, 2010), which calls for motivating factors to be employed in order to motivate individuals to share their knowledge. The literature identifies a number of motivators that influence the success of knowledge sharing and it divides them into intrinsic and extrinsic. Among the significant intrinsic motivators is trust and among the significant extrinsic motivators is reputation. According to the theory of planned behavior, attitude is based on behavioral beliefs linked to attributes or outcomes (Ajzen, 1991). Some of the attributes include trust and confidence (Al-Adaileh & Al-Atawi, 2011), while some of the outcomes include rewards and reputation (Chennamaneni, 2006; Kankanhalli, Tan & Wei, 2005).

Trust is a mutual understanding in others' good intentions and desires (Hsu, Ju, Yen & Chang, 2007) and belief that an intended action -such as knowledge sharing- of some people is appropriate from others' viewpoint. This happens due to the confidence in those people's good intentions, concern, competence, capability, and perceived openness. Davenport and Prusak (1998) also assert that trust is actually one of the major keys of successful knowledge management initiatives and creating a knowledge sharing culture, regardless of how thoroughly they are supported by technology and organization. Indeed, Tan et al. (2010) believe that trust determines the success of knowledge sharing.

In UAE, Seba et al. (2012) studied the influence of trust on employees' attitude toward knowledge sharing and found that trust is a significant motivator of attitude. Rad, Alizadeh, Miandashti and Fami (2011) examined the relationship between trust and employees' attitude and found a significant relationship between them. In Taiwan, Hung, Lai and Chou (2010) studied the influence of trust on attitude towards knowledge sharing, and the latter influence on intention to share knowledge. They found that intention to share knowledge is significantly affected by attitude, which itself is affected by trust.

Reputation refers to the potential of obtaining self-concept gains such as receiving praise, recognition, prestige, or respect, and improving status or image. (Kankanhalli et al., 2005). It positively influences an individual to share knowledge with others. Hung et al. (2010) emphasize that professional reputation encourages people to share their knowledge. Individuals must believe that knowledge sharing is worth the cost and that they will not only benefit others but they themselves will gain benefit. Therefore, in order to motivate people to share knowledge, the best means is showing them that knowledge sharing is for their own personal interest and benefit.

In two American universities and Chinese, Chennamaneni (2006) and Wu and Zhu (2012) respectively explored the influence of reputation on MBA employee-students' attitude towards knowledge sharing. Their findings showed that reputation has a significant positive effect on individuals' attitude. In a different study in Iran, Abzari, Barzaki and Abbasi (2011) studied the influence of reputation on employees' attitude towards knowledge sharing. The results showed that a higher level of perceived reputation enhancement has a significant positive influence on the employees' attitude.

Hindering Factors

On the other hand, knowledge sharing process is hindered by many barriers, and identifying those barriers may contribute in succeeding knowledge sharing strategies. Among the major factors that have been identified as barriers to the individuals' knowledge sharing are lack of time and poor communication (Tan et al., 2010). According to the theory of planned behavior, controllability is based on control beliefs about the availability of certain opportunities and resources (Ajzen, 2002). Examples of such opportunities include time and space availability (Rowely et al., 2012; Seba et al., 2012), while examples of resources include ICT tools and communication skills (Chennamaneni, 2006; Minbaeva & Pedersen, 2010).

Lack of time is understood in terms of lack of available time to perform tasks and actions (Riege, 2005). In this regard, Haas and Hansen (2007) state that individuals' willingness to share knowledge is largely affected by the amount of time allocated to perform their responsibilities for which knowledge sharing can be useful. People may be hindered to share knowledge because they are too busy and do not have sufficient time to communicate and exchange their knowledge or expertise. Rowley et al. (2012) state that time allocation is very essential for individuals in same groups to learn new things, collaborate with each other, share and create knowledge.

In their studies on the factors affecting knowledge sharing in UAE, Rowely et al. (2012) and Seba et al. (2012) found that lack of time is one of the main barriers hindering knowledge sharing among employees. Santos, Soares and Carvalho (2012) explored barriers to knowledge sharing in R&D projects in 6 European countries. The results revealed that lack of time was one of the significant barriers to knowledge sharing, as they pointed out to that absence of balance between work load and available time prevents them to have extra time to share knowledge with others in the organization.

Poor communication refers to weak interaction and interpersonal skills either orally, written, or in body language. Poor communication is seen as a major barrier in the process of knowledge sharing since individuals' ability to share knowledge depends largely on their communication skills (Riege, 2005). The ability of people to effectively share knowledge depends largely on their communication and interaction. Therefore, poor communication and lack of interaction are perceived as significant inhibitors of knowledge sharing (ibid). Also, poor communication negatively influences the success of knowledge sharing in organizations (Sohail & Salina, 2009)

In three different studies in Malaysia, poor communication was found a main barrier to knowledge sharing. Jain, Manjit and Gurvinder (2007) explored different motivators and barriers to knowledge sharing among academic staff, and found that poor communication highly hinders knowledge sharing process. Likewise, Nassuora and Shahizan (2010) investigated barriers to knowledge sharing among academics, and found that poor verbal and written communication skills were key barriers to effective knowledge sharing. In a different context, Ling et al. (2009) examined barriers to knowledge sharing among engineers and managers in a multi-national company. The findings revealed that lack of interaction is a barrier hindering employees from sharing their knowledge.

Knowledge Sharing Practice

Knowledge sharing practice is quite significant for individuals seeking to improve their performance and enhance their career as well as for organizations aiming to achieve their objectives and ensure success and longevity. Organizations invest in different knowledge sharing practices because they are aware of the value of knowledge as a crucial factor for their success. Knowledge sharing practice is manifested in the social interaction among individuals to exchange information, knowledge, expertise, skills, opinions, insights, and ideas (Durmusoglu et al., 2014; Ramlee, 2011). Thus, organizations should adopt the suitable activities and tools in order to ensure best practice of knowledge sharing.

Practising knowledge sharing involves using appropriate activities and tools (Alajmi et al., 2008; Paloti, 2010) that facilitate exchanging, transferring, and utilizing knowledge. Several researchers and practitioners have identified a list of the most popular activities and tools widely employed by many organizations (Ahmad & Daghfous, 2010; Alajmi et al., 2008; Al-Alawi et al., 2007; Skok & Tahir, 2010; Paloti, 2010).

The activities include team meetings, discussions, Communities of Practice, training programs, apprenticeships, workshops, conferences, brainstorming sessions, peer coaching, focus groups, and seminars, while the tools include artifacts, educational materials, manuals, boards, procedures, databases, video screening sessions, decision-support systems, blogs, wikis, portals, and online communication channels like emails, internet and intranet.

Organizations that have survived from knowledge sharing are not those who only implemented technology, but those who developed a knowledge sharing culture (Albres, 2009). Employing technology in addition to applying suitable knowledge sharing practices and strategies can create a knowledge sharing culture to harness organizational development (Mohamed, O'Sullivan & Ribière, 2008)

Previous studies (Ahmad & Daghfous, 2010; Alajmi et al., 2008; Al-Alawi et al., 2007; Haapalainen & Makiranta, 2013; Skok & Tahir, 2010) showed that individuals in various sectors practised knowledge sharing and used several activities and tools to share knowledge. For example, in a case study in the American healthcare sector Alajmi et al. (2008) explored the practice of knowledge sharing and the use of different activities and tools in the practicing process. They found that practice of knowledge sharing was manifested in different ways such as team working, exchanging of knowledge, asking and approaching each other whenever needed. They also found that the participants used different knowledge sharing activities and tools, which has helped in performing daily work processes and in solving and preventing problems. In UAE, Skok and Tahir (2010) explored the practice of knowledge sharing and identified the activities used among managers and employees -Arabs and international- in a private construction company. They found that respondents practice knowledge sharing using certain knowledge sharing activities. In Finland, Haapalainen and Makiranta (2013) investigated the use of IT in acquiring and sharing knowledge in SMEs, and identified various channels to share tacit and explicit knowledge. They found that employees use different channels and tools to share their knowledge and that generally such companies regard IT as a potential tool to share and acquire knowledge.

Knowledge sharing is key component (Alavi & Leidner, 2001) in organizational and individual development (Nahapiet & Ghoshal, 1998; Nonaka, 1994). Therefore, in academic environments such as universities, management should create a knowledge sharing culture and support the practice of knowledge sharing by providing the necessary activities and tools. This would help and encourage academics to share their knowledge, knowing that they play a major role in disseminating knowledge and creating new one.

RESEARCH MODEL AND HYPOTHESES

Based on the theoretical framework and the past researches employing the Theory of Planned Behavior as indicated above, as well as the literature review on the motivating and hindering factors to knowledge sharing, the following research model (Figure 1) is proposed. the current study examines the influence of intention on knowledge sharing behavior as its main determinant, as well as the influence of attitude, subjective norms, self-efficacy, and controllability on intention. Moreover, the study examines the influence of trust and reputation on attitude, and the influence of lack of time and poor communication on controllability.

Based on the theoretical and empirical background of the factors influencing knowledge sharing behavior, the following hypotheses were proposed:

Hypothesis 1: Intention to share knowledge has a significant positive effect on academics' knowledge sharing behavior.

Hypothesis 2: Attitude has a significant positive effect on academics' intention to share knowledge.

Figure 1. Research model

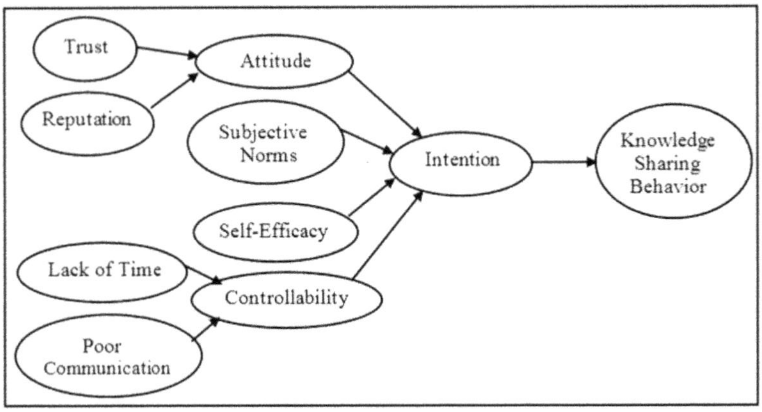

Hypothesis 3: Subjective norms has a significant positive effect on academics' intention to share knowledge.
Hypothesis 4: Self-efficacy has a significant positive effect on academics' intention to share knowledge.
Hypothesis 5: Controllability has a significant positive effect on academics' intention to share knowledge.
Hypothesis 6: Trust has a positive effect on academics' attitude towards knowledge sharing.
Hypothesis 7: Reputation has a positive effect on academics' attitude towards sharing knowledge.
Hypothesis 8: Lack of time has a negative effect on academics' controllability over knowledge sharing.
Hypothesis 9: Poor communication has a negative effect on academics' controllability over knowledge sharing.

METHODOLOGY

Social researches that aim at identifying the factors influencing an outcome or the cause–effect relationships of human behaviors, and testing predetermined hypotheses based on an existing theory require using the quantitative method (Creswell, 2009; Weathington, Cunningham & Pittenger, 2010). Consequently, since this research aims to examine and identify the factors that influence knowledge sharing behavior and test the research hypotheses based on the Theory of Planned Behavior, then it used the quantitative research

A cross-sectional web-based survey was used to collect data from the academics working in the targeted universities in UAE. The cross-sectional design insures collecting data at one point in time from a random sample of participants that represent the larger population and the survey can collect large numbers of data quickly with minimal costs, and it allows to apply correct numerical formulas to test the proposed relationships between the constructs which allows generalizing the findings. The web-based survey has many benefits including (a) overcoming time and space boundaries, (b) ease of data entry (Batinic, Reips & Bosnjak, 2002), (c) cost-effective in developing and distributing through an internet link (Weathington et al., 2010).

Sample and Sampling Technique

The setting of this research composed of 10 public universities in UAE, including 2 federal, 4 governmental, and 4 semi-governmental universities[1]. The targeted population of the research was 2,585 academics, the number of the sample size required to achieve a 95% confidence interval to represent a cross-section of the population was 335. The sample of respondents consisted of 321 academics working in in different faculties in those universities. The sample size was determined using Krejcie and Morgan's (1970) method along with the online sample calculator to ensure determining the exact sample size needed for the research. The sampling technique used for this study was the simple random sampling, which includes samples of whoever would be available at the time of conducting the research (Gay & Airasian, 2003). The simple random sampling maximizes the sample representativeness by eliminating any bias from the process of selection (Weathington et al., 2010), while minimizes the sampling error (Vogt, 2007) and ensures the generalizability (Sekaran, 2003).

Survey Instrument

A comprehensive questionnaire was created to measure the variables in the research model. It comprised of four sections; the first section collected demographic information about the participants. The second section collected data about the participants' knowledge sharing practice. The third section collected data about the participants' knowledge sharing behavior and its predictors, i.e. attitude, subjective norms, self-efficacy, and controllability. The fourth section collected data about the motivating and hindering factors, i.e. trust, reputation, lack of time, and poor communication.

The measurement items for the variables of the theory were developed based on Ajzen's theory of planned behavior (1985, 1991, 2002), and other instruments validated in previous researches conducted on knowledge sharing behavior including (Alajmi, 2011; Bock, Zmud, Kim & Lee, 2005, Chen et al., 2009; Francis et al., 2004 ; Hsu et al., 2007; Ryu et al., 2003; Tohidinia & Mosakhani, 2010). The measurement items for the motivating and hindering variables were developed and validated from previous researches including (Hew & Hara, 2007; Hsu et al., 2007 ; Jain et al., 2007 ; Kankanhalli et al., 2005 ; Ling et al., 2009; Reige, 2005; Sandhu, Jain & Umi Kalthom, 2011; Seba et al., 2012; Wasko & Faraj, 2005). The measurement items for knowledge sharing practice were developed and validated from (Alajmi et al., 2008; Al-Alawi et al., 2007; Hsu et al., 2007; Paloti, 2010; Sandhu et al., 2011; Tohidinia & Mosakhani, 2010).

All items were measured using five-point Likert-scale. The scales are usually used to measure the strength degree of the respondents' attitudes and feelings about a certain subject, and have a score in the middle that allows them to feel neutral (Simonis, 2010). The scale used for the construct of knowledge sharing behavior ranged from 'Never', 'Rarely', 'Sometimes', 'Usually', to 'Always'. Whereas the scale used for knowledge sharing practice as well as the rest of the constructs including the theory variables and the motivating and hindering factors ranged from 'Strongly Disagree', 'Disagree', 'Neutral', 'Agree', to 'Strongly Agree'.

Data Collection

After developing the instrument of the research, a pilot study was conducted to ensure the questionnaire reliability and validity, which led to minor adjustments in the questionnaire. After completion of the pilot study, the data collection stage was conducted. For this purpose, official letters were sent to the presidents, chancellors, or rectors of the targeted universities informing them of the research purpose and seeking their permission and cooperation. Upon receiving the permission, the data collection stage commenced.

The data was collected through a web-based survey. The Internet is seen as one of the fastest techniques to collect data through using an online survey either by sending an email or posting a web page. This method is growing significantly due to being easy, economical, fast, covering a large sample, and targeting the required respondents. For this research, the survey was conducted online by sending an email with a link of the questionnaire to the participants with an Internet-based response mechanism. The email, which was written in both English and Arabic, introduced the study to the participants inviting them to take part in the survey by answering the research questionnaire.

The questionnaire was developed using the Google Drive service in a way that makes it as much easy and user friendly to the respondents as possible. The questionnaire was provided in the email through a URL link that led the participants to a web page where they can answer and submit online. Once the participants submitted their answers, the data were recorded directly into a spreadsheet, which was transferred later into SPSS in order to carry out the required analysis. In order to insure attaining the required sample size, a submission date was arranged to insure that academics would comply with, and reminders were sent to the academics to answer the questionnaire. The data collection period took about five months between 2012 and 2013.

Data Analysis

The data analysis procedure involved a process of analyzing, examining, and interpreting the data in order to test the research hypotheses and highlight useful information based on the findings and suggest recommendations based on the outcomes of the research. Relevant statistical analysis was employed and several steps were involved in the process of analyzing including descriptive statistics and inferential statistics using SPSS, and partial least square structural equation modeling statistics using SmartPLS.

Using SPSS 19.0, the sample descriptive characteristics were assessed based on the demographic information. Inferential statistics were used to (i) measure the exploratory factor analysis to explain the pattern of correlations within a set of observed variables (Battour, 2011) and (ii) test the reliability of the variables. Data analysis for knowledge sharing practice was conducted using descriptive statistics of central tendency, dispersion, and frequencies.

Data analysis for the research model variables was conducted using partial least square path modeling technique (PLS-SEM), which is a variance-based partial least square and is one of the statistical methods for structural equation modeling (Hair, Ringle & Sarstedt, 2011). Indeed the PLS-SEM has increasingly become a more favourable technique in empirical researches and studies in diverse fields. By using SmartPLS 2.0 software (Hansmann & Ringle, 2004), PLS-SEM was applied to assess the measurement and structural models, the mediating relationships, and to test the research hypotheses. Finally, the model was validated using the model fit indices

The assessment of the measurement model involved assessment of indicator reliability, internal consistency reliability, convergent validity, and discriminant validity at indicator and construct levels (Chin, 2010). The assessment of the structural model involved assessment of the coefficient of determination, path coefficient, effect size, and predictive relevance (ibid). Assessment of the mediating relationship was applied using PLS algorithm test on the relationship between the dependent and independent variables with and without the presence of the mediating variable. In case both tests yielded significant relationships, then the mediating variable is partial mediator, while if the tests show that after the inclusion of the mediating variable the direct relationship is no longer significant, then the mediating variable is full mediator (Guenzi, Goerges & Pardo, 2009; Ida, Roshayati & Fazli, 2012). The research model in PLS path modeling has three fit indices: the average communality index, the average redundancy index, and the Goodness of Fit index (Vinzi, Trinchera & Amato, 2010; Vinzi & Russolillo, 2013).

RESEARCH FINDINGS

Demographics of Respondents

Table 1 displays the descriptive statistics of the demographic characteristics for the respondents according to the demographic categories specified in the questionnaire. These categories included gender, age, nationality, level of education, name of university, faculty, position, years of academic experience, and type of contract.

As seen in the table, the percentage of male respondents surpassed that of female respondents. Their age ranged from 22 years to more than 51 years with the majority over 41 years old. In terms of nationality, almost 60% of the respondents were international, and 32.7% were Middle Eastern. 86% of the respondents hold master and Ph.D. degrees. The respondents were distributed among the 10 universities with majority from the 2 federal universities, i.e. UAEU and ZU. They also represented different faculties. As for the position, the respondents ranged from teachers to professors with majority as assistant professors. The academic experience in teaching ranged from less than 5 years to more than 21 years. Finally, the type of contract the respondents were recruited on was mainly either fixed-term or open-ended.

Knowledge Sharing Practice

The first objective of this research was to investigate knowledge sharing practice among academics in the targeted public universities in UAE. In order to measure the practice of knowledge sharing, the respondents were asked to indicate their degree of agreement to eight statements. By using descriptive statistics of central tendency, dispersion, and frequencies, the answers were analyzed. The results showed that majority of the surveyed academics either agreed or strongly agreed to the given statements indicating a great extent of knowledge sharing practice. The minimums, maximums, means, modes, and standard deviations for these statements are shown in Table 2 below.

In general, the respondents indicated that they accomplish certain tasks through teamwork and collaboration with other colleagues, where almost 88% agreed and strongly agreed to the statement. Meanwhile, only 3% disagreed to the statement.

Table 1. Descriptive of respondents' demographics

Category	Item	Percent
Gender	Male	68.5
	Female	31.5
Age	22-30	2.8
	31-40	26.2
	41-50	35.8
	51 & above	35.2
Nationality	Local	7.5
	Middle Eastern	32.7
	International	59.8
Level of Education	Bachelor	1.6
	Master	29.3
	Ph.D	56.7
	Post Doctorate	12.5
Name of University[2]	United Arab Emirates University	48
	Zayed University	10.3
	Khalifa University of Science, Technology & Research	3.4
	The Petroleum Institute University and Research Center	7.2
	University College for Mother and Family Sciences	0.9
	Paris-Sorbonne University Abu Dhabi	1.2
	Ajman University of Science and Technology	9
	American University of Sharjah	8.1
	Ras Al Khaima Medical and Health Sciences University	8.1
	University of Dubai	3.7
Faculty	Agriculture	1.6
	Arts	5.9
	Business & Economics	10.9
	Communication & Media	3.1
	Education	11.5
	Engineering	9.7
	Humanities & Social Sciences	10.9
	Information Technology	5.3
	Islamic Studies	2.2
	Law	1.9
	Medicine & Health Sciences	11.8
	Sciences	10.6
	Other (Pre-academic program, foundation program, general requirement unit, English language center)	14.6
Position	Professor	15.9
	Associate Professor	16.5
	Assistant Professor	34.9
	Instructor	19.9
	Lecturer	11.2
	Other (teacher, research associate, researcher)	1.6
Years of Academic Experience	0-5	17.4
	6-10	20.9
	11-15	19.6
	16-20	13.7
	21 & above	28.3
Type of Contract	Fixed-Term	64.8
	Open-Ended	33.6
	Other (Job-base, rolling, no contract)	1.6

Table 2. Descriptive of knowledge sharing practice

Statement	Min	Max	Mean	Mode	St. Deviation
I accomplish certain tasks through teamwork and collaboration with other colleagues	2	5	4.10	4	0.679
I exchange my knowledge and experience while working with other colleagues	2	5	4.19	4	0.661
I am willing to share my knowledge with my colleagues freely	2	5	4.46	5	0.642
When I learn something new, I tell my colleagues about it	2	5	4.09	4	0.771
When I need certain knowledge, I ask my colleagues	2	5	4.17	4	0.708
I seek my colleagues' experience when I need to learn something	2	5	4.18	4	0.721
I utilize the available tools to share my knowledge with my colleagues	2	5	4.09	4	0.761
I attend and contribute in different knowledge sharing activities	2	5	4.03	4	0.717

Again, almost 90% of respondents agreed and strongly agreed to the second statement showing that they exchange knowledge and experience while working with others, and only 2% disagreed. Almost all respondents 95% agreed and strongly agreed that they are willing to share their knowledge with other colleagues freely. Only 1.2% of them disagreed to the statement. The fourth statement inspected if they would tell other colleagues the new things they learn. Based on the results, 81.3% of respondents either strongly agreed or agreed that they do so, while 3.4% disagreed. Almost 88% of the respondents either strongly agreed or agreed that they ask their colleagues when needing certain knowledge, and almost 3% disagreed. When asked to indicate their agreement to the statement inspecting seeking colleagues' experience when needing to learn something, 87.3% of the respondents strongly agreed and agreed. Meanwhile, only 3% disagreed to the statement.

In terms of using the different tools and activities involved in knowledge sharing practice, the last two statements inspected that. Utilizing certain tools such as educational materials, manuals, boards, procedures, databases, blogs, wikis, portals, emails, internet, and intranet is an indication of practising knowledge sharing. When inspecting their usage of such tools, 83.2% of respondents strongly agreed and agreed that they use these tools, while 12.8% answered neutral and 4% disagreed. Finally, as an indication of practising knowledge sharing, the respondents were asked if they participate in various knowledge sharing activities such as meetings, discussions, CoPs, trainings, apprenticeships, workshops, conferences, seminars, brainstorming, and focus groups. Total of 83% of them strongly agreed, and agreed to this statement. However, 13.7% answered neutral, while 3.4% disagreed to the statement.

Assessment of Measurement Model

The purpose of assessing the measurement model is to evaluate its validity and reliability and thus evaluate the inner path model estimates (Henseler, Ringle & Sinkovics, 2009). It is conducted through the following tests: (a) Indicator reliability by measuring the factor loading of each of the manifest variables, which should be above 0.4 (Hair, Ringle & Sarstedt, 2013), (b) internal consistency reliability by measuring composite reliability and Cronbach's alpha, which should be 0.7 and above (Hair, Black, Babin & Anderson, 2010), (c) convergent validity by measuring the AVE, which should be more than 0.5 (Fornell & Larcker, 1981), and (d) discriminant validity by using Fornell-Larcker's (1981) criterion where the square root of the AVE for each construct exceeds the correlations between the construct and all other constructs (Henesler et al., 2009).

The results of analyzing the measurement model demonstrated reliable and valid measurement model. All factors loaded above the recommended value of 0.7 ranging between 0.791 to 0.943 demonstrating satisfactory indicator reliability. The constructs composite reliability and Cronbach's alpha values exceeded the recommended value of 0.7 indicating satisfactory internal consistency reliability as displayed in Table 3.

The constructs AVE exceeded the recommended value of 0.5 demonstrating adequate convergent validity. The square root of the constructs AVE values exceeded

Table 3. Internal consistency reliability

Construct	Composite Reliability	Cronbach's Alpha
Knowledge Sharing Behavior	0.9055	0.8612
Intention	0.9462	0.9289
Attitude	0.9475	0.9307
Subjective Norms	0.9082	0.8658
Self-Efficacy	0.9549	0.9409
Controllability	0.9500	0.9311
Trust	0.9560	0.9463
Reputation	0.9598	0.9479
Lack of Time	0.9078	0.8661
Poor Communication	0.8370	0.7380

the correlations between the constructs and all indicators loaded higher on their own constructs indicating satisfactory discriminant validity as seen in Table 4.

Assessment of Structural Model

The purpose of assessing the structural model is to evaluate its validity (Kamarul, 2012) and path estimates (Henseler et al., 2009) and thus test the proposed hypotheses (Ida et al., 2012; Kamarul, 2012). This is achieved through the following tests: (a) the coefficient of determination (R^2) by measuring the amount of explained variance of each latent variable, which should be 0.01, 0.09, and 0.25 indicating small, medium and large exploratory power (Mitchell & Jolley, 2013); (b) path coefficient by measuring the path estimates and t-statistics, which should be 0.02, 0.15, and 0.35 indicating small, medium and large relationships (Cohen, 1988; Henseler et al., 2009); (c) effect size (f^2) by measuring the relative impact of a particular exogenous latent variable on an endogenous latent variable by means of changes in the R^2 of the latent variable, which should be 0.02, 0.15, and 0.35 indicating small, medium and large effect (Ida et al., 2012), and (d) predictive relevance (Q^2) by measuring how well observed values are reconstructed by the model and its parameter estimates, which should be higher than zero (Chin, 2010).

The results of analyzing the structural model demonstrated an adequate and valid model. The R^2 values for knowledge sharing behavior and intention were large demonstrating strong explanatory power. Meanwhile the R^2 value for attitude was moderate demonstrating modest explanatory power, and the R^2 value for controllability was small demonstrating weak explanatory power as displayed in Tables 5.

Table 4. Convergent and discriminant validity

Construct	AVE	√AVE
Knowledge Sharing Behavior	0.706	0.840
Intention	0.779	0.882
Attitude	0.783	0.885
Subjective Norms	0.712	0.845
Self-Efficacy	0.809	0.900
Controllability	0.826	0.909
Trust	0.7568	0.870
Reputation	0.8270	0.909
Lack of Time	0.7114	0.843
Poor Communication	0.6372	0.798

Table 5. Coefficient of determination

Construct	R^2	Power
Knowledge Sharing Behavior	0.260	Large
Intention	0.471	Large
Attitude	0.243	Moderate
Controllability	0.063	Small

Table 6 shows that all path coefficient estimates, except one, ranged from -0.164 to 0.510 establishing small, medium, and large relationships. The controllability–intention relationship was -0.010 demonstrating a weak relationship below the recommended value of 0.02, which indicates being too weak to be considered relevant from a practical point of view, that may occur with large sample sizes (Kock, 2012).

The effect size (f^2) values were within the recommended values ranging from 0.002 to 0.219 demonstrating small and medium effect sizes of the independent variables as seen in Table 7.

Table 8 shows the predictive relevance (Q^2) values of the dependent variables. All values were above the recommended value zero indicating an adequate predictive relevance of the model.

Assessment of Mediating Relationship

Mediation assessment provides accurate information whether a mediating variable actually mediates the relation between two other variables (MacKinnon & Fairchild, 2009). An assessment of the significance of the mediation relationship gives a

Table 6. Path coefficients and magnitude

Path	Path Coefficient	Path Magnitude
Intention→ Knowledge sharing behavior	0.510	Large
Attitude → Intention	0.410	Large
Subjective Norms→ Intention	0.232	Medium
Self-Efficacy→ Intention	0.217	Medium
Controllability→ Intention	-0.010	Weak
Trust→ Attitude	0.333	Medium
Reputation → Attitude	0.248	Medium
Lack of Time→ Controllability	-0.128	Small
Poor Communication→ Controllability	-0.164	Medium

Table 7. Effect size

Path	f^2	Effect Size
Attitude→ Intention	0.219	Medium
Subjective Norms → Intention	0.079	Small
Self-Efficacy →Intention	0.049	Small
Controllability → Intention	0.002	Small
Trust→Attitude	0.120	Small
Reputation →Attitude	0.067	Small
Lack of Time →Controllability	0.015	Small
Poor Communication →Controllability	0.023	Small

Table 8. Predictive relevance

Construct	Q^2
Knowledge Sharing Behavior	0.180
Intention	0.364
Attitude	0.183
Controllability	0.044

complete picture of the roles of the mediating variables and provides practical results regarding cause-effect relationships (Hair et al., 2013). The assessment is done by running the PLS algorithm test on the relationship between the independent and dependent variables without the mediator and then with the mediator. Based on the results, it can be judged whether the mediator is full or partial (Guenzi et al., 2009; Ida et al., 2012).

In this research the model is characterized by its complexity for containing three mediators: intention, attitude, and controllability. Table 9 shows the results for the mediator intention indicating that while intention fully mediates between attitude, subjective norms, controllability and knowledge sharing behavior, it partially mediates between self-efficacy and knowledge sharing behavior.

Table 10 shows the results for the mediator attitude indicating that it fully mediates between trust, reputation and intention.

Table 11 shows the results for the mediator controllability indicating that it partially mediates between lack of time and intention, while it has no mediation effect between poor communication and intention.

Table 9. Results of mediator intention

IV	DV	β/t-value wzout mediator	β/t-value wz mediator	Mediating Effect
Attitude	Knowledge Sharing Behavior	β: 0.147 t: 2.640	β:-0.002 t: 0.056	Full
Subjective Norms	Knowledge Sharing Behavior	β: 0.187 t: 3.057	β: 0.099 t: 1.634	Full
Self-Efficacy	Knowledge Sharing Behavior	β: 0.211 t: 2.860	β: 0.129 t: 1.851	Partial
Controllability	Knowledge Sharing Behavior	β: 0.088 t: 1.539	β: 0.095 t: 1.657	Full

Table 10. Results of mediator attitude

IV	DV	β & t-values wzout mediator	β & t-values wz mediator	Mediating Effect
Trust	Intention	β: 0.214 t: 3.479	β:0.021 t: 0.599	Full
Reputation	Intention	β: 0.226 t: 3.655	β: 0.086 t: 1.618	Full

Table 11. Results of mediator controllability

IV	DV	β & t-values wzout mediator	β & t-values wz mediator	Mediating Effect
Lack of time	Intention	β: -0.199 t: 3.231	β: -0.145 t: 2.254	Partial
Poor communication	Intention	β: -0.043 t: 0.896	β: 0.001 t: 0.035	No mediation

Hypotheses Testing

The research hypotheses were tested using the results of the path coefficient. Both path estimates and their t-statistics with p-values are used to support or refute the hypotheses. Path coefficient values were calculated using PLS algorithm test, while t-statistics values were calculated using bootstrapping test, and p-values were calculated using an online free calculator. Path coefficient values of 0.02, 0.15, and 0.35 indicate small, medium, and large relationships (Cohen, 1988). Meanwhile, significant t-statistic values for a two-tailed test are 1.65, 1.96, and 2.59 at P-values

0.1, 0.05, and 0.01 respectively (Hair et al., 2013). Table 12 shows the results of the hypotheses testing.

As seen in the table, the analysis results show that the research hypotheses (H1 – H4) and (H6 – H9) are supported, whereas H5 is not supported. Moreover, the t-statistics values ranged from 2.044 to 11.667 demonstrating significant levels. Figure 2 shows the results of the assessment of the measurement and structural models displaying the path coefficients on top of the arrows, the t-statistics below the arrows, and the R^2 values inside the circles.

MODEL FIT

The model validation in PLS-SEM focuses on the model predictive capability due to being a variance-based one that is strongly oriented to prediction (Vinzi et al., 2010). Because all parts of the model need to be validated; the measurement model, the structural model and the overall model, PLS path modeling has three fit indices: the average communality index, the average redundancy index, and the Goodness of Fit index (ibid).

The average communality index (H^2) is used to measure the global quality of the measurement model. The index values should be positive for all blocks of variables (Tenenhaus, Vinzi, Chatelin & Lauro, 2005), and should be greater than the threshold of Zero (Fornell & Cha, 1994) to indicate an acceptable level of the measurement model quality. By using the PLS algorithm test, the average communality index is

Table 12. Hypotheses testing

No	Hypothesis	Path Coefficients	T-statistics	P-value	Result
1	Intention → Knowledge sharing behaviour	0.510	11.667**	0.0000	Supported
2	Attitude → Intention	0.410	7.857**	0.0000	Supported
3	Subjective Norms → Intention	0.232	4.103**	0.0001	Supported
4	Self-efficacy → Intention	0.217	3.614**	0.0004	Supported
5	Controllability → Intention	-0.010	0.211	-	Not supported
6	Trust →Attitude	0.333	6.306**	0.0000	Supported
7	Reputation →Attitude	0.248	4.424**	0.00001	Supported
8	Lack of Time →Controllability	-0.128	2.044*	0.0418	Supported
9	Poor Communication →Controllability	-0.164	2.661**	0.0082	Supported

(* Significance at t value $\geq$1.96 with $p \leq 0.05$, **Significance at t value $\geq$ 2.59 with $p \leq 0.01$)

Figure 2: Results of Path Coefficients & T-statistics

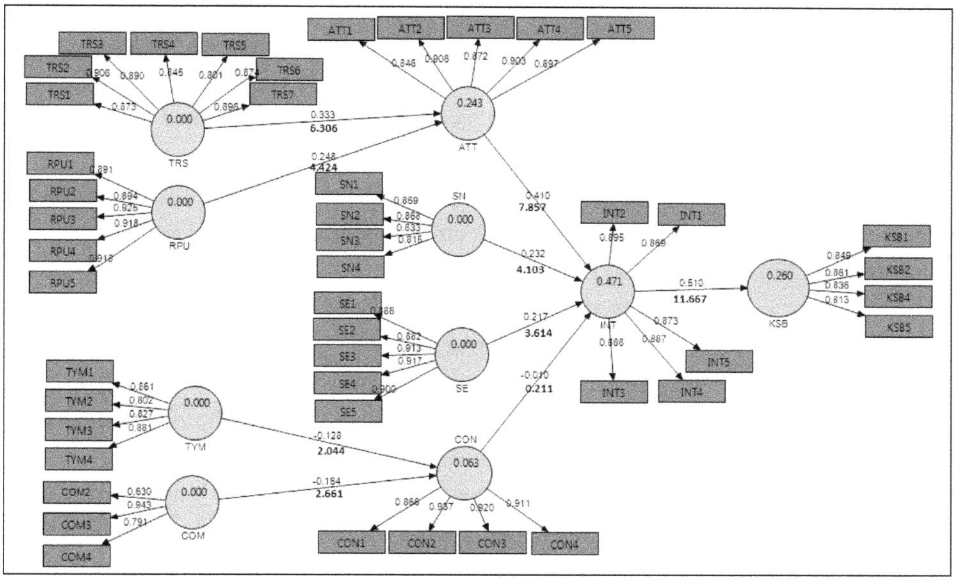

calculated. Table 13 below displays the results of the average communality index of the model, showing that the (H^2) value of the model is 0.755 indicating an acceptable level of the measurement model quality.

The average redundancy index (F^2) is used to measure the global quality of the structural model. The index values should be positive for all the endogenous blocks

Table 13. Average communality index

Construct	H^2
ATT	0.783
COM	0.637
CON	0.826
INT	0.779
KSB	0.706
RPU	0.827
SE	0.809
SN	0.712
TRS	0.757
TYM	0.711
Average	**0.755**

of variables (Tenenhaus et al., 2005), and should be greater than the threshold of 0 (Fornell & Cha, 1994) to indicate an acceptable level of the structural model quality. By using the PLS algorithm test, the redundancy index is calculated. Table 14 below displays the results of the average redundancy index of the model, showing that the (F^2) value of the model is 0.144 indicating an acceptable level of the structural model quality.

The GoF index represents an operational method that verifies and confirms the PLS model performance at the measurement and structural models with focus on the overall performance of the model (Chin, 2010; Tenenhaus et al., 2005; Vinzi et al., 2010). The GoF validates the PLS model globally and is the geometric mean of the average communality H^2 and the average R^2 (Garson, 2012; Henseler & Sarstedt, 2013; Vinzi et al., 2010). The average communality is an index measuring the predictive performance of the measurement model, while the average R^2 is an index measuring the predictive performance of the structural model.

The GoF value varies from 0 to 1 (Vinzi & Russolillo, 2013; Vinzi et al., 2010) with the threshold values of 0.1, 0.25, and 0.36 indicating small, medium, and large goodness of fit respectively (Latan & Ghozali, 2012; Wetzels, Odekerken-Schroder & Van Oppen, 2009). The GoF value is calculated by applying the below formula. So by using the H^2 and R^2 values, the GoF of the research model was measured, which was 0.443.

$$\text{GoF} = \sqrt{\left(\text{Com X R}^2\right)}$$

$$\text{GoF} = \sqrt{\left(0.755 \text{ X}0.259\right)}$$

$$\text{GoF} = \sqrt{.196}$$

Table 14. Average redundancy index

Construct	F^2
ATT	0.102
CON	0.036
INT	0.258
KSB	0.181
Average	**0.144**

$$\mathbf{GoF} = 0.443$$

The result shows that the model goodness of fit is adequately large indicating appropriate global model validity.

DISCUSSION

The results showed a great extent of knowledge sharing practice among academics in UAE public universities. Majority of the respondents agreed or strongly agreed with high percentages to the provided statements investigating their practice of knowledge sharing. Generally the results were similar to those of previous studies (Ahmad & Daghfous, 2010; Alajmi et al., 2008; Al-Alawi et al., 2007; Haapalainen & Makiranta, 2013; Skok & Tahir, 2010). Moreover, one may argue that the results reflect the Arab culture where socializing and interacting with others are prevailing in workplace and society.

Moreover, the results revealed that academics' intention to share knowledge was a significant predictor of their knowledge sharing behavior, where intention presented significantly strong positive influence explaining 26% of the variance in knowledge sharing behavior. The effect of intention indicates that due to the nature of the academics' job, they normally have the intention, willingness, and readiness to share their knowledge and experience with others knowing it could help them to gain knowledge in return. The finding is consistent with the Theory of Planned Behavior (Ajzen, 1995, 1991, 2002), and with previous studies (Alajmi, 2011; Babalhavaeji & Kermani, 2011; Chen et al., 2009; Minbaeva & Pedersen, 2010; Tohidinia & Mosakhani, 2010). The results also showed that academics' intention is significantly influenced by attitude towards knowledge sharing, subjective norms, and self-efficacy. The results demonstrated that attitude and self-efficacy had strong positive effect while subjective norms had moderate positive effect on intention. Collectively, attitude, subjective norms, and self-efficacy explained 47% of the variance in intention. The effect of attitude indicates that academics regard knowledge sharing as a valuable and useful act, the effect of subjective norms indicates that academics care about the opinions of their seniors, peers, and colleagues of knowledge sharing and view these opinions as important, meanwhile the effect of self-efficacy indicates that academics have high level of confidence in their abilities to perform knowledge sharing. This finding is also consistent with the Theory of Planned Behavior (Ajzen, 1995, 1991, 2002), as well as with previous studies (Alajmi, 2011; Babalhavaeji & Kermani, 2011; Chennamaneni, 2006; Elogie & Asemota; 2013; Kuang et al., 2012; Lin & Lee, 2004; Minbaeva & Pedersen, 2010; Ryu et al., 2003).

However, contrary to the theory, the results found that controllability does not have any influence on academics' intention to share knowledge. It can be argued that the reason behind this result is that individuals' intention to perform or not perform certain behavior depends somehow on non-motivational factors as availability of requisite opportunities and resources such as technology, time, money, skills, or cooperation (Ajzen, 1991, Zack, 1999). Also it can be argued that perceived controllability is interrelated with self-efficacy (Chiou, 1998). Yet, the result in this research is consistent with that of Alajmi (2011), who found that individuals' controllability is not a predictor of their intention. It is worth mentioning that in her study, Alajmi examined individuals' intention to share knowledge in online communities and was conducted in an American context.

In addition, the results proved that attitude is significantly and positively influenced by trust and reputation as motivators of knowledge sharing behavior. Trust and reputation explained more than 24% of the variance in attitude. The effect of trust indicates that there are mutual understanding and belief in each other's good wills and competencies, wich helps academics to build strong relationships and thus share knowledge. Meanwhile, the effect of reputation indicates that academics tend to share knowledge as a way of enhancing professional image and impressing others and hence improving their prestige. The results are similar to those of (Abzari et al., 2011; Chennamaneni, 2006; Hung et al., 2010; Rad et al., 2011; Seba et al., 2012; Wu & Zhu, 2012). Finally, the results showed that controllability is significantly and negatively influenced by lack of time and poor communication as barriers of knowledge sharing behavior. Lack of time and poor communication explained 6.3% only of the variance in controllability. The negative effect of lack of time indicates that the work pressure and load prevents academics from allocating time to share knowledge, while the negative effect of poor communication indicates that academics do not have the proper communication skills to engage in verbal and written discussions and that there is lack of interaction among academics. The results are similar to those of (Jain et l., 2007; Ling et al., 2009; Nassuora & Shahizan, 2010; Rowely et al., 2012; Santos et al., 2012; Seba et al., 2012).

SIGNIFICANCE AND CONTRIBUTION

The significance of this research lies in the fact that it is the first to address knowledge sharing in higher education sector in UAE with particular emphasis on academics' knowledge sharing behavior. Considering their important role in creating and disseminating knowledge and the importance of knowledge sharing in achieving universities' goals, the research findings provide significant practical implications for higher education institutions and decision makers in UAE.

Although the findings of this research are somehow consistent with those of previous studies, despite the different setting, they do contribute to the literature of knowledge sharing by addressing knowledge sharing behavior among academics in a new setting and environment i.e. UAE and universities. Majority of previous researches have either studied intention to share knowledge or knowledge sharing, and the few that studied knowledge sharing behavior targeted employees and managers and were conducted in Western or Southern-Asian contexts.

Furthermore, the research extends prior researches on the Theory of Planned Behavior by providing empirical evidence of the determinants of knowledge sharing behavior in new context and new setting, i.e. higher education in UAE. Of equal importance, the research sheds light on the new construct in the theory, i.e. controllability and its role and effect on the individual's intention. This research is the second to adopt the amended Theory of Planned Behavior (2002) after Alajmi's research (2011), while majority of researches still adopt the theory in its earlier versions (Ajzen, 1985, 1991).

FUTURE RESEARCH DIRECTIONS

This research investigated the practice of knowledge sharing and identified the predictors influencing knowledge sharing behavior from a psychological behavioral perspective. Based on the findings, some recommended future research opportunities include:

1. Other individual, organizational, and technological factors should be explored to determine their influence on knowledge sharing behavior. Some of these factors include enjoyment in helping others, peers' influence, loss of knowledge power, lack of social network, organizational structure and culture, commitment, leadership, incentives, ICT, technical support, IT literacy, and knowledge management systems.
2. The result that controllability does not affect intention certainly calls for further researches on this particular construct in the context of knowledge sharing behavior.
3. Replicating the current research in other countries in the Arab world in different sectors is essential to yield comparable results and explore knowledge sharing process in the Arab world.
4. Replicating the current research in other academic institutions such as private universities or technical colleges in other Arab world countries would help exploring any differences between public and private institutions and help validating the research findings.

5. Taking into account, the existence of people from different cultures and non-Middle Eastern origin in UAE and the Arab World, more research is required to explore the influence of cultural attributes and principles on knowledge sharing behavior.
6. Using some qualitative research methods such as interviews, case studies, or longitudinal studies, as well as collecting longitudinal data in future researches may help support the results of the current research

CONCLUSION

This research succeeded in filling the gap in literature on knowledge sharing behavior where the findings support previous researches that have explained the complicated nature of knowledge sharing behavior; however, the findings contribute due to being conducted a new setting, i.e. UAE with particular emphasis on a newly explored context there; higher education, which has not been addressed in previous researches about knowledge sharing in the Arab world.

The need to explore knowledge sharing in higher education is reinforced by the important role of universities in creating and providing knowledge, as well as by the major role of academics as valuable resources of knowledge and wisdom, where knowledge sharing can help them in their scholarly works. At the same time, the findings encourage higher education institutions to adopt knowledge sharing practices through providing necessary activities and tools. This would create a knowledge sharing culture.

REFERENCES

Abu-Ghazaleh, T. (2012). *Abu-Ghazaleh announces the UAE as knowledge and connectivity hub for the Arab region at e-AGE 2012*. Academic Press.

Abzari, M., Barzaki, A. S., & Abbasi, R. (2011). Knowledge sharing behavior: Organizational reputation or losing organizational power: Case study. *International Journal of Business and Social Science*, *2*(17), 144–149.

Ahmad, N., & Daghfous, A. (2010). Knowledge sharing through inter-organizational knowledge networks: Challenges and opportunities in the United Arab Emirates. *European Business Review*, *22*(2), 153–174. doi:10.1108/09555341011023506

Ajzen, I. (1985). From intentions to actions: A theory of planned behavior. In J. Kuhl & J. Beckman (Eds.), *Action control: From cognition to behavior* (pp. 11–39). Berlin, Germany: Springer-Verlag. doi:10.1007/978-3-642-69746-3_2

Ajzen, I. (1991). The theory of planned behavior. *Organizational Behavior and Human Decision Processes, 50*(2), 179–211. doi:10.1016/0749-5978(91)90020-T

Ajzen, I. (2002). Perceived behavioral control, self-efficacy, locus of control, and the theory of planned behavior. *Journal of Applied Social Psychology, 32*(4), 665–683. doi:10.1111/j.1559-1816.2002.tb00236.x

Ajzen, I. (2006). *Constructing a theory of planned behavior questionnaire*. Retrieved December 1, 2012, from http://people.umass.edu/aizen/pdf/tpb.measurement.pdf

Akhavan, P., Ghojavand, S., & Abdali, R. (2012). Knowledge sharing and its impact on knowledge creation. *Journal of Information & Knowledge Management, 11*(2), 1–12. doi:10.1142/S0219649212500128

Al-Adaileh, R., & Al-Atawi, M. (2011). Organizational culture impact on knowledge: Saudi telecom context. *Journal of Knowledge Management, 15*(2), 212–230. doi:10.1108/13673271111119664

Al-Alawi, A. I., Al-Marzooqi, N. Y., & Mohammed, Y. F. (2007). Organizational culture and knowledge sharing: Critical success factors. *Journal of Knowledge Management, 11*(2), 22–42. doi:10.1108/13673270710738898

Al-Nahyan, N. b. M. (2012b). *Education in the United Arab Emirates (UAE)*. The Prospect Group.

Al-Nahyan, N. b. M. (2012a). *Executive focus: Government commitment and investment in education, research initiatives, and improving the quality of education.* UAE Minister of Higher Education & Scientific Research HH Sheikh Nahyan bin Mubarak Al Nahyan.

Al-Shammari, M. (2008). Toward a knowledge management strategic framework in the Arab region. *International Journal of Knowledge Management, 4*(3), 44–63. doi:10.4018/jkm.2008070104

Alajmi, B. M. (2011). *The intention to share: Professional's knowledge sharing behaviors in online communities* (Doctoral thesis). The State University of New Jersey.

Alajmi, B. M., McInerney, C. R., Orzano, A. J., Tallia, A. F., Meese, A., & Vamanu, I. (2008). Knowledge sharing processes and tools in U.S.A. primary health care: Analysis of four case studies. In *Proceedings of the International Conference on Knowledge Management 2008*. Hackensack, NJ: World Scientific doi:10.1142/9789812837578_0009

Alavi, M., & Leidner, D. (2001). Knowledge management and knowledge management systems: Conceptual foundations and research issues. *Management Information Systems Quarterly*, *25*(1), 107–136. doi:10.2307/3250961

Albres, J. A. (2009). A practical approach to implementing knowledge management. *Journal of Knowledge Management Practice*, *10*(1).

Babalhavaeji, F., & Kermani, Z. J. (2011). Knowledge sharing behaviour influences: A case of library and information science faculties in Iran. *Malaysian Journal of Library and Information Science*, *16*(1), 1–14.

Badaracco, J. (2010). Effective knowledge sharing and its benefits. *The Point Newspaper Online*.

Bandura, A. (1997). Self-efficacy: The exercise of control. New York: W.H. Freeman. British Journal of Social Psychology, 44, 479-496.

Batinic, B., Reips, U., & Bosnjak, M. (2002). *Online social sciences*. Seattle, WA: Hogrefe & Hubber Publisher.

Battour, M. (2011). *The impact of the Islamic attributes of destination on tourists' motivation, satisfaction and destination loyalty* (Unpublished doctoral dissertation). University of Malaya.

Biygautane, M., & Al-Yahya, K. (2011). *Knowledge management in the UAE's public sector: The case of Dubai*. Paper presented at the Gulf Research Meeting Conference at the University of Cambridge.

Bock, G. W., Zmud, R. W., Kim, Y. G., & Lee, J. N. (2005). Behavioral intention formation in knowledge sharing: Examining the roles of extrinsic motivators, social-psychological forces, and organizational climate. *Management Information Systems Quarterly*, *29*(1), 87–111.

Bordia, P., Irmer, B. E., Garden, M., Phair, K., & Abusah, D. (2004). *Knowledge sharing in response to a supportive work environment: evidence from an Australian engineering firm*. Paper presented at the iKMS International Conference on Knowledge Management, Singapore.

Burgess, D. (2005). What motivates employees to transfer knowledge outside their work unit? *Journal of Business Communication*, *42*(4), 324–348. doi:10.1177/0021943605279485

Chaudhry, A. S., & Sivakamasundari, B. (2004). *Perceptions of teachers about knowledge sharing in schools.* Paper presented at the Information Resources Management Association International Conference 2004, New Orleans, LA.

Chen, I., & Chen, N.S., & Kinshuk. (2009). Examining the factors influencing participants' knowledge sharing behavior in virtual learning communities. *Journal of Educational Technology & Society*, *12*(1), 134–148.

Chennamaneni, A. (2006). *Determinants of knowledge sharing behaviors: Developing and testing an integrated theoretical model* (Doctoral dissertation). The University of Texas at Arlington.

Chikoore, L., & Ragsdell, G. (2013, March). Knowledge sharing in higher education: A study of students preparing assessed group work. *Journal of Knowledge Management Practice*, *14*(1).

Chin, W. W. (2010). How to write up and report PLS analyses. In V. E. Vinzi, W. W. Chin, J. Henseler, & H. Wang (Eds.), *Handbook of partial least squares: Concepts, methods and applications* (pp. 655–690). Berlin, Germany: Springer-Verlag. doi:10.1007/978-3-540-32827-8_29

Chiou, J.-S. (1998). The effects of attitude, subjective norm, and perceived behavioral control on consumers' purchase intentions: The moderating effects of product knowledge and attention to social comparison information. *Proceedings of the National Science Council*, *9*(2), 298–308.

Chugh, R. (2012). Knowledge sharing with enhanced learning and development opportunities. *IEEEXplore*, 100-104.

Cohen, J. (1988). *Statistical power analysis for the behavioral sciences.* Hillsdale, NJ: Lawrence Erlbaum Associates.

Creswell, J. (2009). *Research design: Qualitative, quantitative, and mixed methods approaches.* Thousand Oaks, CA: Sage Publications.

Cummings, J. N. (2004). Work groups, structural diversity and knowledge sharing in a global organization. *Management Science*, *50*(3), 352–364. doi:10.1287/mnsc.1030.0134

Dokhtesmati, M., & Bousari, R. G. (2013). Knowledge sharing in Iranian academic institutions: Meta-analysis approach. *Procedia: Social and Behavioral Sciences, 73*, 383–387. doi:10.1016/j.sbspro.2013.02.065

Durmusoglu, S., Jacobs, M., Nayir, D., Khilji, S., & Wang, X. (2014). The quasi-moderating role of organizational culture in the relationship between rewards and knowledge shared and gained. *Journal of Knowledge Management, 18*(1), 19–37. doi:10.1108/JKM-05-2013-0183

Earl, M. (2001). Knowledge management strategies: Toward a taxonomy. *Journal of Management Information Systems, 18*(1), 215–233. doi:10.1080/07421222.2001.11045670

Egger, U.K. (2013). *Cross Organisational Knowledge Sharing*. Academic Press.

Elogie, A. A., & Asemota, A. E. (2013). Knowledge sharing behaviour of academics in the University of Ibadan, Nigeria. *PNLA Quarterly, 77*(4), 1–16.

Ensign, P. C., & Hebert, L. (2010, Winter). How reputation affects knowledge sharing among colleagues. *MIT Sloan Management Review, 51*(2), 78–82.

Federal Authority for Government Human Resources (FAHR). (2017). *Guide of Knowledge Management in the Federal Government.* Retrieved from https://www.fahr.gov.ae/Portal/Userfiles/Assets/Documents/a3f395eb.pdf

Fornell, C., & Cha, J. (1994). Partial least squares. In R. P. Bagozzi (Ed.), *Advanced methods of marketing research* (pp. 52–78). Cambridge, UK: Blackwell.

Fornell, C. G., & Larcker, D. F. (1981). Evaluating structural equation models with unobservable variables and measurement error. *JMR, Journal of Marketing Research, 18*(1), 39–50. doi:10.2307/3151312

Francis, J. J., Eccles, M. P., Johnston, M., Walker, A., Grimshaw, J., Foy, R., & Bonetti, D. et al. (2004). *Constructing questionnaires based on the theory of planned behavior: Manual for health services researchers*. University of Newcastle.

Fullwood, R., Rowley, J., & Delbridge, R. (2013). Knowledge sharing amongst academics in UK universities. *Journal of Knowledge Management, 17*(1), 123–136. doi:10.1108/13673271311300831

Garson, G. D. (2012). *Partial least squares: Regression and path modeling*. Statistical Associates Publishing.

Gay, L., & Airasian, P. (2003). *Educational research: Competencies for analysis and applications*. Pearson Education.

Guenzi, P., Georges, L., & Pardo, C. (2009). The impact of strategic account managers' behaviors on relational outcomes: An empirical study. *Industrial Marketing Management, 38*(3), 300–311. doi:10.1016/j.indmarman.2007.09.011

Gurbuz, T. (2008). *Knowledge management in education and research: An elective course at METU*. Paper presented at the 8th International Educational Technology Conference (IETC2008).

Haapalainen, P., & Makiranta, A. (2013, March). Acquiring and sharing knowledge in SMEs: A case study in the manufacturing industry. *Journal of Knowledge Management Practice, 14*(1).

Haas, M. R., & Hansen, M. T. (2005). Different knowledge, different benefits: Toward a productivity perspective on knowledge sharing in organizations. *Strategic Management Journal, 28*(11), 1133–1153. doi:10.1002/smj.631

Hair, J. F., Black, W. C., Babin, B. J., & Anderson, R. E. (2010). *Multivariate data analysis*. Pearson education.

Hair, J. F., Ringle, C. M., & Sarstedt, M. (2011, Spring). PLS-SEM: Indeed a silver bullet. *Journal of Marketing Theory and Practice, 19*(2), 139–151. doi:10.2753/MTP1069-6679190202

Hair, J. F. Jr, Ringle, C. M., & Sarstedt, M. (2013). Editorial: Partial least squares structural equation modeling: Rigorous applications, better results and higher acceptance. *Long Range Planning, 46*(1-2), 1–12. doi:10.1016/j.lrp.2013.01.001

Hansmann, K. W., & Ringle, C. M. (2005). Enterprise-networks and strategic success- An empirical analysis. In T. Theurl & E. C. Meyer (Eds.), *Strategies for cooperation* (pp. 133–152). Shaker Verlag.

Hassanzadeh, M. (2012). *Factors that affect scientists' behavior to share scientific knowledge.* Paper presented at the 7th International Conference on Webometrics, Informetrics and Scientometrics (WIS) and 12th COLLNET Meeting, Istanbul, Turkey. doi:10.1142/9789812702081_0013

Henseler, J., Ringle, C. M., & Sinkovics, R. R. (2009). The use of partial least squares path modeling in international marketing. *Advances in International Marketing, 20*, 277–319.

Henseler, J., & Sarstedt, M. (2013). Goodness-of-fit indices for partial least squares path modeling. *Computational Statistics, 28*(2), 565–580. doi:10.1007/s00180-012-0317-1

Hew, K. F., & Hara, N. (2007). Empirical study of motivators and barriers of teacher online knowledge sharing. *Educational Technology Research and Development, 55*(6), 573–595. doi:10.1007/s11423-007-9049-2

Hsu, I. (2008). Knowledge sharing practices as facilitating factor for improving organizational performance through human capital: A preliminary study. *Expert Systems with Applications, 35*(3), 1316–1326. doi:10.1016/j.eswa.2007.08.012

Hsu, M. H., Ju, T. L., Yen, C. H., & Chang, C. M. (2007). Knowledge sharing behavior in virtual communities: The relationship between trust, self-efficacy, and outcome expectations. *International Journal of Human-Computer Studies, 65*(2), 153–169. doi:10.1016/j.ijhcs.2006.09.003

Hung, S. Y., Lai, H. M., & Chou, Y. C. (2010). *The determinants of knowledge sharing intention in professional virtual communities: An integrative model.* Paper presented at the 14th Pacific Asia Conference in Information Systems, Taipei, Taiwan.

Ida, R. I., Roshayati, H., & Fazli, I. (2012). *PLS application in journal of operations management: A review.* Paper presented at the Global Conference on Operations and Supply Chain Management, Bandung, Indonesia.

Jain, K. K., Sandhu, M. S., & Sidhu, G. K. (2007). Knowledge sharing among academic staff: A case study of business schools in Klang Valley, Malaysia. *JASA, 2*, 23–28.

Kamarul Faizal Hashim. (2012). *Understanding the determinants of continuous knowledge sharing intention within business online communities* (Doctoral dissertation). Auckland University of Technology.

Kankanhalli, A., Tan, B. C., & Wei, K. K. (2005). Contributing knowledge to electronic knowledge repositories: An empirical investigation. *Management Information Systems Quarterly, 29*(1), 113–143.

Kock, N. (2013). *Easily calculating the GoF fit index with WarpPLS outputs.* Retrieved from http://warppls.blogspot.com/2013/01/easily-calculating-gof-fit-index-with.html

Kraft, P., Rise, J., Sutton, S., & Roysamb, E. (2005). Perceived difficulty in the theory of planned behaviour: Perceived behavioural control or affective attitude? *British Journal of Social Psychology, 44*(3), 479–496. doi:10.1348/014466604X17533 PMID:16238850

Krejcie, R. V., & Morgan, D. W. (1970). Determining sample size for research activities. *Educational and Psychological Measurement*, *30*(3), 607–610. doi:10.1177/001316447003000308

Kuang, L. C., Davidson, E., & Yao, L. (2012). *A planned behavior-based investigation of knowledge sharing in construction industry*. Paper presented at the International Management and Engineering Conference.

Latan, H., & Ghozali, I. (2012). Partial least squares: Concept, technique and application using program SmartPLS for empirical research. In V. Pez, R. Butori, & G. C. de Kerviler (Eds.), *Because I'm worth it: The impact of given versus perceived status on preferential treatment efficiency. Academic Press.*

Lilleoere, A. M., & Hansen, E. H. (2011). Knowledge sharing enablers and barriers in pharmaceutical research and development. *Journal of Knowledge Management*, *15*(1), 53–70. doi:10.1108/13673271111108693

Lin, H. F., & Lee, G. G. (2004). Perceptions of senior managers toward knowledge sharing behaviour. *Journal of Management Decision*, *42*(1), 108–125. doi:10.1108/00251740410510181

Ling, C. W., Sandhu, M., & Jain, K. K. (2009). Knowledge sharing in an American multinational company based in Malaysia. *Journal of Workplace Learning*, *21*(2), 125–142. doi:10.1108/13665620910934825

MacKinnon, D. P., & Fairchild, A. J. (2009). Current directions in mediation analysis. *Current Directions in Psychological Science*, *18*(1), 16–20. doi:10.1111/j.1467-8721.2009.01598.x PMID:20157637

Madden, T. J., Ellen, P. S., & Ajzen, I. (1992). A comparison of the theory of planned behavior and the theory of reasoned action. *Personality and Social Psychology Bulletin*, *18*(1), 3–9. doi:10.1177/0146167292181001

Maktoum, A. M. b. R. (2007). *Mohammed addresses the knowledge conference*. Retrieved on February 17, 2011 from http://www.sheikhmohammed.co.ae/vgn-ext-

McDermott, R., & O'Dell, C. (2001). Overcoming cultural barriers to sharing knowledge. *Journal of Knowledge Management*, *5*(1), 76–85. doi:10.1108/13673270110384428

McNeil, C. M. (2003). Line managers: Facilitators of knowledge sharing in teams. *Employee Relations*, *25*(3), 294–307. doi:10.1108/01425450310475874

Minbaeva, D., & Pedersen, T. (2010). Governing individual knowledge-sharing behaviour. *International Journal of Strategic Change Management, 2*(2/3), 200–222. doi:10.1504/IJSCM.2010.034414

Mitchell, M. L., & Jolley, M. J. (2013). *Research design explained* (8th ed.). Wadsworth Cengage Learning.

Mohamed, M., O'Sullivan, K. J., & Ribière, V. (2008). A paradigm shift in the Arab region knowledge evolution. *Journal of Knowledge Management, 12*(5), 107–120. doi:10.1108/13673270810902975

Nahapiet, J., & Ghoshal, S. (1998). Social capital, intellectual capital, and the organizational advantage. *Academy of Management Review, 23*(2), 242–266.

Nassuora, A.B., & Hassan, S. (2010). *Knowledge sharing among academics in institutions of higher learning*. Retrieved October 27, 2013 from http://www.kmice.cms.net.my/ProcKMICe/KMICe2010/Paper/PG164_173.pdf

Nonaka, I. (1994). A dynamic theory of organizational knowledge creation. *Organization Science, 5*(1), 14–37. doi:10.1287/orsc.5.1.14

Nonaka, I., & Takeuchi, H. (1995). *The knowledge-creating company: How Japanese companies create the dynamics of innovation*. Oxford University Press.

Paloti, R. G. (2010). *Knowledge sharing*. Retrieved January 17, 2011. http://eprints.rclis.org/bitstream/10760/15179/3/KnowledgeSharinFinal.pdf

Pangil, F., & Chan, J. M. (2014). The mediating effect of knowledge sharing on the relationship between trust and virtual team effectiveness. *Journal of Knowledge Management, 18*(1), 1–24. doi:10.1108/JKM-09-2013-0341

Patel, M., & Ragsdell, G. (2011, June). To share or not to share knowledge: An ethical dilemma for UK academics? *Journal of Knowledge Management Practice, 12*(2).

Rad, G. P., Alizadeh, N., Miandashti, N. Z., & Fami, H. S. (2011). Factors influencing knowledge sharing among personnel of agricultural extension and education organization in Iranian Ministry of Jihad-e Agriculture. *Journal of Agricultural Science and Technology, 13*(4), 491–501.

Ramlee, A. R. (2011). *Knowledge sharing practices among researchers and officers of the national institutes of health Malaysia* (Unpublished doctoral dissertation). International Islamic University Malaysia, Kuala Lumpur, Malaysia.

Reychav, I., & Weisberg, J. (2009). Good for workers, good for companies: How knowledge sharing benefits individual employees. *Knowledge and Process Management, 16*(4), 186–197. doi:10.1002/kpm.335

Riege, A. (2005). Three-dozen knowledge-sharing barriers managers should consider. *Journal of Knowledge Management, 9*(3), 18–35. doi:10.1108/13673270510602746

Rowley, J., Seba, I., & Delbridge, R. (2012). Knowledge sharing in the Dubai Police Force. *Journal of Knowledge Management, 16*(1), 1–24.

Ryu, S., Ho, S. H., & Han, I. (2003). Knowledge sharing behavior of physicians in hospitals. *Expert Systems with Applications, 25*(1), 113–122. doi:10.1016/S0957-4174(03)00011-3

Sallis, E., & Jones, G. (2002). *Knowledge Management in Education: Enhancing learning and education*. London: Kogan Page.

Sandhawalia, B. S., & Dalcher, D. (2011). Effective knowledge sharing for competitive advantage. *Journal of Knowledge Management, 15*(2).

Sandhu, M. S., Jain, K. K., & Umi Kalthom, A. (2011). Knowledge sharing among public sector employees: Evidence from Malaysia. *International Journal of Public Sector Management, 24*(3), 206–226. doi:10.1108/09513551111121347

Santos, V. R., Soares, A. L., & Carvalho, J. A. (2012). Knowledge sharing barriers in complex research and development projects: An exploratory study on the perceptions of project managers. *Knowledge and Process Management, 19*(1), 27–38. doi:10.1002/kpm.1379

Seba, I., Rowley, J., & Lambert, S. (2012). Factors affecting attitudes and intentions towards knowledge sharing in the Dubai Police Force. *International Journal of Information Management, 1120*, 1–9.

Sekaran, U. (2003). *Research methods for business: A skill building approach (4th ed.)*. New York: John Wiley & Sons.

Sharma, A. (2010). *Enabling knowledge management of organizational memory for groups through shared topic maps* (Master dissertation). Iowa State University.

Siddique, C. M. (2012). Knowledge management initiatives in the United Arab Emirates: A baseline study. *Journal of Knowledge Management, 16*(5).

Simonis, J. (2010). *A guide to question types: An art and a science*. Academic Press.

Singer, P., & Hurley, J. E. (2005). The Importance of Knowledge Management Today. *ALA-APA Library Worklife Home, 2*(6), 1–3.

Skok, W., & Tahir, S. (2010). Developing a knowledge management strategy for the Arab World. *The Electronic Journal on Information Systems in Developing Countries*, *41*(7), 1–11.

Sohail, S., & Daud, S. (2009). Knowledge sharing in higher education institutions: Perspectives from Malaysia. *The Journal of Information and Knowledge Management Systems*, *39*(2), 125–142.

Sveiby, K. E. (2002). A knowledge-based theory of the firm to guide strategy formulation. *Journal of Intellectual Capital*, *2*(4), 344–358. doi:10.1108/14691930110409651

Syed, S. A., Zaini, A., Noormala, A. I., & Zahariah, M. Z. (2009). Assessing knowledge sharing behaviour among employees in SMEs: An empirical study. *International Business Research*, *2*(2), 115–122.

Tan, C.N., Lye, Y.H., Ng, T.H., & Lim, Y.S. (2010). Motivational factors in influencing knowledge sharing among banks in Malaysia. *International Research Journal of Finance and Economics*, (44), 191-201.

Tenenhaus, M., Vinzi, V. E., Chatelin, Y. M., & Lauro, C. (2005). PLS path modeling. *Computational Statistics & Data Analysis*, *48*(1), 159–205. doi:10.1016/j.csda.2004.03.005

Tohidinia, Z., & Mosakhani, M. (2010). Knowledge sharing behavior and its predictors. *Industrial Management & Data Systems*, *110*(4), 611–631. doi:10.1108/02635571011039052

Vinzi, V. E., & Russolillo, G. (2013). Partial least squares algorithms and methods. *Wiley Interdisciplinary Reviews: Computational Statistics*, *5*(1), 1–19. doi:10.1002/wics.1239

Vinzi, V. E., Trinchera, L., & Amato, S. (2010). PLS path modeling: From foundations to recent developments and open issues for model assessment and improvement. In V. E. Vinzi, W. W. Chin, J. Henseler, & H. Wang (Eds.), *Handbook of partial least squares: Concepts, methods and applications* (pp. 47–82). Berlin, Germany: Springer-Verlag. doi:10.1007/978-3-540-32827-8_3

Vogt, W. P. (2007). *Quantitative research methods for professionals*. Boston, MA: Pearson Allyn and Bacon.

Wang, S., & Noe, R. A. (2010). Knowledge sharing: A review and directions for future research. *Human Resource Management Review*, *20*(2), 115–131. doi:10.1016/j.hrmr.2009.10.001

Wasko, M. M., & Faraj, S. (2005). Why should I share? Examining social capital and knowledge contribution in electronic networks of practice. *Management Information Systems Quarterly*, *29*(1), 35–57.

Weathington, B. L., Cunningham, C. J., & Pittenger, D. J. (2010). *Research methods for the behavioral and social sciences*. Hoboken, NJ: John Wiley & Sons, Inc.

Wetzels, M., Odekerken-Schroder, G., & Van Oppen, C. (2009). Using PLS path modeling for assessing hierarchical construct models: Guidelines and empirical illustration. *Management Information Systems Quarterly*, *33*(1), 177–196.

Wu, Y., & Zhu, W. (2012). An integrated theoretical model for determinants of knowledge sharing behaviours. *Kybernetes*, *41*(10), 1462–1482. doi:10.1108/03684921211276675

Xiao, Y. C., & Jin, Y. H. (2010). The hierarchical linear modeling of shared mental model on virtual team effectiveness. *Kybernetes*, *39*(8), 1322–1329. doi:10.1108/03684921011063619

Zack, M. H. (1999, Summer). Managing codified knowledge. *Sloan Management Review*, *40*(4), 45–58.

Zhang, P., & Ng, F. (2013). Explaining knowledge-sharing intention in construction teams in Hong Kong. *Journal of Construction Engineering and Management*, *139*(3), 280–293. doi:10.1061/(ASCE)CO.1943-7862.0000607

ENDNOTES

1 Federal universities are owned and financed by the federal government, where the Minister serves as the president/chancellor, while the provost performs the management tasks. Governmental universities are owned and financed by the government of the emirate. Semi-governmental universities are owned in partnership by the government of the emirate and the private sector, which acts as financer and composes majority of the board of trustees.

2 Federal universities: United Arab Emirates University and Zayed University. Governmental universities: Khalifa University of Science, Technology & Research, The Petroleum Institute University and Research Center, University College for Mother and Family Sciences, and Paris-Sorbonne University Abu Dhabi. Semi-governmental universities: Ajman University of Science and Technology, American University of Sharjah, Ras Al Khaima Medical and Health Sciences University, and University of Dubai.

Chapter 8
Knowledge-Intensive Evolutionary Algorithms for Solving a Healthcare Fleet Optimization Problem:
An Ontological Approach

Carlos Adrian Catania
National University of Cuyo, Argentina

Cecilia Zanni-Merk
INSA Rouen Normandie, France

François de Bertrand de Beuvron
INSA Strasbourg, France

Pierre Collet
Strasbourg University, France

ABSTRACT

In this chapter, the authors show how knowledge engineering techniques can be used to guide the definition of evolutionary algorithms (EA) for problems involving a large amount of structured data, through the resolution of a real problem. Various representations of the fitness functions, the genome, and mutation/crossover operators adapted to different types of problems (routing, scheduling, etc.) have been proposed in the literature. However, real problems including specific constraints (legal restrictions, specific usages, etc.) are often overlooked by the proposed generic models. To ensure that these constraints are effectively considered, the authors propose a methodology based on the structuring of the conceptual model underlying the problem, as a labelled domain ontology suitable for optimization by EA. The authors show that a precise definition of the knowledge model with a labelled domain ontology can be used to describe the chromosome, the evaluation functions, and the crossover and mutation operators. The authors show the details for a real implementation and some experimental results.

DOI: 10.4018/978-1-5225-5655-8.ch008

INTRODUCTION

In this article, we will show that the use of knowledge engineering can greatly enhance the definition of an evolutionary algorithm for a real case. This study is the result of a collaboration between our team and an Alsatian Patient Transportation Service (PTS), which has provided the real data.

The project is developed under the health care system environment, specifically oriented to the transportation of patients, normally from or to some health care center. The need to develop a specific application arises from the fact that companies that take care of the logistics of the trips must manage many requirements and restrictions when making an itinerary. This logistics affects many enterprise resources, like employees and vehicles, which should be assigned in an efficient way to guarantee, among other things, the satisfaction of the patient and the conformity of numerous law regulations.

The problem consists in satisfying the daily requests of the patients while maximizing the benefits, minimizing the costs and fulfilling different constraints. The requests are basically for pick-ups and/or deliveries of the patients to or from their house to some health care center.There are different types of vehicles than can be used for a journey and each of them has an associated cost. There are also the costs for the assignment of a crew, composed by a set of one or two employees, to a given vehicle or for a given patient.

This case of Vehicle Routing Problem (VRP) can be defined as a Rich Vehicle Routing Problem (RVRP). According to a recent literature review (Lahyani et al. 2014), RVRP is defined as a problem which simultaneously includes several types of challenging and complicating features. It is associated with the complexity of real-life routing problems.

Many studies concern the Vehicle Routing Problem (Raghavan&Wasil, 2008). These studies can be categorized along two axes:

- The solving approach; mainly exact algorithms or metaheuristics (stochastic or nature inspired algorithms) (Russell &Norvig, 2009).
- The variants of the problem; including time windows constraints (Giraldez et al. 2005) or multiple heterogeneous vehicles with pick-up and delivery (Tasan& Gen, 2012), among others.

We have chosen to follow an approach based on Evolutionary Algorithms (EAs) for this specific problem to solve, since we will show in this article that some families of EA and Genetic Algorithms (GA) in particular are suitable for a knowledge driven definition.Genetic algorithms are stochastic search algorithms based on abstraction of the process of Darwinian evolution. They are based on a simple principle: better individuals survive and reproduce more often than other individuals. A classic genetic algorithm maintains a population of individuals, each of them encodes a candidate solution to a given problem. Each individual is evaluated by a fitness function, which measures the quality of its corresponding candidate solution. Then, a selection mechanism identifies the fittest individuals of the current population to serve as parents of the next generation. Hence, the better the quality of an individual, the higher the probability it reproduces and breed to form a next generation. This process is repeated until some stopping criterion. Usual stopping criteria are that the fitness value is beyond a threshold or that a number ofgenerationshave been reached. To explore the search space, different genetic operators are used during the evolutionary process. Classic genetic algorithms use two operators: mutation and crossover. However, several other operators can be chosen according to the kind of problem.

Even if this problem belongs to the family of Rich Vehicle Routing Problems (RVRP), it soon became clear that the solutions proposed in the literature, as specific as they are (Parragh et al.,2008; Kilcher&Calvo, 2013; Paquette et al. 2013) were not intended to take into account all the specific constraints of this particular VRP problem. In addition, we believe that this situation is found in many other optimization problems when the parameters are numerous and varied in nature. Thus, in our example, legal constraints such as the working time of ambulance attendants, or medical constraints such as the disinfection of the vehicles, or the personnel qualification, cannot be overlooked while minimizing costs or distances.

In addition, even if the evolutionary optimization algorithm can be seen as the heart of this application, it is clear that it needs to be integrated into a system with more traditional management modules, such as, customer billing, calculation of working hours and remuneration of ambulance attendants according to the legislation in force, customer support, among others.

This entire legacy software environment, particularly the conceptual model of the information system, represents a body of knowledge and skills within the company, which should be used in the implementation of the optimization module. But strangely enough, we did not find in the literature any framework taking these needs into account. In fact, in the early days of genetic algorithms, much emphasis was put on the domain-independent nature of the basic crossover and mutation algorithms, working on a standard, binary coded, fixed length chromosome (Holland, 1975). The foundation of genetic algorithms relies on the exploitation of similarities (building blocks or schemas) in the chromosome. Using a specific chromosome representation and operators that emphasize meaningful building blocks in the application domain should improve the efficiency, as highlighted by C. Janikow (1993). This general idea has led to numerous specific genetic algorithms based on domain knowledge. For example, in our problem, the distribution of tasks between ambulances is a special case of a set partition problem. This kind of problem led to the definition of specialized so-called Grouping Genetic Algorithms (Falkenauer, 1998), and it is not surprising that some further specialization has produced very efficient genetic algorithms for the pickup and delivery problem (Pankratz, 2005). In addition to the coding of the chromosome, knowledge may be used in other parts of a genetic algorithm: in the fitness function or the initial population, or to keep collective knowledge among individuals (Cultural Algorithms) (Nguyen, 2007). A survey about such use of knowledge for the development of highly specialized EAs can be found in Giraldez et al. (2005).

Our goal is different: we want to use domain knowledge to assist in the definition of the algorithm. That is why we propose a methodology centered on an extended domain ontology for the definition of evolutionary algorithms whose parameters and constraints represent a huge volume of structured data (this is what we call *Knowledge-Intensive Evolutionary Algorithms*).

The rest of the paper is structured as follows. In *Methodology Overviews*, we present the main steps of the methodology we propose. In Methodology Overviews, we show the way to build an extended conceptual model to link the domain ontology with the EA own constructions. We then see how this structuring of the conceptual model can be used to define the evaluation functions (*Multi-Objective Optimization and Evaluation Functions*), the chromosome and the associated mutation/crossover operators (*Definition of the Chromosome and the Evolutionary Operators*), ensuring that all the specific constraints of the problem are taken into account. Finally, we present in *Experiment Setup* the preliminary results on the health care routing problem, and we give some conclusions and perspectives in the *Conclusion*

METHODOLOGY OVERVIEWS

We propose a methodology in three phases (Figure 1):

- **Analysis**: The main goal of this phase is to understand the domain in need of optimization.
    - *Input:* The data model associated to the problem and the optimization goals.
    - *Output:* A labeled domain ontology; that is, a classical domain ontology where some of the concepts are labeled according to the type of entity they represent (parameter entities, evaluation entities or entities to optimize) It is important to remark that an entity of the domain can be an evaluation entity and an entity to optimize at the same time.

- **Definition**: The goal of this phase is to define the genetic operators and the fitness functions
    - *Input:* The labeled domain ontology.
    - *Output:* The specification of the chromosome, the definition of the genetic operators, and several fitness functions.
- **Implementation**: here the goal is the production of an operational prototype. In the case we are describing in this article, two different implementations were produced: a first sequential version, developed classically with an object-oriented language; and a parallel version, to comply with constraints about the maximum time allowed to reformulate an existing schedule when an urgent trip appears.

Analysis

This phase consists in defining precisely the function to be optimized through a specific labeling of the domain ontology. The project objectives are identified by the domain experts, and a comprehensive list of constraints and costs related to the realization of these objectives is identified. A team around the domain experts,

Figure 1. Methodology overview

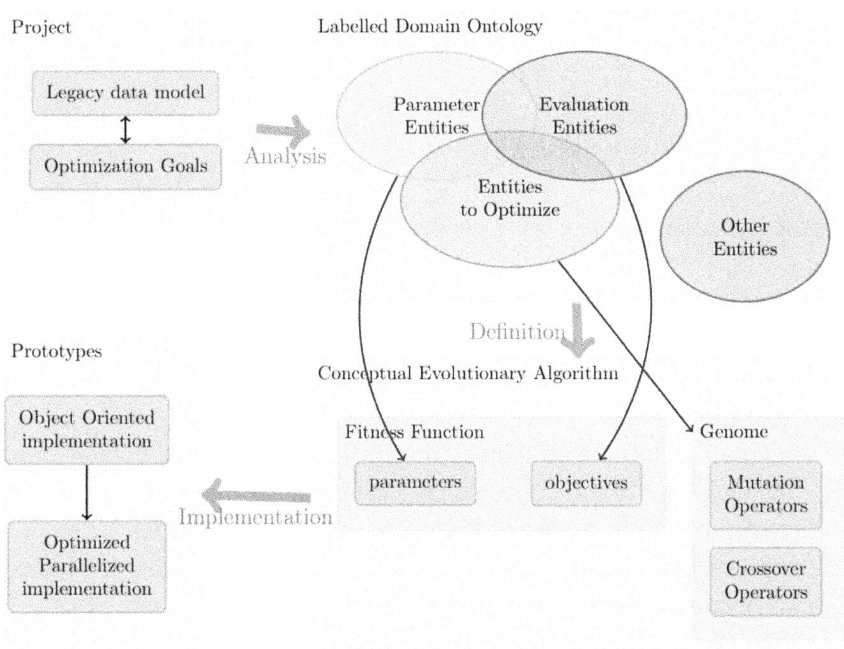

incorporating expertise in Knowledge Engineering and in Evolutionary Algorithms is then formed. The aim is to link the goals, the costs and the constraints in the project with the generic concepts of EAs (evaluation functions, representation of the genome, mutation, and crossover) through a specific labeling of the domain entities. Four general, overlapping categories of domain entities are defined:

1. **Entities to Optimize:** Entities whose values are to be determined by the optimization algorithm (e.g. which ambulance will take care of each client). The structure of the chromosome will largely depend on these entities and their relationships. It prefigures the output data structure of the evolutionary algorithm.
2. **Parameter Entities:** Entities whose values act as costs or constraints (e.g. cost/km for a vehicle, legal maximum number of daily driving for an ambulance attendant). The calculation of the fitness functions depends on these entities. They prefigure the structure of the input data.
3. **Evaluation Entities:** Entities whose values are objectives to optimize (e.g. minimize the total number of kilometers traveled by all ambulances or maximize the benefits). It must be ensured that each of these goals is represented by one or more fitness functions in the evolutionary algorithm.
4. **Other Entities:** Entities that are not involved in the optimization (e.g. history of customer calls to the customer call center).

The first three categories overlap. In fact, the membership of an entity to a category depends mainly on the usage of the values of its properties in the optimization problem. Some of the entities in the application domain may be "heterogeneous" when not all their properties serve the same purpose.

In the frequent case where the function to be optimized must be integrated into a larger existing system, most entities already exist in its data model. Most often, however, new entities will have to be introduced to have a more fine-grained representation of the entities to optimize.

We describe in *Structuring Knowledge for Evolutiionary Algorithms* a specific labeling of the domain ontology to specify these categories.

Definition

From the labeled domain ontology defined in the analysis phase, the fitness functions, the structure of the chromosome and the associated evolutionary operators must now be defined. This process will be described in *Multi-Objective Optimization and Evaluation Functions* for the fitness functions and in *Definition of the Chromosome and the Evolutionary Operators* for the genome and operators of the genetic operators.

Implementation

A first prototype using the EASEA platform (Maitre et al. 2012), (Collet et al., 2000) was developed to test the feasibility of the approach. However, it soon became clear that a sequential implementation of the algorithm will not be sufficient to solve our particular real problem. Limitations come from:

- The total memory footprint (*genomeSize*× *populationSize*) is too large to be handled by a single computer.
- The convergence time to an acceptable solution is too long.

For our application, the convergence time is critical: ambulance itineraries cannot be fully calculated off-line, because waiting times and consultation durations are largely unpredictable, and new journeys may be requested at any time, some of them urgent. The system must be able to take these changes into account and propose a new schedule in a time less than two minutes. Only a massively parallel implementation as proposed also by the EASEA platform can achieve such efficiency.

Experiment Setup presents the evaluation of the first prototype. Notice that the highly specialized technical details concerning the parallel implementation are not discussed in this article.

STRUCTURING KNOWLEDGE FOR EVOLUTIONARY ALGORITHMS

Given the large volume of information needed to solve the problem, the generation of its data model can be very complex. This is why we have decided to formalize it as an ontology, to lead the definition of the evolutionary algorithm thanks to the formal relationships that appear among the entities in it.

An ontology is a knowledge representation of a domain or a field that provides conceptual resources for knowledge-based systems (KBS). It gathers and defines the set of objects that are known as belonging to the domain (Stabd&Studer 2004).

In general, an ontology is composed of entities sometimes called *concepts* or *classes* and relationships between these entities usually called *roles*, *properties*, or *attributes* if they are mono-valued (notice that within this paper we use *entity* and *property*). In fact, ontologies provide the terminological and conceptual resources needed for knowledge formulation and for making knowledge explicit.

Our ontology covers two distinct aspects. On the one hand, there is a domain ontology, which formalizes the main concepts concerning the problem, such as vehicles, crews, patients, addresses, journeys and so on (see entities below

DomainEntity in Figure 2). And on the other hand, there are the entities associated to the genetic algorithm itself *(EAEntity* in Figure 2). It is evident that the concepts below *DomainEntity*need to be reformulated for every new problem, as well as the properties linking them with the concepts below *EAEntity*.

One of the essential components of the routing problem is *PlannedElement*. It is an event to be held at a given location (*Address*) which should start at a desired time (*requestedDate*), and should last a known or estimated time (*duration*). There are many subtypes of *PlannedElement*, some of which are shown in Figure 2:

- **PlannedEmployeeElement:** Events related to an employee. For example, such employee shall be home by noon. On the same principle, there are also *PlannedVehicleElement* events related to a vehicle: the vehicle must be revised or disinfected.
- **PlannedFleetElement:** An event that is not directly dependent on a vehicle or on a particular employee. The optimization algorithm will have to determine which vehicle driven by which employees (*Crew*) will be assigned to the event. There are again two subtypes of *PlannedFleetElement*:
    - **BusinessFleetElement:** This is the most classical event: a patient must be collected or deposited somewhere. These events are paired within a *Journey* which includes the collection, the ride and the deposit of

Figure 2. Excerpt of the domain EA ontology

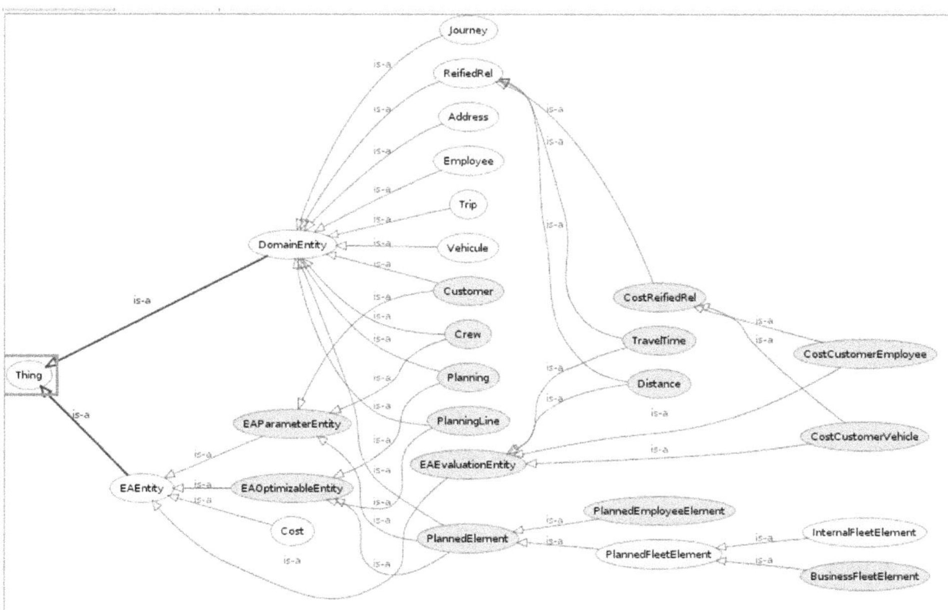

the patient. Of course, a patient who was picked up by an ambulance must be deposited by the same one. The optimization will obviously have to take this basic constraint into account. The *Journey* is associated to *PlannedElement* and not directly to a *BusinessFleetElement*since constraints related to the vehicle may be associated with *Journey*: for example, for some infectious patients, the ambulance must be disinfected immediately after the ride.

  - **InternalFleetElement:** These are constraints that are not directly related to the patients (e.g. fetching a document in a hospital).

The assignment of the events to individual vehicles is the main goal of the optimization. A *PlanningLine* represents the list, ordered by increasing time, of the events supported by a given vehicle. A complete *Planning* is simply the set of the *PlanningLine* for all the vehicles.

To optimize itineraries, notions of distance and travel time are crucial. They are represented by the *Distances* and *TFN* (Time for the Next) entities. This data is huge: the ambulances of a large company may have to visit several thousands of addresses per day. As evolutionary algorithms may randomly test any path, the distance between any two addresses must be known or estimated. This is even more critical for the travel time which usually depends on the time of day (more details will be shown in *Multi-Objective Optimization and Evaluation Functions*).

Finally, the algorithm must consider all the constraints that should be necessary for generating the itineraries. These constraints can be differentiated between the so-called *soft constraints* and *hard constraints*. Soft constraints are the requirements it is desirable to satisfy, whereas hard constraints are the requirements that must be satisfied for an itinerary to be valid. As an example of a soft constraint we can mention the limit in the working hours of certain employee. There are situations in which the working hours limit could be exceeded when transporting a patient to hospital. In those cases, there will be no choice but ignoring the limit and finishing the trip to the hospital and assuming the associated cost. Even if a constraint has been violated, the trip was still possible. On the other hand, the maximum occupant capacity of a vehicle is a physical limitation that cannot be overlooked. In other words, an ambulance that is prepared to receive just one patient cannot under any circumstances deal with two or more patients. The latter can certainly be considered as a hard constraint. Nevertheless, it is important to mention that the transport company will be the ultimately responsible of differentiating between hard and soft constraints. For the sake of simplicity, only two appear in Figure 2:

- **CostCustomerVehicule:** Vehicles are more or less suitable for patients. A patient in a wheelchair is easier to take care of (soft constraint) in an adapted vehicle. A patient who must travel lying down requires (hard constraint) a real ambulance.

- **CostCustomerEmployee:** Employees speaking the same language as the patient or personal preferences of the patient, among others.

The ontology for Evolutionary Algorithms (EA ontology) provides generic, domain independent entities and properties, which facilitate the definition of what use will be made of entities and properties of the domain in the genetic algorithm. The EA ontology is therefore a characterization of the entities and properties of the domain ontology seen as individuals. A precise modeling would require for the EA ontology to be defined at a meta-model level over the domain ontology. The separation of entities into four categories as proposed in *Methodology Overviews* should be represented as specializations of a meta-entity "Entity" itself. Similarly, a parameter used in the fitness function should be explicitly linked to some entities or properties in the application domain. For example, in the application domain of ambulances, the specific evaluation function "minimizing travel distances" should be connected (among others) to the property "address" of the "patient" entity. Unfortunately, not many formalisms allow the representation of relationships between different semantic levels (model and meta-model). In the ontology description language OWL 2 (Cuenca et al., 2007) for example, the possibility for a single object to be seen both as an entity and as an individual is reserved to the OWL-full expressibility which is undecidable, and although theoretical studies have been conducted (Motik, 2007), few tools or reasoners, if any, allow to take into account this kind of structures.

We therefore chose a simpler model for the current version of our methodology in OWL-DL:

- Two generic entities, *DomainEntity* and *EAEntity* are defined. All the domain entities are defined as subclasses of *DomainEntity*.
- We define a generic object property *EAProperty* and three mutually exclusive sub-Properties *EAEvaluationProperty*, *EAOptimizableProperty*, *EAParameterProperty*. All object properties of the domain ontology involved in the genetic algorithm must be sub properties of one of these three specific properties depending on whether they are involved for optimization, in the evaluation, or as a parameter.
- For the same purpose, but for atomic domain knowledge, we also define a generic data property *EADataProperty* and three mutually exclusive sub-data properties. The top level EA object and data properties, and their specialization for the ambulance example can be seen in Figure 3.
- The entity *EAEvaluationEntity* is defined as a subclass of *EAEntity* having at least one evaluation property:

    - EAEvaluationEntity $\sqsubseteq$ EAEntity.
    - EAEvaluationEntity $\equiv$ (EAEvaluationProperty some Thing) or (EAEvaluationDataProperty some (boolean or dateTime or integer or real or ...)).
- *EAParameterEntity* and *EAOptimizableEntity* are defined similarly.

The use of domain entities in the evolutionary algorithm is shown at two levels of granularity in the EA ontology: properties are classified according to their use within the EA, and every entity involved in the realization of the EA will be automatically classified into one or more sub-classes of *EAEntity* based on its properties. Thus, it can be seen in Figure 2 that *PlannedElement* plays a role both as a parameter and as an entity to optimize.

It is more difficult to visualize the use of properties. In the absence of appropriate visualization tools, we manually recreated a UML diagram (see Figure 4) corresponding to our ontology with the following conventions:

- The background color of the UML classes represents their category: red for subclasses of *EAOptimizableEntity*, green for *EAParameterEntity*, blue for *EAEvaluationEntity*, and black for other entities[1]. As already noted in

Figure 3. Object and data properties

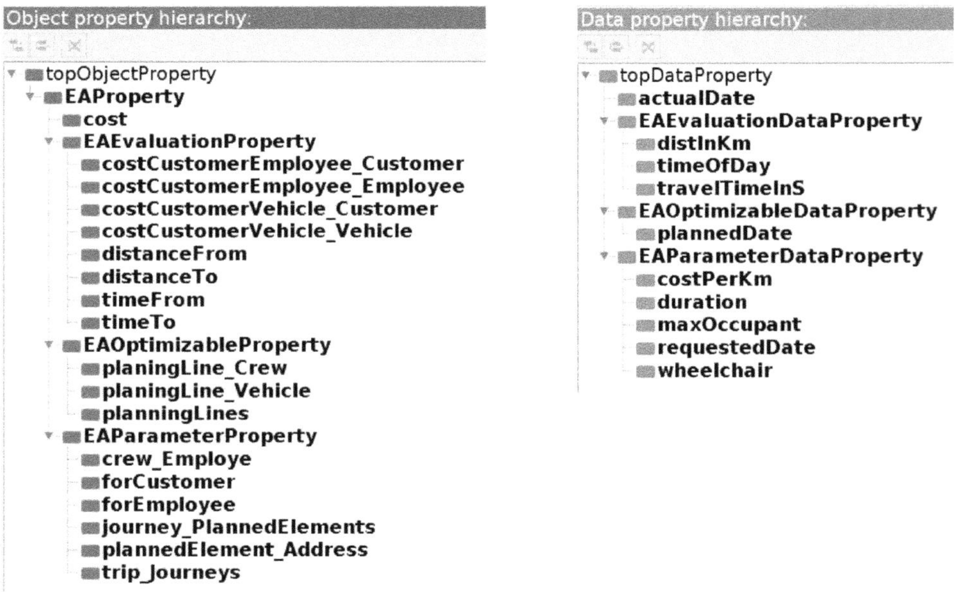

Analysis, some entities belong to more than one category. The background color of these heterogeneous entities is yellow.
- In heterogeneous entities (yellow background UML classes), each property is colored with the same color convention used for the classes.

The data corresponding to the conceptual model of Figure 4 is stored in a relational database in the information system of the company. Some tables are very large such as *timeForTheNext* which includes cached information about the time to travel between two addresses at a given day and time.

We will see in more detail the use of some entities in the EA in the following sections.

Figure 4. Simplified extract of the ontology

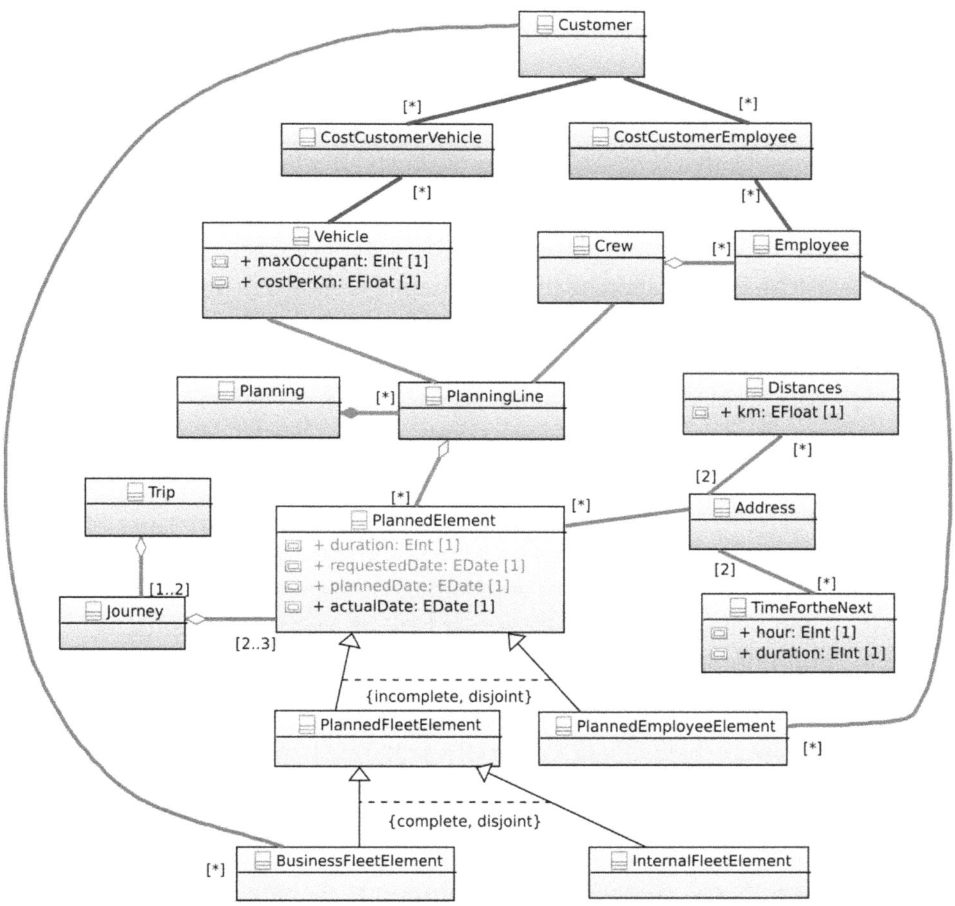

** For a more accurate representation see the electronic version.*

MULTI-OBJECTIVE OPTIMIZATION AND EVALUATION FUNCTIONS

In agreement with the entities from the data model detailed in *Structuring Knowledge for Evolutionary Algorithms*, the problem can be formalized as follows:

Let C = {0, 1, ..., c} be the set of *Customers* who need transportation and H $=\{0,1,\ \dots,h\}$ be the set of health centers. We have a complete directed graph $G=(P,A)$, where $P=\{0,1,\ \dots,v\}$ are the different *Addresses* of the *Customers* and health centers and $A=\{(i,j),i,j\in V,i\neq j\}$ is the set of arcs connecting the *Addresses*. For each $arc(i,j)\in$ A, a benefit b_{ij}, a travel time $t_{ij}(\geq 0)$ and a *Distance* $d_{ij}(\geq 0)$ are associated. $V=\{0,1,\ \dots,v\}$ is the set of *Vehicles* and $E=\{0,1,\ \dots,e\}$ is the set of *Employees* required to operate each *Vehicle*. Each *Vehicle* $k\in V$ has a capacity $q\leq max_q$ of passengers that it can transport at the same time. Each *Employee* r has a maximal number W of working hours and a mandatory pause every P hours. Each *Customer* $l\in C$ has specific demands regarding the appointment time S_i at some given element of *Addresses* i_h, a service duration SD_i and the required type of *Vehicle* v_l .

The macro objective is to design a set of routes $R=\{r_1\ ,\ \dots r_k\}$ such that each route begins at ends at a health centers in H and such that each costumer is transported by exactly one *Vehicle*. Routing design is performed considering several constraints/ requirements. An enumeration of such constraints is shown in Table 1 categorized by the *Customer* personal needs, the availability of the *Employees* and the *Vehicle* fleet. It is important to mention these are just a subset of the complete set of constraints observed in the problem.

All the constrains and requirements of the problem can be analyzed according to three axes:

- The **Economical** axis includes all the requirements and constraints that have some associated monetary cost or income. For instance, the cost per kilometer associated to each vehicle, the cost per hour of each employee, etc.
- The **Quality of Service (QoS) axis** includes all the requirements and constraints related to patients. For instance, respecting the time for the pickup and delivery of a patient, providing the patient the crew capable to deal with his specific needs, etc.
- The **Physical** axis includes all the requirements and constraints related to physical aspects of the problem. For instance, the maximum number of passengers that a given vehicle can transport at the same time.

Table 1. Problem constrains associated to the different entities

Employee	***Customer***	***Vehicle***
Number of Employees $\|E\|$ available.	Required Type of Vehicle $e_l \in E$	Number of Vehicles available |V|.
Max. W number of hours for Employee E_i	Respect the appointment time S_i.	Number of *E* for each Vehicle $v_i \in V$.
Mandatory break every P hours.	Possibility of sharing the Vehicle	Max. number *q* of Passengers per vehicle v_i
	Specificities: Wheeling chair, oxygen tank, etc.	Max. number of Km d before next mechanical service for vehicle v_i
	Preferred Crew.	Vehicle Disinfection required.

These axes are, of course, in relation with the entities in the ontology in Figure 2 (see for example the *CostCustomerVehicle* entity or the relationships between *Crew* and *Vehicle* in the ontology).From the analysis of such axes it is possible to infer that Multi-objective approaches appear clearly as a possibility to solve our problem after a careful analysis of the constraints coming from the underlying data model. For instance, the economical axis arises as a clear objective function, where all constraints and requirements regarding the vehicles and employees' costs can be included. Similar is the situation of the QoS axis, which will include all the information about the service and the patient. On the other hand, the so called *hard constraints*, such as the maximum number of occupants, are also modeled as an objective function and included in the multi-objective algorithm. A similar approach was described in (Mezura-Montes &Coello, 2006). Here, the main idea is to have *m+n* objectives, where *n* is the number of constraints and *m* the number of actual objectives. Then, we can apply a multi objective optimization technique to the vector $v = \left(f, f_{1,} \ldots, f_n, f_{n+1}, \ldots, f_m\right)$ where $f_1, \ldots, f_n$ are the constraints. Then, a valid solution *X* would have $F_i\left(X\right) = 0$ for $n+1 \leq i \leq m$. It is interesting to remark that the inclusion of hard constraints as part of the fitness functions, implies easy inclusion of newer hard constraints to the algorithm without producing major modifications to the chromosome. At the end of the evolution process, only those individuals that have satisfied the constraints (e.g. considered as valid individuals) will be available for the vehicle fleet management application.

It is important to mention that the scalar concept of "optimality" does not apply directly in the multi-objective framework. Here the notion of *Pareto optimality* must be introduced. A point x^* is said to be a *Pareto optimum* or an *efficient solution* for the multi-objective problem if and only if there is x∈ S such that for all i∈1,...,n (Zhou et al., 2011).

The image of the efficient set, i.e., the image of all the efficient solutions, is called *Pareto front* or *Pareto curve or surface*. The shape of the Pareto surface indicates the nature of the trade-off between the different objective functions.

A detailed description of the implemented fitness functions is shown below.

Calculating the Economical Axis

The economical axis is one of the most important aspects to be considered when calculating an itinerary. In the current implementation we have found that the calculation of the benefits arises as the most natural approach to deal with most of the economical details of the problem. Most of the fine economical details related to an itinerary can be included in a fitness function, which consider on the one hand the Billings and on the other hand the associated Cost.

In general, for computing the billing for a *Journeyj*, we have to consider the distance *D* between the *PlannedElementsp*$_j$ and p_{j+1}. Given the value of $D(p_j, p_{j+1})$, a price per kilometer (P_{km}) must be considered for computing the billing price. The billing price could be increased when the kilometers traveled are under specific ranges. On the other hand, the billing price is decreased if a particular customer has to share the vehicle with someone else. This way the final function for calculating the billing of a given journey *j* could be expressed according to equation (1).

$$Billing\left(j\right) = \left(D\left(p_i, p_{i+1}\right) P_{km} + I\left(D\left(p_j, p_{j+1}\right)\right), CNO_j\right) \quad (1)$$

where $I\left(D\left(p_j, p_{j+1}\right)\right)$ is the increase that is applied when the specific number of kilometers is traveled and CNO_j refers to the concurrent number of occupants.

The distance $D\left(p_i, p_{i+1}\right)$ is also considered for calculating the cost associated to *Journey j*. In this case, we have to consider the cost per kilometer of the given vehicle CKm_v. In addition, we need to calculate the cost for the *Crew* related to the vehicle. The cost of a *Crew* will depend on the number of employees and the cost per hour CHs_e of each one of them. We can express the cost function more formally according to equation (2)

$$Cost\left(j\right) = D\left(p_j, p_{j+1}\right) CKm + \sum_{e \in E} CHs_e \tag{2}$$

Finally, equation (3) shows de process for calculating the objective function.

$$maximize \sum_{v \in V} \sum_{p \in Pv} Billing\left(inJ\left(p\right)\right) - Cost\left(inJ\left(p\right)\right) \tag{3}$$

Where we denote by V the set of vehicle, by P the set of *PlannedElements*, and for each vehicle v , we associate the ordered list $P_v = \left[p_1^v, \ldots, p_{n_v}^v\right], p_j^i$ assigned to the vehicle *inJ*(*p*) refers to the set containing *p* and all the *PlannedElements* in the same *Journey* as *p*.

Calculating the Quality of Service

In the current version of the Quality of Service function we have focused on the cost associated with arrivals to a certain point in delay or in advance. The idea is to minimize the waiting time of the patient (in the case of a late arrival) as well as the waiting time of the vehicle (in the case of an early arrival). Each *Journey* between two *PlannedElementsp*$_j$and p_{j+1} is represented in a temporal line (Figure 5) considering the *RequestedDate* (*RD*) of both *PlannedElement*s. Then, *Duration*(*D*) and the estimated time to arrive to the next point *TimeForTheNext* (*TFN*) are computed. The *Duration(D)* estimates the time requiredfor taking care of the patient. This time depends on multiple factors, for example, the size of the wheelchair if the patient needs one or the fact that the patient is in a stretcher or not. On the other hand TFN data is represented as a three-dimensional matrix (called *cubeTFN*), where each element represents the travel times between two addresses for a given (discrete) time in the day. Therefore, $cubeTFN_{ijt}$ is the time in seconds it takes to go from the point *i* to point *j* at time *t*.

The cost is represented through a piecewise function $f_o\left(x\right)$, where $x = RD_{j+1} - \left(RD_j + D_j + TFN_j\right)$ is the temporal difference shown in Figure 5. As it can be seen in equation (4), if this difference is negative (arrival in delay), the cost is quadratic; otherwise (arrival in advance), the cost is linear.

$$f_0\left(x\right) = \begin{cases} x^2 & if x < 0 \\ x & if x \geq 0 \end{cases} \tag{4}$$

Figure 5. Temporal line between RequestedDatevalues of point p_j and p_{j+1}

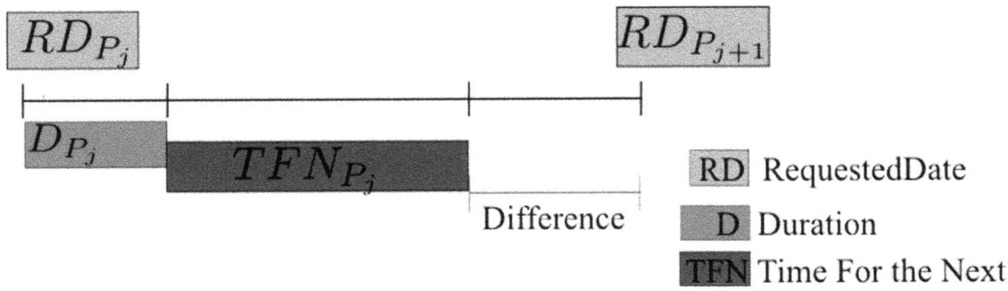

Therefore, the objective can be formalized according to equation (5):

$$minimize \sum_{v \in V} \sum_{p \in P_v} f_0 \left(RD_{j+1} - \left(RD_j + D_j + cubeTFN\left(p_j, p_{j+1}, RD_{j+1}\right)\right)\right) \quad (5)$$

Calculating the Physical Axis

In the current implementation of the algorithm we deal with the maximum number of passengers a given vehicle can transport at the same time (denoted as $maxOccup_v$). In a similar way to the QoS axis calculation, this axis is represented through the piecewise function *offset*(x), where x is the current number of occupants for a given *PlannedElement*. If x is greater than $maxOccup_v$ the offset is the difference between x and $maxOccup_v$. If x is less than $maxOccup_v$, 0 is returned.

$$offset\left(x\right) = \max\left(0,\ x - maxOccup_v\right) \quad (6)$$

Therefore, the objective can be formalized according to equation (7)

$$\text{minimize} \quad \sum_{v \in V} \sum_{p \in P_i} offset\left(occupant_p\right) \quad (7)$$

DEFINITION OF THE CHROMOSOME AND THE EVOLUTIONARY OPERATORS

The structure of the chromosome is presented in Figure 6. As it can be seen, the chromosome represents a complete *Planning* where all the vehicles and all the *PlannedElements*are considered. For every vehicle available we associate a *Crew* and the list of points (*PlannedElements*) that the vehicle needs to attend. Such a list of *PlannedElement* is the *PlanningLine* associated to each vehicle. The*PlanningLine* is sorted in ascending order according to the requested time for the event.

Variation Operators

For evolving the population, it is necessary to define a set of evolution operators. These operators should take into account all the requirements and constraints that are necessary for the planning to tend towards a possible final solution. There are two major axes that differ in the structure of a chromosome, the distribution of the *Crews* and the *PlannedElements* on the set of vehicles. Any changes in each one of the two axes will affect the values associated to the objective functions and will provide the means for evolving populations with individuals of better quality. Several operators for the two axes have been defined, including Crossover, Mutation and Swap. For space reason just two of them are described below.

PlannedElement Crossover

When considering the Crossover Operator implementation, it should be noticed that *PlannedElements* may be regrouped in a *Journey* and therefore should be supported by the same vehicle. This constraint must be considered during the implementation of the operator which certainly increases the algorithm complexity. It has also led us to implement the Crossover operator following a classic one-point approach instead of some other multi-point approaches, which could add unnecessary complexity to the operator. Nevertheless, the simple one-point approach has proved to be very effective as will be shown in *Experiment Setup*.

A graphical description of the crossover operator is shown in Figure 7. To perform the crossover between two individuals, named mother and father in the figure, a *PlannedElement* p_r is randomly chosen in this list. For each *PlanningLineI*, we have the list p_i^f (resp. p_i^m) of *PlannedElements* affected to vehicle *i* in the father (resp. mother) individual. The set p_i^s of *PlannedElement* affected to vehicle *i* in the new son individual is defined by equation (8).

Figure 6. Structure of the chromosome

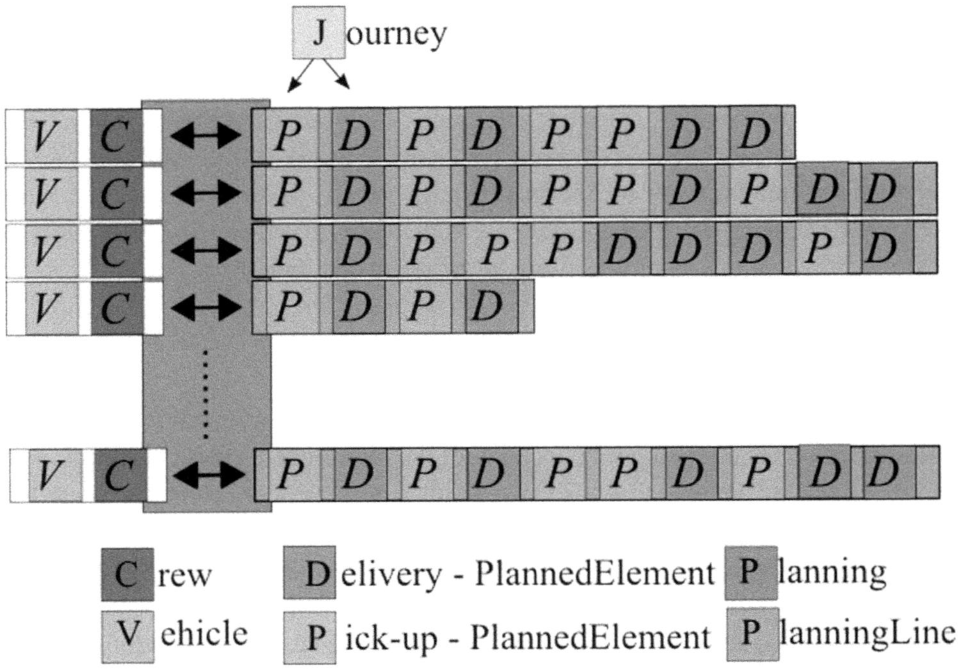

$$[l]\, P_i^s = \left\{ p^f \in P_i^f;\ \exists\, p \in inJ\left(p^f\right) RD\left(p\right) \le RD(p_r \right\} \cup$$
$$\left\{ p^m \in P_i^m;\ \forall\, p \in inJ\left(p^m\right) RD\left(p\right) > RD(p_r \right\} \quad (8)$$

In other words, the son gets assigned the earliest *PlannedElements* of its mother and the latest ones of its father.

PlannedElement Mutation

Like the *PlannedElement* crossover operator, we have to deal with the constraint regarding those*PlannedElement*belonging to a *Journey*. A graphical representation of the mutation operator is illustrated in Figure 8. To perform the mutation on one previously selected individual, we randomly select a *PlanningLinei* and a *PlannedElementp*$_r$. Then, all the *PlannedElements*that belong to the *Journey* (e.g. *inJ(p*$_r$*)*) are removed from *PlanningLinei*and reinserted in *PlanningLine j*. It is important to mention that the insertion in the *PlanningLine j* is done maintaining the proper temporal order.

Figure 7. The PlannedElement Crossover operator

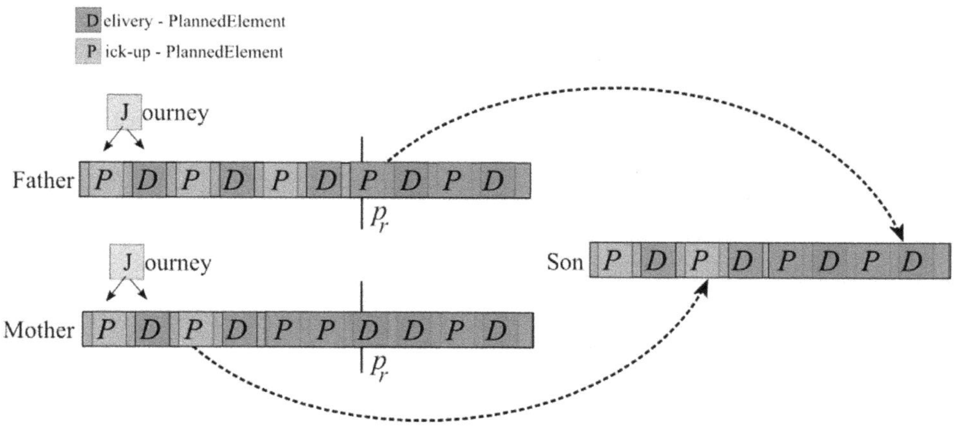

EXPERIMENT SETUP

A first prototype of the algorithm was developed using EASEA (Collet et al. 2000), (Maitre et al. 2012) a high-level framework for easily developing evolutionary approaches. The framework was extended for supporting multi-objectives algorithms. An adaptation of the NSGA-II algorithm (Deb et al., 2002) was included in it.

Figure 8. The PlannedElement Mutation operator

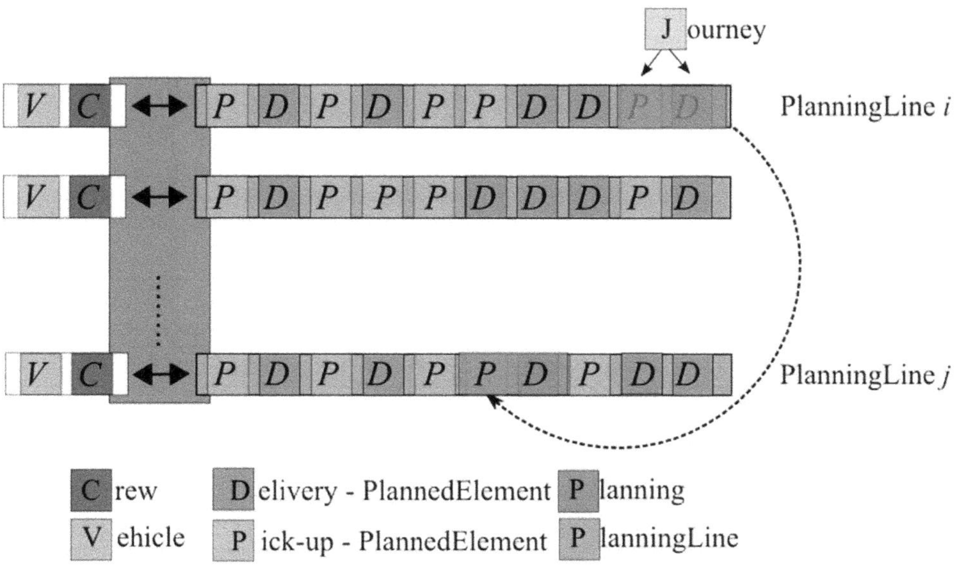

For evaluating the behavior of the algorithm, we carried out two different experiments on an average size problem. Real data comes from a small transport company with a fleet of 23 vehicles. The vehicle fleet is composed of VSL (the French acronym for Lightweight Sanitary Vehicle) and Ambulances. In this article, we are using the data corresponding to the vehicle itineraries for six days from two different weeks.

The three first columns of Table 2 show the number of journeys and vehicles used for satisfying the costumer's request for each day on the two considered weeks. The vehicle itineraries for each day as they were planned by the PTS staff have been evaluated using the objective functions described in *Structuring Knowledge for Evolutionary Algorithms*. The results of that evaluation are shown in columns four (Delay function from QoS Axis) and five (Benefits Function from Economical Axis) of Table 2. Given that the result of the evaluation of the objective function from the physical axis of *Calculating the Physical Axis* is zero, we exclude it from the table. Notice that the negative values for the Benefits function indicates losses for each one of the evaluated days. A situation that was confirmed by the PTS staff.

The algorithm is executed 35 times for each vehicle number scenario using a probability of 0.4 for the crossover operator and a probability of 0.1 for the mutation operator. In other words, after an individual has been selected, the crossover operator will be executed on it with a probability of 0.4 and mutation will be executed with a probability of 0.1.

First Experimentation

Here, we focus on evaluating the capacity of the algorithm to provide efficient solutions along the evolutionary process. During a first stage of this experiment, the algorithm will be evaluated by reducing number of vehicles from 16 to 12 and

Table 2. Evaluation in terms of the proposed fitness functions for the five days of real planning provided by the PTS company

	Day of the Week	Number of Journeys	Number of Vehicles	Fitness Delay(Minutes)	Fitness Benefits(k€)
First week	Friday	108	16	6.54	-4.3
Second week	Monday	159	21	13.7	-4.8
	Tuesday	132	21	8.28	-1.32
	Wednesday	126	21	4.85	-1.41
	Thursday	146	23	4.86	-2.01
	Friday	115	21	11.7	- 1.15

8 for a specific day of the first week (i.e. Friday). Here, we will analyze the fitness values of the population and discuss possible strategies for choosing the right solution according to the company needs. Then, in a second stage, we analyze the quality of the chosen solutions provided by the algorithm by comparing them against the solution that was scheduled by the transport company. This comparison is made in terms of values for the benefits and QoS fitness functions.

Selecting the Solutions After the Evolutionary Process

In Figure 9 we show the distribution of the fitness values only for the valid solutions (e.g. those respecting the constraints about maximum number of occupants of each vehicle) obtained after 35 executions of the algorithm on each one of the different vehicle setups. Regarding the Delay fitness function, we can observe that the scenarios where the number of vehicles is larger are those showing the minimum delays. This is a result that certainly can be expected. In the case of the Benefit fitness the behavior is different. The first observable situation is that there are no benefits but losses in all the scenarios. The second is that there is not an important difference in their average fitness values. The difference is more observable in the border of the box plot, where the case of 16 vehicles seems to be capable to achieve the better results.

As mentioned in *Multi-Objective Optimization and Evaluation Functions*, the final goal of the algorithm is to provide a set of valid possibilities to the end-user. These solutions are those observed along the Pareto front. In Figure 10, we plot all the valid found solutions in terms of their fitness values and we analyze in detail

Figure 9. Distribution of Fitness function values of valid solutions

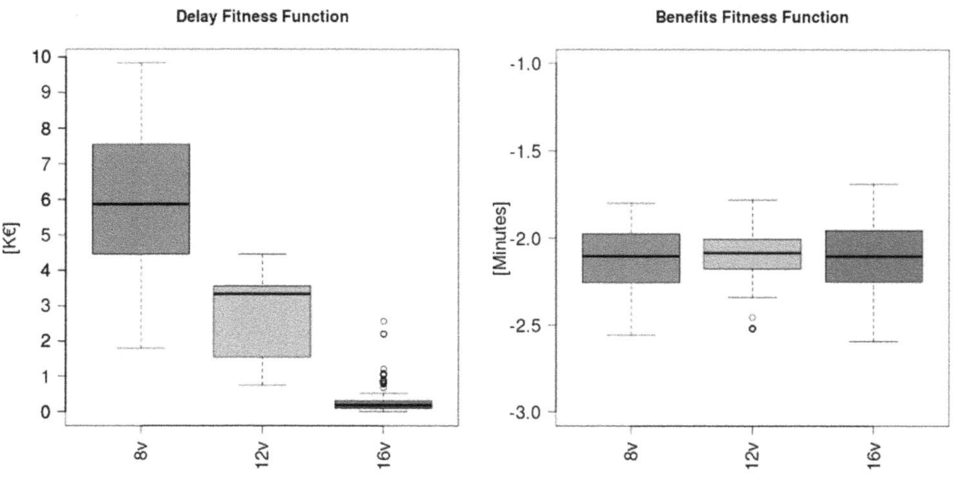

those in the Pareto front. Even if under a real-life situation, the final decision should be made by the end-user, in order to be able to proceed with our analysis we decide to follow two simple (and somewhat extreme) approaches. A first approach could be a solution where the ultimate goal is to reduce the losses without considering the delay (denoted as Solution A). Those points that are acceptable according to this criterion are the rightmost points inside the Pareto front. Alternatively, as part of the decision process, the end-user could select a valid solution considering a completely opposite approach. That is, reducing the delay without considering the benefits (Solution B). For that case, the points that are acceptable according to this criterion are the leftmost points inside the Pareto front. These points are marked in blue in Figure 10 on every possible scenario.

In Table 3, we show the different fitness values for each one of the possible vehicle setups. As we can see for both solutions, the Economical axis (e.g. Benefits fitness function)values increases as the number of vehicles is increased. Remarkably, the algorithm has shown that sometimes the use of fewer vehicles could be less efficient in terms of benefits; a situation already observed in Figure 9. This behavior could

Figure 10. Pareto Front on valid solutions obtained after 35 executions

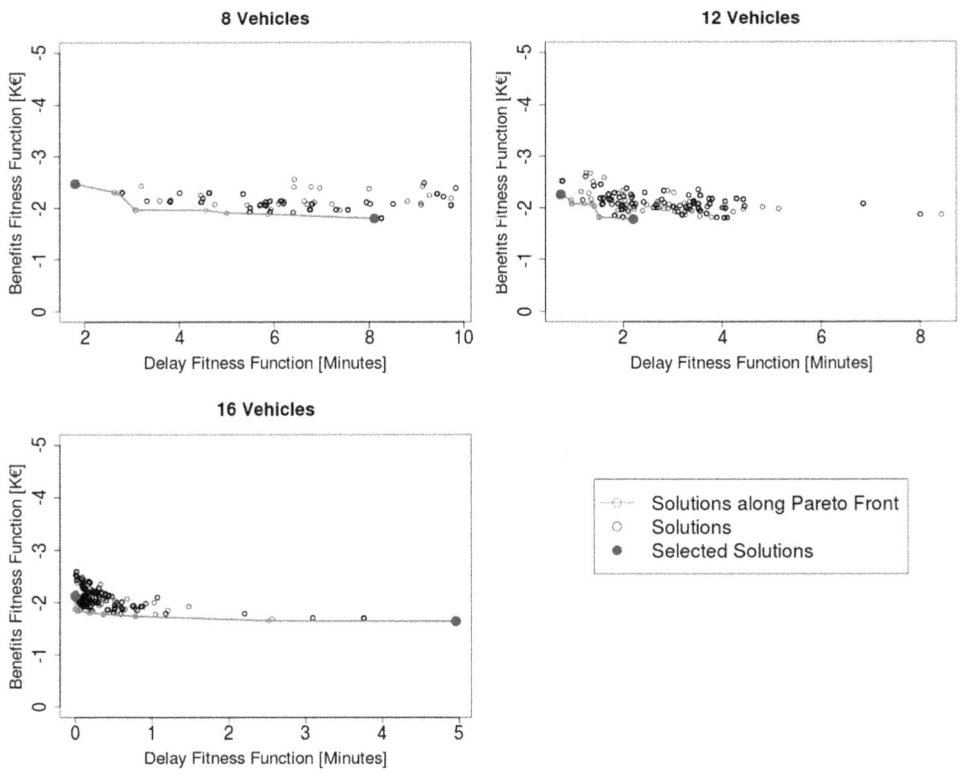

be explained because of the cost of keeping the same crew during a working day, which could be more expensive than using a new vehicle with a new crew.

Regarding the QoS axis (e.g. Delay fitness function), the values decrease if the number of vehicles increments; showing that if the end user wants to find a solution that satisfies the customer, adding more vehicles appears as the most convenient option.

To conclude with this section, we can remark that for this case even when we have chosen a solution from both extremes of the Pareto front, the Benefits function shows a difference that is of the same order of magnitude for both solutions. This is not the case if we compare the delay Fitness function, which observes a difference from 1 to 3 orders of magnitude between Solution A and B.

Quality of the Solutions

In this section we compare the solutions scheduled by the algorithm with the original solution as it was scheduled by the transport company. For space reasons we will only analyze Solution A (i.e. the solution that maximizes the benefits)

The original solution requires 16 vehicles for satisfying all the patient requests. In Figure 11 we provide a detailed view of the actual schedule chosen by the transport company. Each bar plot in Figure 11 shows the *PlannedElement* managed by a given vehicle. On the x-axes, we show the sequence of *PlannedElements* and we indicate its type (e.g. Pickup and Delivery) while in the y-axes we show the time in seconds. For each *PlannedElement* we show the duration (in blue), the time to the next in green and finally in red the delay for that *PlannedElement*.

In general terms, the distribution of journeys among the fleet seems to be well balanced. However, we can observe that in the fifth row, the second and third vehicles are responsible for just one journey, which seems clearly an inefficient use of the available fleet of vehicles.

A detailed view for the solution that maximizes the benefits (Solution A) as it was scheduled by the proposed algorithm is shown in Figure 12. In this case, we can observe that there are some VSL vehicles whose delays values are considerable bigger than in the original solution in Figure 11.

Table 3. Fitness values for the solution along the Pareto front

	Solution A			**Solution B**		
	8v	**12v**	**16v**	**8v**	**12v**	**16v**
Economical Axis (k€)	-1.79	-1.77	-1.64	-2.47	-2.26	-2.12
QoS Axis (Minutes)	8.10	2.19	4.96	1.79	0.738	0.0004

Figure 11. Original solution considering a vehicle fleet with 16 vehicles

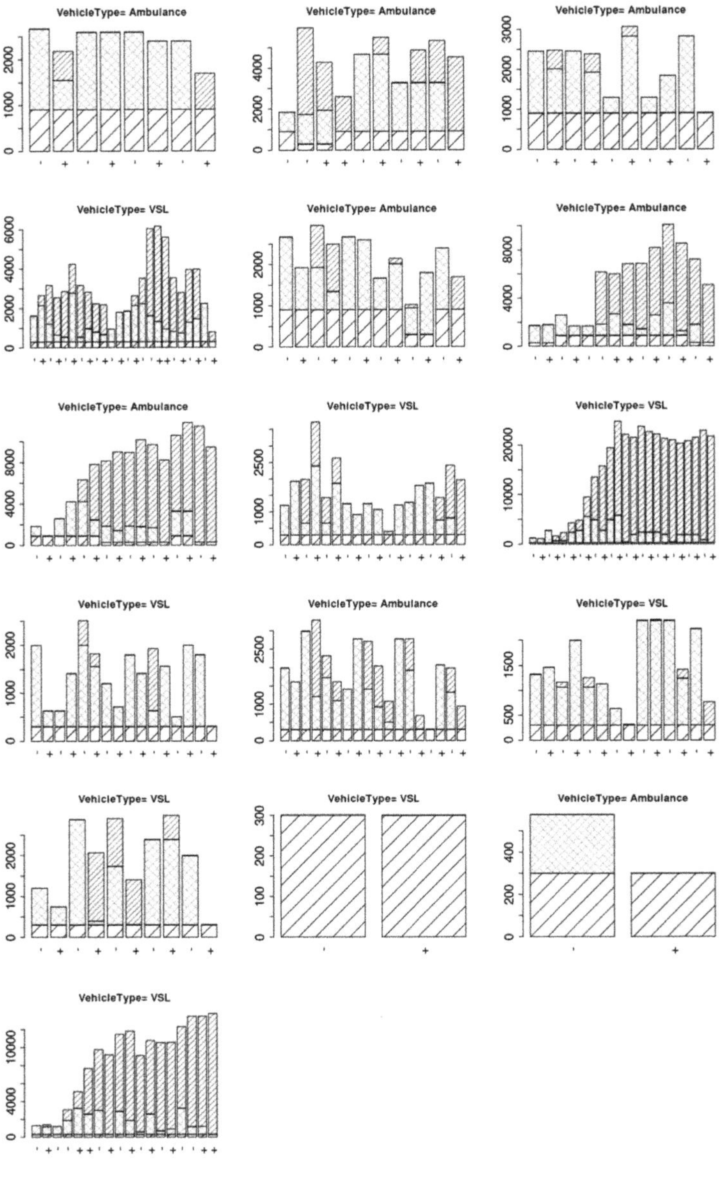

This situation is a consequence of the choice made in *First Experimentation* where we have preferred to increase the benefits without considering the delays. This way, given the VSL capacity for carrying multiple occupants, the algorithm prefers to include multiple customers in the same vehicle at the same time to improve the benefits. The consequence is that some vehicles could overcharge their capacity

Figure 12. Selected solution A considering a vehicle fleet with 14 vehicles

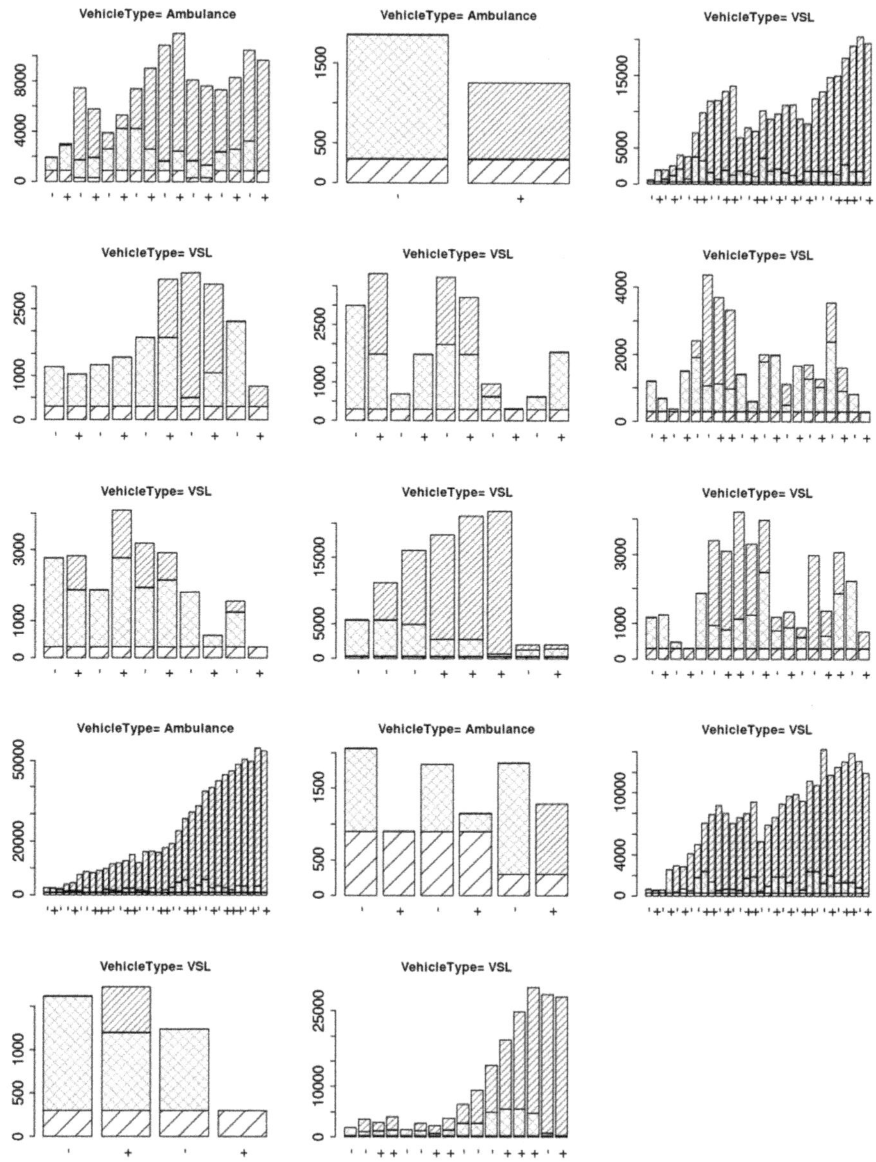

resulting in high delay values. Such situation can be observed in the first and third vehicles of the fourth row of Figure 12. Additionally, we can observe that the algorithm has been more efficient by using fewer vehicles for dealing with the same number of journeys. The original solution found during the real working journey used all the 16 vehicles while the solution got by the EA has used only 14 vehicles.

Results provided by the PTS and Solution A are summarized in Table 4. As it can be observed, the fitness values for the Economical axis provided by solution A have been improved. For this scenario, the algorithm has been able to find a combination of *Vehicles*, *Journeys* and *Crews* that minimizes the cost and maximizes the benefits. Notice that the algorithm has also been capable to find a solution that reduces the number of vehicles from 16 to 14. On the other hand, the Quality of Service axis has been also improved. However, in this case fitness value remains closer to the original solution.

Finally, we can conclude that even when the final solution will always depend on the user selection over the Pareto front, there will be solutions provided by the algorithm that can potentially show improvements over both axes. Moreover, as it can be seen in the box plot in Figure 13, all the solutions provided by the algorithm have been able to further improve the fitness values provided by the solution as it was scheduled by the transport company.

Second Experimentation

Here, we aim at providing a more general result by evaluating the MOEA using the data corresponding to the vehicle itineraries for five days (from Monday to Friday) of a second week.

In the Table 5, we present the average and standard deviation (σ) of the fitness values obtained by the MOEA on the five days of data provided by the PTS company. As stated in *Calculating the Physical Axis*, only solutions satisfying the vehicle maximum number of occupants are considered for calculating the average values. That is, we consider only those solutions whose values for the objective function from the physical axis of *Calculating the Physical Axis* equals to zero. As it can be seen, the average values for the delay and benefits objective functions are considerable better than the ones observed in Table 2. Moreover, in some cases the values show a difference of one order of magnitude. This is the case of the first 4 days in the Benefits fitness function and the first and the fifth days in the case of the Delay fitness function.

Table 4. Fitness values for the original solution and the solution obtained at the end of the evolutionary process

	Original Solution	EA Solution A
Economical Axis (k€)	-4.34	-1.64
QoS Axis	6.54	4.96
Number Of Vehicles	16	14

Figure 13. Distribution of the fitness function values for valid solutions for 16 vehicles and the fitness values for the original solution provided by PTS

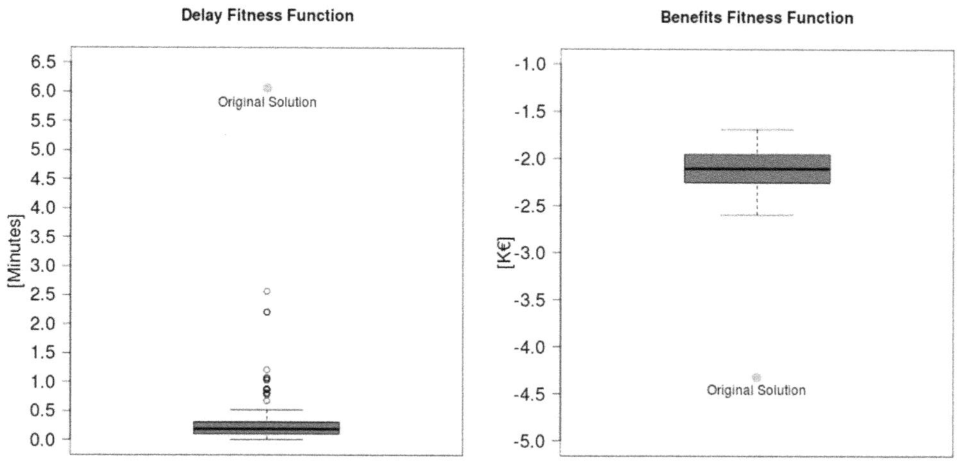

A graphical representation of the solutions for the five days is shown in the box plots in Figure 14. Regarding the Delay fitness function, we can observe the planning provided by the PTS company for Monday, Tuesday and Friday shows a lower performance than the worst solution found by the MOEMA. In the two other cases the solution provided by PTS are closer to the upper limit of the error box, but performing considerable worse than the solution included in the IQR (interquartile range). In the case of the Benefits function, the solutions provided by the PTS company presents bigger losses than the worst solution provided by the MOEA represented by the upper whisker of the box plot.

Table 5. Average fitness values and standard deviations for the five days of real planning provided after 35 executions

Day of the week	Fitness Delay (minutes)	sd	Fitness Benefits(k€)	sd
Monday	2.4	1.2	-0.64	0.24
Tuesday	0.9	0.9	-0.68	0.19
Wednesday	1.2	1.3	-0.49	0.15
Thursday	1.0	0.7	-0.55	0.27
Friday	0.6	0.9	-0.59	0.21

Figure 14. Comparison between the quality of the solutions provided by the PTS staff and all the solution found by the MOEA. Benefits and Delay functions are considered.

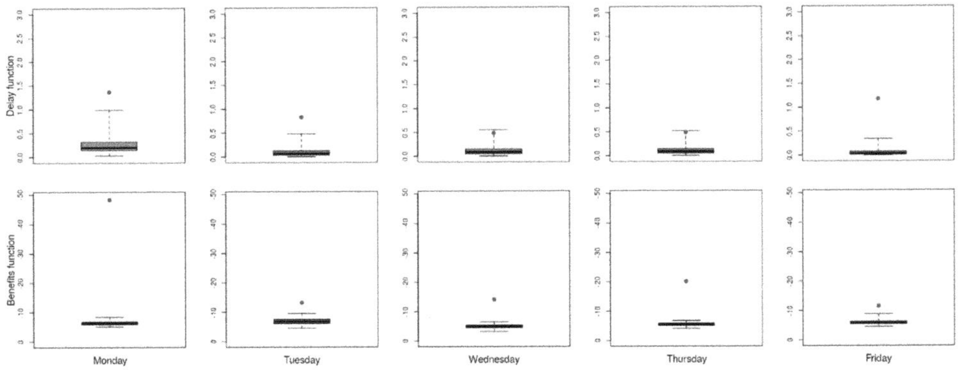

CONCLUSION

We have shown here, through the resolution of a real problem, how knowledge engineering techniques can be used to guide the definition of Evolutionary Algorithms (EA) for problems involving a large amount of structured data.

Even if the use of Evolutionary Algorithms on VRP-like problems has been deeply analyzed in the past years, the proposed solutions do not deal with a high number of constraints and requirements. Modeling the problem through a domain ontology is an effective way to cope with the problem complexity. Moreover, the inclusion in the ontology of concepts that are specific to evolutionary algorithms has allowed to systematize the coding of the genome, the evaluation functions and the genetic operators.

Nevertheless, the presented method linking ontologies to genetic algorithms does not benefit enough from the reasoning capabilities provided by ontologies. A more precise and formal definition of this process is under study. In particular, apart from only encoding domain knowledge into the genetic algorithm, we are now working on expressing semantic information in the chromosome representation, to preserve it when applying genetic operators in what we call “semantic evolution”.

Regarding the real-world problem we needed to solve, by the development of a multi objective genetic algorithm we have been able to include most of the constraints (hard and soft) into different fitness functions. The feasibility of the algorithm has been verified not only on a relatively small but real problem with a fleet from 8 to 16 vehicles and 108 Journeys but also in a larger size problem consisting of five days of vehicle itineraries with 23 vehicles and more than 115 journeys in average. The results have shown that the algorithm is able to calculate a set of valid solutions considering different values for the Economical and QoS axis and outperforming the solutions provided by the PTS staff.

The final goal of the algorithm is to be able to provide a solution in less than two minutes; to deal with the urgent calls in a realistic time. This is a goal that we expect to reach by means of the parallelization of the algorithm on GP-GPU platforms.

To increase the efficiency of the EA, we are also considering the use of machine learning techniques. In fact, most of the *PlannedElements* are recurrent in the same geographical area; the development of a module considering the past experience is being undertaken to attempt guiding the population of the EA to more accurate and close-to-reality itineraries.

REFERENCES

Collet, P., Lutton, E., Schoenauer, M., & Louchet, J. (2000). Take it EASEA. In Parallel Problem Solving from Nature PPSN VI (vol. 1971, pp. 891–901). Springer Berlin Heidelberg. doi:10.1007/3-540-45356-3_87

Deb, K., Pratap, A., Agarwal, S., & Meyarivan, T. (2002, April). A fast and elitist multiobjective genetic algorithm: NSGA-II. *Trans. Evol. Comp*, *6*(2), 182–197. doi:10.1109/4235.996017

Falkenauer, E. (1998). *Genetic Algorithms and Grouping Problems*. New York, NY: John Wiley & Sons, Inc.

Garcia-Najera, A., & Bullinaria, J. A. (2011). An improved multi-objective evolutionary algorithm for the vehicle routing problem with time windows. *Computers & Operations Research*, *38*(1), 287–300. doi:10.1016/j.cor.2010.05.004

Giraldez, R., Aguilar-Ruiz, J. S., & Riquelme, J. C. (2005, May). Knowledge-based fast evaluation for evolutionary learning. *Systems, Man, and Cybernetics, Part C: Applications and Reviews. IEEE Transactions on*, *35*(2), 254–261.

Golden, B., Raghavan, R., & Wasil, E. (Eds.). (2008). The Vehicle Routing Problem: Latest Advances and New Challenges. Springer. doi:10.1007/978-0-387-77778-8

Grau, B.C., Horrocks, I., Motik, B., Parsia, B., Patel-Schneider, P., & Sattler, U. (2008). OWL 2: The next step for OWL. *Web Semantics: Science, Services and Agents on the World Wide Web*, *6*(4), 309 – 322.

Holland, J. H. (1975). *Adaptation in natural and artificial systems: An introductory analysis with applications to biology, control, and artificial intelligence*. University of Michigan Press.

Janikow, C. Z. (1993). A knowledge-intensive genetic algorithm for supervised learning. *Machine Learning, 13*(2-3), 189–228. doi:10.1007/BF00993043

Kirchler, D., & Wolfler Calvo, R. (2013). A Granular Tabu Search algorithm for the Dial-a-Ride Problem. *Transportation Research Part B: Methodological, 56*(October), 120–135. doi:10.1016/j.trb.2013.07.014

Maitre, O., Kruger, F., Querry, S., Lachiche, N., & Collet, P. (2012). EASEA: Specification and execution of evolutionary algorithms on GPGPU. *Soft Computing, 16*(2), 261–279. doi:10.1007/s00500-011-0718-z

Mezura-Montes, E., & Coello, C. A. (2006). A survey of constraint-handling techniques based on evolutionary multiobjective optimization. Workshop paper at PPSN.

Motik, B. (2007). On the properties of metamodeling in owl. *Journal of Logic and Computation, 17*(4), 617–637. doi:10.1093/logcom/exm027

Nguyen, N. T. (2007). *Advanced Methods for Inconsistent Knowledge Management (Advanced Information and Knowledge Processing)*. Secaucus, NJ: Springer-Verlag.

Pankratz, G. (2005). A grouping genetic algorithm for the pickup and delivery problem with time windows. *OR Spectrum, 27*(1).

Paquette, J., Cordeau, J., Laporte, G., & Pascoal, M. M. B. (2013). Combining multicriteria analysis and tabu search for dial-a-ride problems. *Transportation Research Part B: Methodological, 52*, 1–16. doi:10.1016/j.trb.2013.02.007

Parragh, S.N., Doerner, K.F., & Hartl, R.F. (2008). A survey on pickup and delivery problems. *Journal fürBetriebswirtschaft, 58*(1), 21-51.

Rahma, L., Khemakhem, M., & Semet, F. (2015). Rich vehicle routing problems: From a taxonomy to a definition. *European Journal of Operational Research, 241*(1), 1–14. doi:10.1016/j.ejor.2014.07.048

Russell, S. J., & Norvig, P. (2009). *Artificial Intelligence: A Modern Approach* (3rd ed.). Pearson Education.

Saar-Tsechansky, M., & Provost, F. (2007). Handling missing values when applying classification models. *Journal of Machine Learning Research, 8*, 1217–1250.

Staab, S., & Studer, R. (Eds.). (2004). *Handbook on Ontologies.* Springer.

Tasan, A. S., & Gen, M. (2012). A genetic algorithm based approach to vehicle routing problem with simultaneous pick-up and deliveries. *Computers & Industrial Engineering, 62*(3), 755–761. doi:10.1016/j.cie.2011.11.025

Witten, I. H., Frankc, E., & Hall, M. A. (2011). *Data Mining: Practical Machine Learning Tools and Techniques (3rd ed.)*. Morgan Kaufmann Series in Data Management Systems.

Zhou, A., Qu, B., Li, H., Zhao, S., Suganthan, P. N., & Zhang, Q. (2011). Ponnuthurai NagaratnamSuganthan, and Qingfu Zhang. Multiobjective evolutionary algorithms: A survey of the state of the art. *Swarm and Evolutionary Computation, 1*(1), 32–49. doi:10.1016/j.swevo.2011.03.001

ENDNOTE

1 Not appearing in Figure 4 since only the most significant entities in relation with the problem are represented.

Chapter 9
Assessing Knowledge Management Performance in Organisations Based on the Criteria of Total Quality Management

Kit Fai Pun
University of the West Indies, Trinidad and Tobago

Man Yin Rebecca Yiu
University of the West Indies, Trinidad and Tobago

ABSTRACT

Knowledge management (KM) has been gaining momentum as the means toward organisational growth. Recent literature also gives much prominence to emerging performance measurement (PM) systems for assessing performance. This chapter comprehends the concepts of integrating KM with PM, and in short, KM performance measures in organisations. It reviews the issues that surround KM/PM initiatives, and discusses the assessment criteria that integrate the philosophy and the principles of total quality management (TQM) with KM performance measures. Five KM/PM enablers/criteria are identified, namely 1) senior management leadership, 2) management by processes, 3) people development, 4) continuous improvement, and 5) results orientation. These concepts are constituted the development of an integrated knowledge management (IKM) model for assessing organisational KM performance. Future research could evaluate the efficacy of the IKM model, validate the key KM/PM criteria, and develop an implementation framework to foster the KM/PM efforts in organisations in different sectors.

DOI: 10.4018/978-1-5225-5655-8.ch009

INTRODUCTION

Interests on performance measurement (PM) in organisations have notably increased in the last two decades (Pun and White, 2005; de Lima et al., 2009; Yiu et al., 2013). Measurement systems incorporating financial and non-financial measures have been a topic of considerable interest to both business practitioners and academics (Taticchi and Balachandran, 2008; Taticchi et al., 2010). It has been reported that knowledge and its management has linked to organisational performance and strategy (Pun and White, 2005; Yiu and Sankat, 2012). There has also been an increasing recognition on the importance of intangible assets (expertise, experiences and patents) to the performance of any organisation (Wong and Aspinwall, 2005; Migdadi, 2009; Ma and Yu, 2010). Over the past two decades, knowledge management (KM) has progressed as a strategic management concept drawing from various disciplinary areas (Pillania, 2009). It has also emerged as a phenomenon with wide-ranging implications for organisational performance and competitiveness (Heisig, 2009; Serenko and Bontis, 2009; Yiu et al., 2013).

Knowledge is recognised as a vital resource and source of competitive advantage (Maqsood et al., 2007; Migdadi 2009), and knowledge management (KM) is a key differentiator (Sujatha, 2007; Ho 2009; Wei, Choy, and Yew 2009). KM involves the design, review and implementation of both social and technological activities and processes to improve creating, sharing and applying/using knowledge (Standards Australia, 2005). The challenges for today's organisations are to 1) match and align performance measures with business strategy, structures and corporate culture, 2) deploy the measures so that the results are used and acted upon, and 3) integrate KM with performance measurement (PM) to attain sustainable competitive performance (del-Rey-Chamorro et al., 2003; Pun and White, 2005). Without measurable success, enthusiasm and support for KM is unlikely to continue. Bose (2004) contends that the best and most logical approach to measuring the impact of KM on an organisation's performance is to tie-in measurement of KM with the organisation's overall PM systems. Meanwhile, many organisations have been adopting the philosophies of total quality management (TQM) and business excellence (BE) to foster their performance goals (Chin and Pun, 2002; Lyons et al., 2008).

De Gooijer (2000) argues that the resolution of PM problems lays in two areas: linking KM to the overall business performance framework and organisational culture change. Besides, it is widely recognised that using performance metrics would help management and stakeholders to quantify KM capabilities and embed them in the policy, strategy and implementation processes of their organisations (Bose 2004; Yiu, 2012). Development of performance metrics for KM began in recent years and various metrics are being applied by forward-looking organisations. Many researchers (e.g. Ho (2009) and Zack et al. (2009)) argue that a viable KM approach with well-defined metrics would help organisations to measure and monitor their performance. Various models, frameworks and approaches had been postulated pertinent to KM, PM and their integration (e.g. Sasson and Douglas, 2006; Heisig, 2009; Khatibian et al., 2010; Yiu et al., 2013).

There has been an increasing need to assess their KM competence and integrate KM with PM in the pursuit of continuous improvement and organisational goals (Yiu et al., 2013). This chapter comprehends the concepts of integrating KM with PM, and in short, KM performance measures in organisations. It begins with a review of the PM concepts and the development, of PM systems since the 1980s. The chapter then discusses the needs of, and explores the scope of recent studies on measuring KM performance. Common enablers and performance metrics of KM are reviewed, and five (5) categories of KM criteria are identified. The integration between KM and PM is addressed, with respect to the total quality management (TQM) philosophy and the principles of business excellence (BE) models (e.g. the Malcolm Baldrige National Quality Award (MBNQA) (NIST 2017) and the European Quality Award (EQA) (EFQM 2017). An integrated knowledge management (IKM) model incorporating the TQM-BE philosophies is derived, along with a results-oriented scoring method that facilitates the self-assessment of KM performance in organisations. This paper concludes by discussing the applicability of the IKM model in amalgamating performance measure efforts for organisational goals.

PM SYSTEMS IN CONTEXT

Performance measurement can be defined as "a process of quantifying the efficiency and effectiveness of action that leads to performance" (Kennerley and Neely, 2002), whereas performance measures are "the numerical or quantitative indicators that show how well each objective is being met" (Pun and White, 2005). PM is thus composed of various performance measures through the setting of goals, standards and targets for improving an enterprise's performance. del-Fey-Chamorro et al. (2003) regard PM as:

1. a philosophy of continuous learning, using the performance measures to make adjustments to the course of an organisation in order to achieve the vision of the company;
2. a continuous and ongoing process that begins with the setting of objectives and the development of the vision and mission; and
3. a structure in which strategic, tactical and operational actions are linked to processes to provide the information required to improve the programme or service on a systematic basis.

Before the 1980s, the financial-oriented systems and tools had dominated the PM practices in almost all industry sectors (Simons 2000). PM in organisations had been primarily based on management accounting systems (Kaplan and Norton, 2000). This has resulted in most measures focusing on return on investment, return on sales, price variances, sales per employee, productivity and profit per unit production. Many recent studies (e.g., Kennerley and Neely, 2002; Kanji, 2009) indicated that these financial data have the advantage of being precise and objective, but they do not match entirely with the competencies and skills required by companies for today's changing business environment. Chin et al. (2003) add that several performance indicators (such as quality, customer satisfaction, innovation and market share) always reflect an organisation's economic condition and growth prospects better than its reported earnings can. Since the 1980s, many PM systems of non-financial orientation have been emerging.

Many organisations measure their performance systematically and thoroughly, and others conduct it on an *ad hoc* basis. This attracts much cynicism and scepticism over why, how and when performance measures are used (Neely, 2005). Pun and White (2005) discussed the strategy/measurement initiatives and compared ten (10) selected PM systems from the late 1980s to early 2000s. It was found that the majority of these systems covered most dimensions of performance. Many exhibit properties could be mapped to the characteristics of performance measures and to the requirements of development process. Moreover, Taticchi et al. (2010) provided an update of Neely (2005)'s work on the evolution of PM research under two dimensions: the research field evolution from 2005 to 2008 and the investigation of the PM dimension. Findings showed that the evolution of PM systems focuses performance from a financial perspective to a non-financial perspective. The state of art that emerges appears different for large companies versus small and medium-sized enterprises (SMEs). There are the needs to 1) solve the difficulty of those large companies in effectively translating information coming from the measurement of processes into effective tasks, and 2) focus on the creation of early conditions for PM implementation and on the development of PM systems tailored for specific characteristics and needs of SMEs (Taticchi et al., 2010). Table 1 depicts a consolidated list of representative PM systems (including models and tools) from literature in chronological order.

NEEDS OF MEASURING KM PERFORMANCE IN ORGANISATIONS

Measuring organisational performance, and in particular KM performance, plays a very important part in translating corporate strategy into results (Yiu and Sankat,

Table 1. List of selected PM systems, models and frameworks in chronological order since the 2000s

Year	Name of Performance Measurement Systems, Models and Frameworks	References
2000	The Dynamic Performance Measurement System (DPMS)	Bititci et al. (2000)
2000	OPM®: A System for Organisational Performance Measurement	Chennell at al (2000)
2001	The Action-Profit Linkage Model (APL)	Epstein and Westbrook (2001)
2001	The Manufacturing System Design Decomposition (MSDD)	Cochran et al. (2001)
2001	The Performance Prism (PP)	Neely et al. (2001, 2002)
2002	Dynamic Integrated Performance Measurement System (DIPMS)	Laitinen (2002)
2003	The Strategy Formulation and Performance Measurement Integration (SPI) Model	Pun (2003)
2004	The Performance Planning Value Chain (PPVC)	Neely and Jarrar (2004)
2004	The Capability Economic Value of Intangible and Tangible Assets Model (CEVITA™)	Ratnatunga et al. (2004)
2006	The Performance, Development, Growth Benchmarking System (PDGBS)	St-Pierre and Delisle (2006)
2007	The Unused Capacity Decomposition Framework (UCDF)	Balachandran et al. (2007)
2007	A Business Process Improvement (BPI) Framework and Performance Assessment Methodology (PAM) for SMEs	Khan et al. (2007)
2008	A Performance Measurement Model based on the Grounded Theory (GT) Approach	Chong (2008)
2011	A Holistic Model for Assessing Organisational Learning Performance	Nathai-Balkissoon and Pun (2011)
2012	An Integrated Model of Measuring Knowledge Management Performance	Yiu (2012)

Sources: Modified from Taticchi et al. (2010) and Yiu et al. (2013)

2008, 2012). Malhotra (2005) argues that knowledge becomes the accelerator of organisational performance when identified with execution of business strategy rather than with the choices of tools and technologies that keep changing with time. It embodies organisational processes that seek synergistic combination of data and information-processing capacity of information technologies, and the creative and innovative capacity of human beings.

Demarest (1997) contends that the degree to which KM is beneficial to an organisation depends on: 1) management culture pertaining to the value and purpose of knowledge, 2) how knowledge is created, transferred, and used in the company, 3) the benefits expected to come about from KM, 4) the existing level of KM systems at the company, 5) the methods used to institute KM, and 6) the level to which IT will be applied within the KM system. Gupta et al. (2000) add that KM performance (benefits) should be measured through customer satisfaction, financial indices and innovation metrics, among others. The challenges are thus on how to match and align KM performance measures with business strategies, structures and corporate culture, the type and number of measures to use, the balance between the merits and costs of introducing these measures, and how to deploy the measures so that the results are used and acted upon (Kennerley and Neely, 2002; Bose, 2004; Yiu et al., 2013).

The necessity of PM becomes evident to control the contribution of KM solutions to business operations and performance (del-Rey-Chamorro et al., 2003; Yiu and Sankat, 2008). Much specific work has recently been done in the particular domain of PM related to KM. For instance, the KM Performance Scorecard advocated by de Gooijer (2000) adapted the balanced scorecard approach in which an organisation could measure its performance in four key result areas, namely: 1) financial performance; 2) internal business processes; 3) customers; and 4) growth. The next step is to map the key result areas for KM across the scorecard. Hence, del-Rey-Chamorro et al. (2003) advocated that KM results obtained could be categorised into two groups: objective and subjective measures. The objective measures were related to issues, which provided a value that could be compared in order to follow the evolution of KM solutions and the improvement on business processes. The subjective ones provided information about the KM solution performance from the viewpoint of business process workers. Moreover, Mohamed et al. (2004) proposed a systematic approach for combining the principles of KM and cross-functional teaming in ways that purposefully enhanced knowledge flows and resulted in significant improvements in organisational performance.

Furthermore, Paulzen et al. (2002) proposed a Knowledge Process Quality Model (KPQM) to control knowledge processes and improve the KM structures for supporting systematic KM learning. Lyon et al. (2008) investigated into integrating KM and quality management as a seamless operational model to sustain performance in organisations. Yiu and Sankat (2008, 2012) also advocated a self-assessment initiative to measure KM performance along with the total quality management (TQM) concepts and an Analytical Hierarchical Process (AHP) approach in industrial enterprises. The integration of KM and PM would improve the control and accountability mechanisms in organisations (Yiu, 2012; Yiu and Pun, 2014).

KM/PM ENABLERS AND TQM PHILOSOPHIES

Wong and Aspinwall (2005) contended that KM processes capture fundamental KM activities that cannot distinguish the performance differences of one organisation from another. It would be the way these processes interact and embed within the organisational enablers that define performance outcomes (Malhotra, 2005; Wang and Ahmed, 2005). Anantatmula and Kanungo (2010) adopted Yu et al. (2004)'s classification and identified three main dimensions for enablers (including organisational characteristics, IT and managerial support) and nine manifest enablers (incorporating learning orientation, communication, knowledge sharing, flexibility, KM system quality, functionality, top management support, KM reward and KM team activity). Moreover, according to Wang and Ahmed (2005), KM/PM enablers are capability-building items and involve infra-structural elements that are the underlining factors that enable KM processes to arrive at the anticipated results. Pillania (2009) suggested that there are five main infra-structural elements of KM, namely 1) Managing the knowledge system, 2) Fostering knowledge culture, 3) Developing organisational memory, 4) Promoting knowledge sharing, and 5) Knowledge benchmarking.

Malhotra (2005) argued that the KM ability of an organisation would hinge significantly on a host of determinants such as company's mission and objectives, senior management leadership, people involvement and organisational resources and constraints (e.g. human, financial and technology). Anantatmula and Kanungo (2010) added that well-defined performance attributes and dimensions would help develop specific measures to monitor KM performance. Development of performance metrics for KM has begun in recent years and these metrics are being applied by some organisations. However, more research is needed to define these measures and to make them universal.

Despite the diversity of specific KM/PM metrics in organisations across different industry sectors, Bose (2004) argued that it should be possible to define general types of metrics. Kanji and Moura e Sá (2002) contended that these metrics or factors could be grouped into some principles that have been systematically proven to be universally valid. For instance, various balanced scorecard techniques (Kaplan and Norton, 2000), quality and excellence awards approaches (EFQM, 2017; NIST, 2017), and the ISO type of processes (ISO, 2015) are examples that incorporate the performance/quality management principles into KM and have been empirically tested and validated in different contexts.

Many researchers (e.g. Kanji and Moura e Sá 2002; Wongrassamee et al., 2003; Wang and Ahmed, 2005; Lyons et al., 2008) suggested that efforts on KM performance measures must be effectively directed towards building and improving organisational capabilities, such as responsiveness to customers, new product development, organisational learning, and strategic flexibility. This is consistent with the TQM philosophies. Moreover, various quality and business excellence (BE) models/awards (such as Malcolm Baldrige National Quality Award (MBNQA) and European Quality Award (EQA)) propagate the TQM principles and have their respective requirements that could be served as evaluation criteria for assessing a company's performance (Yiu, 2012; Yiu and Pun, 2014).

In essence, the TQM philosophies demand supportive leadership patterns and stress self-assessments. Policy and strategy would be based on TQM principles, systematically deployed and transformed into operative plans and actions (Chin et al., 2003; Yiu and Sankat, 2012). Self-assessment is a comprehensive, systematic and regular review of an organisation's activities that ultimately result in planned improvement actions (EFQM 2010). According to Pun (2002, 2003), a reliable self-assessment should satisfy two cardinal conditions. First, it should include what it is supposed to measure, and the dimensions of measurement that are deemed to have impact on overall organisational performance. Second, it should be able to measure them correctly. An organisation has to identify and permanently review its processes. It must credibly come up with the interests of all stakeholders and of the society as a whole. The philosophies also demand a structured process to determine the corporate vision and objectives. Besides, the philosophies assess the organisation's proceeding on people empowerment and process management particularly in the context of KM processes. The importance to process control and to review and improvement cycles is also stressed (EFQM, 2017; NIST, 2017).

Table 2 shows the relevance of TQM principles to performance measures with respect to the quality awards and BE models. These principles stress the importance of measurement for identifying and monitoring improvement, and are thus considered interchangeably as KM/PM criteria (or enablers) for this study. Besides, these awards advocate a point-scoring system on assessing performance, and are similar in the sense that both of them give a greatest weight to the results. One key benefit of using these awards and models is the opportunity for self-assessment and benchmarking (Chin et al., 2003; Yiu and Sankat, 2012; Yiu and Pun, 2014).

INTEGRATING TQM PRINCIPLES INTO KM PERFORMANCE ASSESSMENT

Model Constructs and Criteria

Recent literature has identified various principles and attributes in connection with KM (e.g. Wang and Ahmed 2005; Dufour and Steane 2007; Pillania 2009; Pun and Yiu, 2017) and PM (e.g. Kanji 1998, 2009; Kaplan and Norton 1992, 2000; Neely

Table 2. Relevance of TQM principles to performance measures

TQM Principles	Relevance to Performance Measures
1. Senior management leadership	• Top management recognises its roles and responsibilities to set directions, management principles and vision, and develops strategies and policies. • Management should exercise its involvement and commitment in developing the management structure and environment in which the organisation and its people can excel to achieve the organisation's objectives.
2. Management by processes (business/ operations and KM)	• Using reliable information and analysis of data make effective decisions for the current operations and planned improvements. • Vital components of the KM processes are the acquisition, manipulation, distribution, storage and use of knowledge. • KM processes stress benchmarking (sensing and acquiring new knowledge), cross-functional process improvement teams (modifying and using knowledge), training (distributing knowledge), and procedures and records (storing knowledge). • More predicable results can be obtained and achieved more efficiently when the inter-related activities are managed as a process. • Improvements are made though sharing of information and knowledge and effective implementation of organisational strategies and policies.
3. People development	• Through shared values, trust and empowerment, which encourages the involvement of people in all levels in the organisation to best release their full potential to be used for organisation's benefit. • Achieving the highest levels of employee performance requires well-developed people education and training and adoption of ethical approach to promote people well-being and satisfaction.
4. Continuous improvement	• The resources are planned, managed, and improved with continuous review and update of strategies and policies. • The importance of continuous innovation with the emphasis of learning culture should be developed and maintained. • Excellence is dependent upon balancing and satisfying the needs of all relevant stakeholders.
5. Results orientation	• Service quality and customer loyalty, retention and market share gain are best optimised through a clear focus on the needs of current and potential customers. • An organisation works more effectively when it has mutually beneficial relationships with its people and partners focusing on both financial and non-financial results and organisational effectiveness. • The long-term interest of the organisation and its people are best served by exceeding the expectations and regulations of the community at large.

Source: Modified from Pun and Lau (2003) and Yiu and Sankat (2008)

1999, 2005). Many practitioners and researchers have postulated different models, frameworks and approaches pertinent to KM, PM and their integration (e.g. Sasson and Douglas, 2006; Heisig 2009; Khatibian, Hasan gholoi pour, and Jafari 2010). Although most of these planning models stand by themselves empirically and/or theoretically, they have constraints borne with their own application domains.

By measuring the right quantities, an organisation can identify where to improve and how the limited resources can be used more effectively for performance improvement (Pun, 2002, 2003). The assessment process helps organisations identify their strengths, shortcomings and best practices where they exist (Neely 1998, 2005; Pun, Chin, and

Lewis 2003). With the common direction and an increased consistency of purpose, self-assessments can provide organisations with opportunities to build greater unity in pursuit of improvement initiatives (Pun, 2002, 2003). A paradigm is attempted to integrate KM and PM practices in organisations based on the TQM philosophies and the self-assessment method advocated by EFQM (2000, 2017). The paradigm constitutes an integrated knowledge management (IKM) model. Three (3) basic assumptions have been made for the development of the model and accompanied self-assessment system. These are:

1. The model is not in isolation from the organisation's business performance framework;
2. There is direct alignment among individual work plans, team goals, business unit objectives and the organisation's key result areas; and
3. KM performance measures would be embedded in various aspects of the organisation's work.

Recent research studies had developed self-assessment paradigms using the TQM-BE principles for use in industry. For instance, Pun (2003) developed a model for integrating manufacturing strategy formulation and performance measurement using five categories of performance determinants. Chin, Yeung, and Pun (2006) developed a web-based system for assessing supplier quality performance. Lewis, Pun, and Lalla (2007) advocated a self-assessment scale on measuring employees' perception of organisational performance in SMEs. Besides, Pun, Chin, and Yiu (2010) developed a generic measurement model for NPD performance with five categories of performance determinants, including 1) Management and strategy, 2) Customer focus, 3) Information focus, 4) People focus, and 5) Product/process focus.

Based on the self-assessment method advocated by EFQM (2000, 2017) and the rational and political approaches of KM implementation suggested by Dufour and Steane (2007), the proposed IKM model builds largely upon the constructs advocated by Pun (2003) with a stronger focus on senior management leadership and KM processes. It has five (5) categories of self-assessment criteria, including Senior Management Leadership, KM Processes, People Development, Continuous Improvement, and Results Orientation (Yiu, 2012; Yiu and Pun, 2014). Table 3 depicts the self-assessment constructs of the IKM model as compared to the constructs used for two selected models of TQM-BE domains.

Senior management leadership is the driver of the KM/PM integration that leads to the sustained pursuit of customer value and improvement in performance. The integration of KM/PM efforts rests on senior management leadership with the KM processes and continuous improvement to meet the performance requirements. It also stresses the people development on fostering KM practices and performance. The

Table 3. Comparison of the self-assessment constructs among IKM, SPI and NPD success models

IKM Model	SPI Model[1]	NPD Success Model[2]
Constructs and Elements	**Constructs and Elements**	**Constructs and Elements**
1. Senior Management Leadership • Corporate mission and value • Management involvement • Management commitment • Strategy/policy development	***1. Leadership and Constancy of Purposes*** • Corporate mission, vision and values • Management involvement • Management commitment • Strategy and policy Development	**1. Management and Strategy** • Management leadership • Organisational environment • Market and technology strategy • Strategic product planning
2. KM Processes • Identification/acquisition of knowledge • Codification/storage of knowledge • Dissemination/refinement of knowledge • Application/creation of knowledge	**2. Results Orientation** • Customer focus • Financial results • Non-financial results • Organisational effectiveness • Social responsibilities	**2. Customer Focus** • Customer knowledge • Customer relationship
3. People Development • People education and training • People well-being/satisfaction • People involvement • People empowerment	**3. Management by Processes** • Product and service processes • Sharing of information • Sharing of knowledge • Implementation of strategy and policy	**3. Information Focus** • Acquisition of product information • Use of information and data • Feedback of product analysis and review
4. Continuous Improvement • Learning/knowledge culture • Continuous innovation • Review of strategy/policy • Balancing stakeholders' needs	**4. People Development and Involvement** • People education, training and development • People well-being and satisfaction • People involvement • People empowerment	**4. People Focus** • Teamwork • People education and training
5. Results Orientation • Customer focus • Financial results • Non-financial results • Social responsibilities	**5. Continuous Improvement** • Learning culture • Continuous innovation • Review and update of strategy/policy • Balancing and satisfying stakeholders' needs	**5. Product/Process Focus** • Product/process development • Product/process design optimization • Continuous improvement • Resources and technology • Supplier management

Remarks: [1]Pun (2003); [2]Pun, Chin, and Yiu (2010)

results-oriented measures of progress provide a basis for channeling actions to gaining competitive advantage, improving operational efficiency, nurturing cultural changes, and fostering organisational learning (EFQM, 2000). The self-assessment constructs constitute several core enablers and results elements that govern the operations of the IKM model. These form a holistic framework that could be integrated into the PM system of any industrial enterprises, as depicted in Figure 1.

The proposed IKM model adopts the guiding principles of TQM, and promotes structured measurement and self-assessments of KM performance in organisations. Each criterion has 4 sub-criteria. The model has a total of twenty (20) sub-criteria constituting a total score of 1,000 points, as illustrated in Figure 2. An item listing of these criteria and sub-criteria with respective score points is given in Table 4. The point values are derived based on the empirical evidences from industry that was reported in a recent study by Yiu (2012). By managing the KM performance via self-assessments, the results could allow user organisations to examine the dynamic relationships between these criteria/sub-criteria and their resource constraints (Yiu and Sankat, 2012; Yiu and Pun, 2014).

Structured Instrument and Scoring Method

As guided by the European Quality Award (EQA), an organisation can furnish its performance information using the RADAR logic that incorporates four measurement cores, namely Results, Approach, Deployment, and Assessment and Review (EFQM, 2000, 2017). Table 5 depicts the anticipated outcomes and focal areas of measurements with respect to these cores. Since most assessment criteria are non-prescriptive and

Figure 1. The construct of an integrated knowledge management model

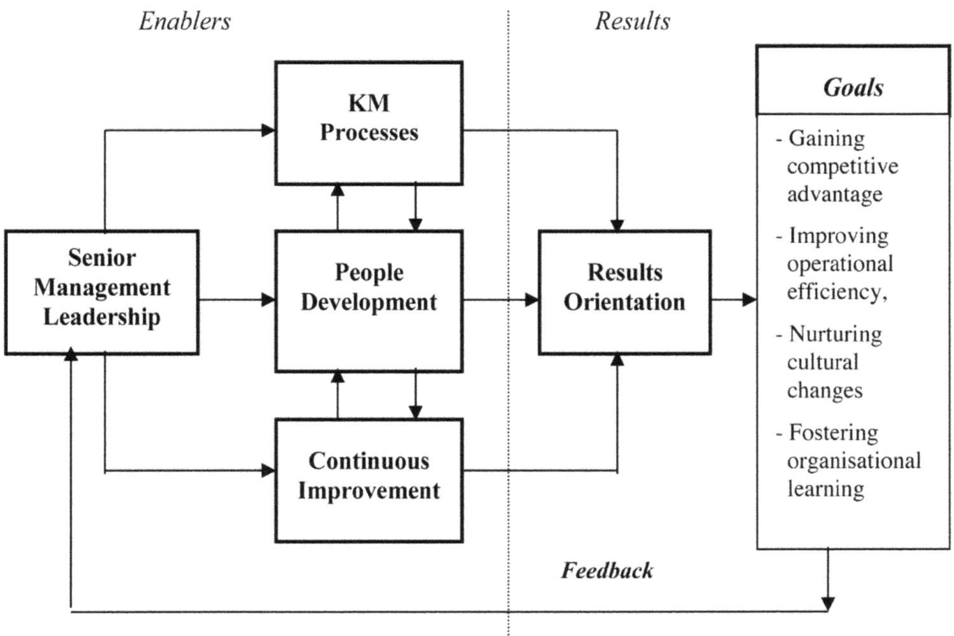

Figure 2. Five KM/PM criteria and scores of the IKM model

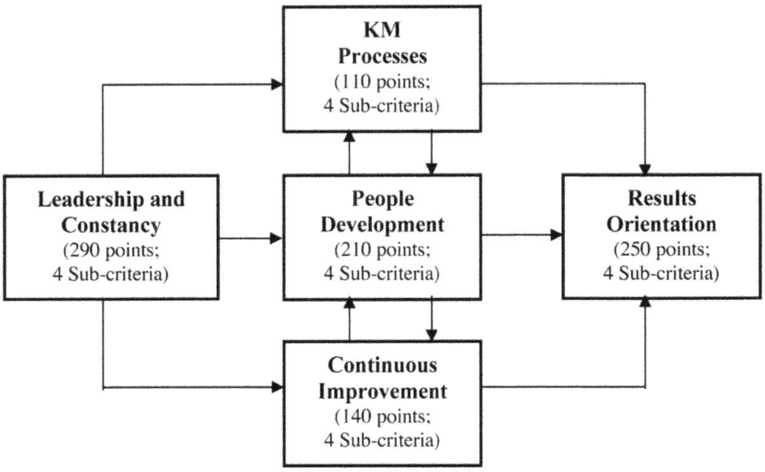

cannot be directly measured, they are translated into a set of performance indicators and measuring items. Pun (2002) argues that a standard score matrix questionnaire for structured measurement and self-assessments should be developed and used on a regular basis, so that organisations could monitor progress over time and anticipate changes. It would also be desirable for organisations to design and adopt their own self-assessment questionnaires.

It is suggested that the self-assessment questionnaire does not query companies about their specific approaches. Instead, companies are asked questions on whether there are approaches in place and they capture data that reflect the performance results of some of the approaches (Pun, 2002). For individual criteria of enablers, measuring items are developed to assess the presence of approaches and the extent of deployment. Similarly, for the criterion of Results Orientation, the extent of positive trend in the results would be assessed (Yiu and Sankat, 2012; EFQM, 2017). Based on the method advocated by Lee and Quazi (2001) and Pun and Chin (2005), the structure for developing items in the self-assessment tool is shown in Figure 3. Basically, each question addresses a practice that the organisation is expected to have in place. An objective score has to be derived from responses entered in the questionnaire.

Senior management is responsible for the design and revision of the score matrix questionnaire (or similar means), taking into consideration various inputs from representatives of various stakeholders (including employees, customers and the public). Organisations need to design their own questionnaire with respect to their performance objectives and business environments within which they operate. If any adjustments need to be made to the questionnaire that was designed and currently

Table 4. An item listing of KM/PM criteria and sub-criteria with score points

Categories/Items		Point Values*	
1	**Senior Management Leadership (SL)**		**290**
	1(a) Corporate Mission and Values (CMV)	50	
	1(b) Management Involvement (MIN)	100	
	1(c) Management Commitment (MAC)	80	
	1(d) Strategy/Policy Development (SPD)	60	
2	**KM Processes (KP)**		**110**
	2(a) Identification/Acquisition of Knowledge (IAK)	20	
	2(b) Codification/Storage of Knowledge (CSK)	30	
	2(c) Dissemination/Refinement of Knowledge (DRK)	20	
	2(d) Application/Creation of Knowledge (ACK)	40	
3	**People Development (PD)**		**210**
	3(a) People Education and Training (PET)	70	
	3(b) People Well-being/Satisfaction (PWS)	40	
	3(c) People Involvement (PIN)	40	
	3(d) People Empowerment (PEM)	60	
4	**Continuous Improvement (CI)**		**140**
	4(a) Learning/Knowledge Culture (LKC)	35	
	4(b) Continuous Innovation (COI)	35	
	4(c) Review of Strategy/Policy (RSP)	30	
	4(d) Balancing Stakeholders' Needs (BSN)	40	
5	**Results Orientation (RO)**		**250**
	5(a) Customer Focus (CUF)	85	
	5(b) Financial Results (FIR)	80	
	5(c) Non-financial Results (NFR)	45	
	5(d) Social Responsibilities (SOR)	40	
Total:			**1,000**

Remarks:
* Point values are derived based on the empirical findings of a recent study by Yiu (2012)

in use, these adjustments must be kept to a reasonable level. Any radical changes might jeopardise the reliability of the measurements and/or self-assessments and the chance of getting meaningful and comparable results (Pun et al., 2010).

For the purpose of this study, a standard self-assessment questionnaire developed by Pun (2003) was adopted and modified, incorporating the KM performance components defined by Wang and Ahmed (2005) and several CSF elements advocated

Table 5. The RADAR, assessment outcomes and focal areas

RADAR	Assessment Outcomes	Focal Areas
Results (R)	• The results should show positive trends and/or sustained performance. • Performance measurement targets should be met or exceeded. • Performance will compare with others and will have been caused by the approaches. • The scope of the results should address the relevant areas.	• Company's current performance • Performance relative to appropriate comparisons and/or benchmarks • Rate, breadth, and importance of the performance improvements • Linkage of results measures to process and action plan
Approach (A)	• A sound approach includes having a clear rationale, defined and developed processes and a clear focus on stakeholder. • The approach supports policy and strategy. • It links to other approaches where appropriate.	• Appropriateness and effectiveness of use of the methods. • Alignment with the organisation's needs • Degree to which the approach is repeatable, integrated, and consistently applied • Reliable information and data • Evidence of innovation
Deployment (D)	• The strategies, policies and actions should be deployed in relevant areas, in systematic manner	• Deployment addressing the evaluation requirements • Adopted by all appropriate work units.
Assessment and Review (AR)	• Performance measurement activities should be undertaken. • The outputs should be used to identify priorities, plan and implement improvement.	• All relevant factors of other three cores (i.e. Results, Approach and Deployment). • Subject to regular measurement

Sources: Based on EFQM (2000) and Pun et al. (2010)

Figure 3. The structure for developing a self-assessment item; Sources: Based on Lee and Quazi (2001) and Pun and Chin (2005)

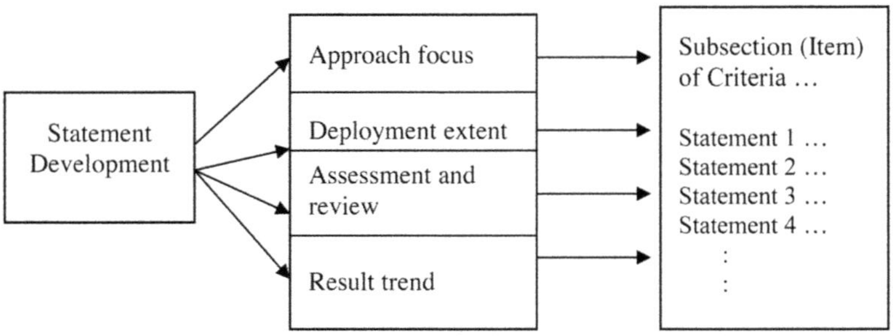

by Wong and Aspinwall (2005). The questionnaire contains 130 self-assessment statements or items in five categories of criteria and twenty sub-criteria. The first four categories are Enablers criteria comprising 95 items, whereas the last Results criterion contains 35 items. This score-matrix instrument serves mainly to explain the concepts and focal areas to be addressed in the design and revision of any self-assessment instrument by respective users. A 10-point numerical scale format is used for all measuring items in the questionnaire. The ratings range from 1 (i.e. 'Not at all' and/or 'least significant') to 10 (i.e. 'To a very large extent' and/or 'most significant'). The score is able to differentiate the overall performance of an organisation with respect to the requirement of both enablers and results criteria of the IKM model.

To illustrate how items are developed and included in the questionnaire, an example from the Category 2(a) criteria of the IKM model is shown in Figure 4. The particular items refer to the "Identification and Acquisition of Knowledge" under the criterion of Knowledge Management Processes. The self-assessments of these items are based on the approach, deployment, review and assessment under the enabler dimensions (EFQM, 2000, 2017).

Figure 5 depicts the construct of EQA's scoring method that is to be adopted for guiding the measurement/self-assessment exercise. In accordance with the EQA guidelines (EFQM, 2000, 2017), the percentage scores (i.e., from 0% to 100%) would be objectively assigned to each sub-criterion using the RADAR logic. In essence, 'Results' would be represented by trends, target comparison, courses and scope. 'Approach' would be examined to determine whether it is a sound one or is integrated in nature. Implemented versus systematic contents would govern the 'Deployment', whereas 'Assessment and Review' would fall into measurement, learning or improvement (Pun, 2002; Chin et al., 2003). The management or authorised

Figure 4. An illustrated example of questionnaire development

I. Criteria Requirement:

Areas to Address:
a. Searching for, and locating new information, ideas and knowledge.
b. Acquiring knowledge identified to be relevant.

II. Questionnaire Version:

A2(a) Identification and Acquisition of Knowledge	
1) The organisation searches for new information, ideas and knowledge which are relevant to the organisation.	1 2 3 4 5 6 7 8 9 10 Not at all … To a very Large extent
2) The organisation identifies the knowledge and associated processes for daily operations.	1 2 3 4 5 6 7 8 9 10

personnel, who would take up the roles of internal assessors and/or auditors, perform the exercise using the questionnaire. During the exercise, these assessors/auditors could assign scores to each matrix element and item for respective criteria and sub-criteria in accordance with the scoring method and guide. Improper measurements/self-assessments would cause organisations to be pushed blind alleys and to take initiative that are not suited to their current state of performance improvement. It is therefore advisable that a proper familiarisation session and relevant training should be provided on the use of the score matrix questionnaire (Pun, 2002).

The maximum possible score of the overall performance index is 1,000. Table 6 depicts a 4-level scheme for assessing the maturity status of KM performance in organisations. This assessment scheme is to be developed based on the EQA guidelines (EFQM, 2000, 2017) and modified according to the similar classification as advocated by Chin and Pun (2002). It is proposed that an organisation attaining an overall performance scores of 800 or above could be considered as an excellent performer (i.e., Class 'A' Achievers), and those attaining a score of 600 to 799 as

Figure 5. The scoring method of the European Quality Award; Source: Abstracted from EFQM (2000)

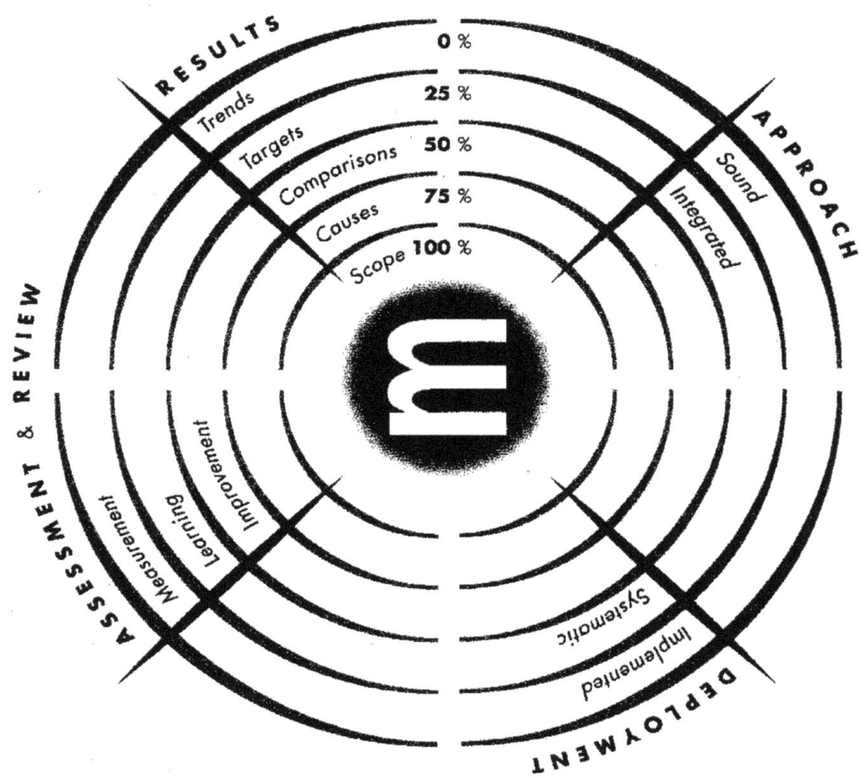

good performers (i.e., Class 'B' Improvers), respectively. For those attaining a score of 400 to 599 would be classified as Class 'C' Initiators and those scoring 399 or less are Class 'D' Starters. The 4 levels of maturity status provide a reference base for organisations to assess and then determine their current status of performance accomplishments. Besides, it could indicate how these KM/PM criteria are interrelated in responding effectively to the mission, goals and requirements of respective organisations (Yiu, 2012; Yiu and Pun, 2014).

A TRIAL EVALUATION OF THE IKM MODEL

In order to explore the creditability of using the KM/PM criteria for self-assessment of KM performance in industrial enterprises, a panel was invited comprising three (3) industry/academic experts. These experts hold senior positions in the respective organisations, including a consultant firm, a learning society (i.e. the Technology and Engineering Management Society Chapter of the IEEE Trinidad and Tobago Section), and the Department of Mechanical and Manufacturing Engineering at The

Table 6. A 4-level scheme of KM performance maturity status

Level/Class	Descriptions of IKM Performance Maturity Status
Level 1: **Class 'D' Starters** (399 scores or less)	These are 'Starter' organisations which are not familiar with the concepts, practices and tools and techniques in KM performance measures. They may have some understanding (or misunderstanding) of the measures and have decided that the principles and practices underpinning the concept are not for them. For instance, they may give an impression that they have adopted KM practices, but no real changes have been made.
Level 2: **Class 'C' Initiators** (400 – 599 scores)	These are 'Initiators' which have become aware of the importance of KM performance measures in their organisations, but they are still in the earlier stages of putting the basic elements of the measures in place. These organisations still need clear guidance of what to do in order to facilitate the adoption of concepts, practices and tools and techniques in fostering the KM/PM process.
Level 3: **Class 'B' Improvers** (600 – 799 scores)	These 'Improvers' are moving in the right direction and have made real progress in KM performance measures. The process of improvement is typically not self-sustaining, and the KM/PM efforts may not be internalised throughout the organisation. These organisations are often vulnerable to short-term pressures and various difficulties related to KM performance measures.
Level 4: **Class 'A' Achievers** (800 scores or above)	These are 'Achievers' which have reached the highest level of KM performance maturity in their organisations. The kind of culture, values, trust, capabilities, relationship and employee involvement required to attain the recognised standards or specific business excellence awards have been developed. Continuous improvement is self-sustaining in nature.

Sources: Modified from Chin and Pun (2002) and Yiu (2012)

University of the West Indies in Trinidad and Tobago. Some basic information of this expert group is given in Table 7. The invited experts have substantial industry experience in the design and implementation of performance measurement systems, related consulting services and teaching in the areas of organisational performance measures. The researcher had made a short presentation on the construct of the IKM model and design of the self-assessment tools and scoring methods. Then, it was followed by the panel discussion on the efficacy of the model and its PM system.

On the Design of the IKM Model and PM System

The potential adoption of the proposed IKM model is subject to objective evaluations and empirical validation. Bititci et al. (2000) suggested a set of nine (9) specifications for a dynamic PM system, whereas Hudson et al. (2001) identified nine (9) requirements for an effective PM development process. Pun and White (2005) also proposed a taxonomy method to evaluate the design and development process of PM systems. A host of 25 evaluation elements was identified and an instrument was developed for facilitating the panel's assessment on the design of IKM model. The design of KM performance measures is concerned with the mapping of existing KM processes and activities to performance measurement requirements. The experts were asked to share views and assess the IKM model on three main areas:

1. The Specifications of the IKM Model (with 9 elements),
2. The Requirements for IKM Model Development (with 9 elements), and
3. The Features of IKM Self-Assessments (with 7 elements).

It is expected that a viable PM system needs to be designed, managed and evaluated periodically to ensure that it yields the desired KM performance results (Yiu and Sankat, 2012). The proposed model is a balanced and dynamic one that enables support of decision-making processes by gathering, elaborating and analysing information. Table 8 summarises the main results of experts' evaluation on the design and development process of the IKM model. Of three categories of evaluation areas, it

Table 7. Basic information of industry experts

Industry Experts	Representing Organisations	Position and Capabilities	Relevant Industry Experiences
A	Consulting Company	Management Consultant	> 10 years
B	Learning Society	Past Chairman	> 20 years
C	University Sector	Professor	> 20 years

was found that the IKM model would have attained a full compliance of 18 elements (i.e. 72 percent) and a partial compliance of another 7 (i.e. 28 percent). If looking into separate categories, the model fits well with all the seven elements compliant for the 'Features of IKM Self-Assessments' (see Sub-section 6.2 below). Comparatively, there are more rooms for improvement in the design of the 'Requirements for IKM Model Development' (with full compliance of 6 elements and partial compliance of 3 elements) and the 'Specifications of the IKM Model' (with full compliance of 5 elements and partial compliance of 4 elements), respectively.

All three experts agreed that the IKM model was theoretically sound. However, both industry experts (i.e. A and B) have queried the compliance status of four elements (i.e., #2, #7, #8 and #9) under the model specifications. These are:

Table 8. Experts' evaluation on the design and development process of the IKM model

Set of Evaluation Elements	Experts' Evaluation*			Overall*
	A	B	C	
I. Specifications of the IKM Model:				
1) Internal control system	∅	√	√	√
2) External control system	∅	∅	∅	∅
3) Review mechanism	∅	√	√	√
4) Deployment system	∅	√	√	√
5) Causal relationships	√	√	√	√
6) Quantifying the criticality	√	√	√	√
7) Gaining maintenance	∅	∅	√	∅
8) Alarm signal	∅	∅	√	∅
9) IT platform	∅	∅	√	∅
II. Requirements for IKM Model Development:				
10) Need evaluation/existing PM audit	√	√	√	√
11) Key user involvement	√	√	√	√
12) Strategic objective identification	√	√	√	√
13) Performance measure development	√	√	√	√
14) Periodic maintenance structure	∅	∅	√	∅
15) Top management support	√	√	√	√
16) Full employee support	√	√	√	√
17) Clear and explicit objectives	∅	∅	√	∅
18) Set timescales	∅	∅	∅	∅
III. Features of IKM Self-Assessments:				
19) Be derived from business/KM strategy	∅	√	√	√
20) Link operations to strategic goals	∅	√	√	√
21) Stimulate continuous improvement	√	√	√	√
22) Provide fast, accurate feedback	√	√	√	√
23) Be of clearly defined/explicit purpose	√	∅	√	√
24) Be relevant and easy to maintain	√	√	√	√
25) Be simple to understand and use	√	√	√	√

* *Keys*: √ - indicates fully compliance or present; ∅ - indicates partially compliance or present

- Element #2: 'External control system' which would use KM performance measures to continuously monitor the critical parameters in the external environment for changes;
- Element #7: 'Gaining maintenance' which ensures that *gains* made as a result of improvement initiatives *maintained* through KM performance measures used by the people who work within activities and processes;
- Element #8: 'Alarm signal' which would facilitate identification and use of performance limits and thresholds to generate *alarm signals* to provide early warning of potential KM performance problems; and
- Element #9: 'IT platform' which would provide an executive information system as a mean of maintaining the PM system.

The academic expert agreed with others on the deficiency in the embodiment of 'External control system' in the design of the IKM model. Hence, Expert A further added that there have been no evidences of presence of Elements #1 and #4 below.

- Element #1: Internal control system which would use KM performance measures to continuously monitor the critical parameters in the internal environment for changes, and Element #4: Deployment system which would deploy the revised objectives and priorities to business units, processes and activities using KM performance measures.

Regarding the requirements for IKM model development, all three experts were inclined to the similar views on the compliance status of most evaluation elements, with the exceptions of Element #14 (i.e., Periodic maintenance structure) and Element #17 (i.e., Clear and explicit objectives). Both industry experts (i.e., A and B) have reservations on their full compliance status that were in line with the deficiencies found in the specifications of the IKM model (as discussed). Besides, all three experts argued that there was insufficient base to justify the full compliance status of 'setting timescales' (i.e. Element #18).

The two industry experts (i.e., A and B) commented that key internal and external KM performance metrics (such as Senior Management Leadership, Result Orientation, and People Development) must be identified including both qualitative and quantitative assessments of the critical KM processes and activities in organisations. The design of KM performance measures would include the technical details of the implementation plan(s), for instance, the completion dates, the owners of implementation items, and those actually executing the KM strategy and plan. The frequency of measurement, responsibilities for data collection, analysis and action must also be determined. Moreover, the academic expert (i.e. C) added that the IKM model incorporated with the PM system would be a balanced and dynamic one that enables support of decision-making processes by gathering, elaborating and analysing information. He added that many factors would determine the success of KM performance measures, including management commitment, management involvement, customer focus, people empowerment, and people education and training. The understanding of the PM system must be promoted by supporting networking and collaboration that ensures cohesion and interconnectivity among important strategic goals and measures.

On the Design of IKM Self-Assessment and Facilitative Tools

As shown in Part III of Table 7, the evaluations from the expert panel were largely consistent and positive on the 'Features of IKM Self-Assessments'. Although there were some reservations as indicated by both industry experts on the full compliance status of three elements (i.e., Elements #19, #20 and #23), the panel generally agreed that the IKM model constituted a set of core elements of KM self-assessments. In short, they argued that the IKM model and accompanied self-assessment system could:

- Be derived from business/KM strategy,
- Be of clearly defined/explicit purpose,
- Be relevant and easy to maintain,
- Be simple to understand and use,
- Link operations to strategic goals,
- Provide fast, accurate feedback, and
- Stimulate continuous improvement.

Being an integral part of the discussion, the expert panel was asked to preview a set of score matrix questionnaire and comment the use of the facilitative tools and their implications on the self-assessments of KM performance. Both industry experts (i.e. A and B) appraised that the handful set of self-assessment questionnaire would be useful for assisting industrial enterprises with their adoption of KM practices and performance measures. The academic expert added that the instrument and a set of facilitative tools help organisations to analyse their performance indices and the scores obtained from individual criteria. This would serve as references for assisting them to design KM self-assessment practices, and devise and guide the implementation of improvement actions. Besides, all three experts commended that the design of the 4-levels of IKM performance maturity status (with respect to four (4) different Classes, namely 'Starters', 'Initiators', 'Improvers' and 'Achievers') could be considerable and which would provide an easy (scoring) reference to facilitating organisations in assessing and/or determining the current status of their KM performance accomplishments.

RECOMMENDATIONS FOR FUTURE WORK

The premise for the need for KM is based on a paradigm shift in the business environment where knowledge is central to organisational performance. According to the 2008 survey conducted by the Asian Productivity Organisation (APO) on the KM status among its member countries, many leading organisations are shifting their focus from excellence in operational productivity and quality management to excellence in strategic innovation and learning through KM (Talisayon, 2008). There is no one "right" way to implement KM. Building organisational capabilities to acquire, create, and disseminate knowledge on a continual basis has become a key challenge for strategy and organisational design experts. Besides, while KM is well entrenched and practiced among large enterprises, this is not the case in the vast majority of SMEs (Wong and Aspinwall, 2005; Ogiwara, 2009; Yiu and Sankat, 2012). In this context, the possibility of enriching the theories and extending the applications of KM performance measures needs to be explored. This paper identifies three venues that warrant future research as follows.

Extended Scope of Empirical Studies

The pursuit of performance excellence as a way of managing businesses has been increasingly recognisable (Chin et al., 2003; Yiu, 2012; EFQM, 2017). Many companies focused on PM would have adopted some kind of quality or BE models in their system. There would be a research venue for investigating into whether these companies should implement a separate KM model or integrate the KM paradigm into their current system. Besides, the future of KM would lie in better integration with the common business processes and a concentration on the human organisation interface (Pun and Yiu, 2017). Comparative evaluations and case studies are suggested to examine the critical KM processes and determinants of KM performance measures among organisations across various industry sectors. Future research could thus validate the key KM/PM criteria identified for large enterprises versus SMEs of varied operations nature, separately and collectively. Moreover, there is a need to investigate the extent to which the employment of the proposed IKM model would contribute towards achieving sustainable competitive advantage in organisations. The adoption of such model could be extended to different industry sectors and disciplines whereby new attributes and elements could be included.

Development of an IKM Implementation Framework

Many practitioners and researchers have stressed the design, implementation and audits of PM systems (e.g. see Bitichi et al. 2000; Pun, 2003; Neely 2005; Neely et al., 2005; Yiu, 2012). In order to encapsulate the requirements of critical KM/PM processes and related activities into the way being managed in organisations, future research work would investigate into the development of an implementation framework that assists organisations to:

- Create learning/knowledge culture conducive to KM/PM integration;
- Align the KM performance measures with corporate missions and objectives;
- Facilitate training of internal assessors/auditors;
- Measure performance with respect to the KM/PM criteria;
- Streamline and monitor the KM/PM processes; and
- Foster people development, organisational learning and benchmarking of KM practices.

Development of IT-Enabled KM/PM Capabilities

The way of determining the KM/PM criteria and sub-elements and the design of self-assessment method(s) and instrument(s) (e.g., score matrix questionnaire) are of methodological focus. These criteria could be readily codified and the data be processed using computing languages and Internet technologies. A computer- or web-based system could enable the management to focus the efforts of everyone across the business on performance measures and improvement activities (Chin et al., 2003; Turner et al., 2004; Pun and Chin, 2005). Aiding tools and means (such as those that are designed in computer- or web-based systems) would facilitate structured measurements and self-assessments of performance, and make it attractive to practitioners of the wider business community (Yiu, 2012). Future research work could be on the development of IT-enabled systems with emphasis on user interface design, intelligent tutoring facilities and organisational learning capabilities. Internal assessors/auditors and those who are involved in KM performance measures would be able to complete the instrument(s) and generate scores real-time based on the data and inputs acquired.

CONCLUSION

The overall global business environment has been changing. Strategic responses to it have changed, in an effort to foster control and performance. Fierce competition had forced a lot of less-performed or weak firms out of business. It has been widely acknowledged that the competitive advantage of firms stems not from market position, but from knowledge assets and the manner in which they are deployed. Businesses that can efficiently capture the knowledge embedded in their organisations and deploy it into their operations, productions and services will have an edge over their competitors (Migdadi 2009; Khatibian, Hasan gholoi pour, and Jafari 2010). Many leading organisations are shying away from strictly technological advancement and are redirecting their resources toward the management of their unique inherent knowledge. Energising KM practices could help these organisations build and strengthen their capacities/competitiveness in terms of leveraging the knowledge; creating new knowledge or promoting innovation; and increasing collaboration and hence enhancing the skills level of people. KM performance measures are also embedded in various aspects of the existing management systems of organisations.

This chapter comprehends the concepts of integrating KM with PM, and proposes an integrated paradigm for assessing performance. The IKM model adopts the TQM principles and stresses five (5) categories of KM/PM criteria, namely *Senior management leadership*, *KM processes*, *People development*, *Continuous improvement*, and *Results orientation*. The scoring method is developed with a total of 1,000 score points. Using the model on a regular basis, organisations could determine their strengths, weaknesses and constraints that they might have in getting performance improvements for those identifiable areas. More importantly, the model is not in isolation from any organisation's business performance framework.

Having regards the design and efficacy of the IKM model, this research initiated a panel discussion with industry experts and practitioners. The findings from the panel discussion showed that the IKM model was theoretically sound, despite having queries on the compliance status of some elements under the specifications of the model and the requirements for model development. Besides, participating experts/practitioners generally agreed that the IKM model constituted a set of core elements for assessing KM practices, and appraised the handful set of self-assessment questionnaire. They also commended that the design of the 4-levels scheme of IKM performance maturity status was considered applicable and which provided an easy (scoring) reference for organisations in assessing/determining their KM performance accomplishments.

Performance improvement is a never-ending process. The ultimate objective of integrating KM and PM is to assist organisations in their quest to attain performance improvement and business excellence. Industry practitioners could adopt the model for formalising their KM/PM efforts, and consistently managing performance. This would promote both internal and external communications, foster customer-focused measurements and allow benchmarking of KM performance to be made. Besides, much information regarding the design and evaluation of the integrated paradigm would be of substantial values to researchers and scholars in the areas.

As business and operational situations vary in these enterprises, managing the KM/PM integration will only succeed if they are implemented as a long-term organisational paradigm shift. Future research is thus to evaluate the efficacy of the IKM model, validate the key KM/PM criteria, and develop an implementation framework with IT-enabled capabilities to foster the KM/PM efforts in organisations across different sectors.

REFERENCES

Anantatmula, V. S., & Kanungo, S. (2010). Modeling enablers for successful KM implementation. *Journal of Knowledge Management*, *14*(1), 100–113. doi:10.1108/13673271011015598

Balachandran, K. R., Shu, H. L., & Suresh, R. (2007, Winter). A framework for unused capacity: Theory and empirical analysis. *Journal of Applied Management Accounting Research*, 21-38.

Bititci, U. S., Turner, T., & Begemann, C. (2000). Dynamics of performance measurement systems. *International Journal of Operations & Production Management*, *20*(6), 692–704. doi:10.1108/01443570010321676

Bose, R. (2004). Knowledge management metrics. *International Journal of Technology Management*, *24*(4), 457–468.

Chennell, A. F., Dransfield, S. B., Field, J. B., Fisher, N. I., Saunders, I. W., & Shaw, D. E. (2000). *OPM®: A system for organisational performance measurement. In Performance Measurement – Past, Present and Future*. Cambridge, UK: University of Cambridge.

Chin, K. S., & Pun, K. F. (2002). A proposed framework for implementing TQM in Chinese organisations. *International Journal of Quality & Reliability Management*, *19*(3), 272–294. doi:10.1108/02656710210415686

Chin, K. S., Pun, K. F., & Lau, H. C. W. (2003). Development of a knowledge-based self-assessment system for organisational performance. *Expert Systems with Applications*, *24*(4), 443–455. doi:10.1016/S0957-4174(02)00192-6

Chin, K. S., Yeung, I. K., & Pun, K. F. (2006). Development of a web-based system for assessing supplier quality performance. *International Journal of Quality & Reliability Management*, *23*(7), 743–765. doi:10.1108/02656710610679806

Chong, H. G. (2008). Measuring performance of small-and-medium sized enterprises: The grounded theory approach. *Journal of Business and Public Affairs*, *2*(1), 10–20.

Cochran, D. S., Arinez, J. F., Duda, J. W., & Linck, J. (2001). A decomposition approach for manufacturing system design. *Journal of Manufacturing Systems, 20*(6), 371–380. doi:10.1016/S0278-6125(01)80058-3

Cooper, R., & Kaplan, R. S. (1988, September). Measure costs right: Make the right decisions. *Harvard Business Review*, 10–20.

Cross, K. F., & Lynch, R. L. (1988). The SMART way to define and sustain success. *National Productivity Review, 8*(1), 23–33. doi:10.1002/npr.4040080105

de Gooijer, J. (2000). Designing a knowledge management performance framework. *Journal of Knowledge Management, 4*(4), 303–310. doi:10.1108/13673270010379858

de Lima, E. P., da Costa, S. E. G., & Angelis, J. J. (2009). Strategic performance measurement systems: A discussion about their roles. *Measuring Business Excellence, 13*(3), 39–48. doi:10.1108/13683040910984310

del-Rey-Chamorro, F. M., Roy, R., van Wegen, B., & Steele, A. (2003). A framework to create key performance indicators for knowledge management solutions. *Journal of Knowledge Management, 7*(2), 46–62. doi:10.1108/13673270310477289

Demarest, M. (1997). Understanding knowledge management. *Long Range Planning, 30*(3), 374–384. doi:10.1016/S0024-6301(97)90250-8

EFQM. (2000). *The EFQM Excellence Model: Assessor Training Modules. MC Modules (Company) V1/E*. Brussels: European Foundation for Quality Management.

EFQM. (2017). *European Foundation for Quality Management.* Retrieved March 1, 2017, from http://www.EFQM.org/

Epstein, M. J., & Westbrook, R. A. (2001). Linking action to profits in strategic decision making. *MIT Sloan Management Review, 42*(3), 39–49.

Gupta, B., Iyer, L. S., & Aronson, J. E. (2000). Knowledge management: Practices and challenges. *Industrial Management & Data Systems, 100*(1), 17–21. doi:10.1108/02635570010273018

Heisig, P. (2009). Harmonisation of knowledge management: Comparing 160 KM frameworks around the globe. *Journal of Knowledge Management, 13*(4), 4–31. doi:10.1108/13673270910971798

Ho, C.-T. (2009). The relationship between knowledge management enablers and performance. *Industrial Management & Data Systems, 109*(1), 98–117. doi:10.1108/02635570910926618

Hudson, M., Smart, A., & Bourne, M. (2001). Theory and practice in SME performance measurement systems. *International Journal of Operations & Production Management, 21*(8), 1096–1115. doi:10.1108/EUM0000000005587

ISO. (2015). *ISO 9001: Quality Management Systems - Requirements.* Geneva: International Organisation of Standardisation.

Kanji, G. K. (2009). *An integrated Approach of Organisational Excellence.* Retrieved February 1, 2009, from http://www.gopal-kanji.com

Kanji, G. K., & Moura e Sá, P. (2002). Kanji's Business Scorecard. *Total Quality Management, 13*(1), 13–27. doi:10.1080/09544120120098537

Kaplan, R. S., & Norton, D. P. (2000). *The Strategy-Focused Organisation: How Balanced Scorecard Companies Thrive in the New Business Environment.* Boston, MA: Harvard Business School Press.

Kennerley, M., & Neely, A. (2002). A framework of the factors affecting the evolution of performance measurement systems. *International Journal of Operations & Production Management, 22*(11), 1222–1245. doi:10.1108/01443570210450293

Khan, Z., Bali, R. K., & Wickramasinghe, N. (2007). Developing a BPI framework and PAM for SMEs. *Industrial Management & Data Systems, 107*(3), 345–360. doi:10.1108/02635570710734262

Khatibian, N., Hasan gholoi pour, T., & Abedi Jafari, H. (2010). Measurement of knowledge management maturity level within organisations. *Business Strategy Series, 11*(1), 54–70. doi:10.1108/17515631011013113

Laitinen, E. K. (2002). A dynamic performance measurement system: Evidence from small Finnish technology companies. *Scandinavian Journal of Management, 18*(1), 65–99. doi:10.1016/S0956-5221(00)00021-X

Lewis, W. G., Pun, K. F., & Lalla, T. R. M. (2007). Measuring employees' perception in small and medium-sized enterprises: A self-assessment scale. *International Journal of Management Science and Engineering Management, 2*(3), 229–238.

Lyons, K., Acsente, D., & van Waesberghe, M. (2008). Integrating knowledge management and quality management to sustain knowledge enabled excellence in performance. *Vine, 38*(2), 241–253. doi:10.1108/03055720810889879

Ma, Z., & Yu, K. H. (2010). Research paradigms of contemporary knowledge management studies: 1998-2007. *Journal of Knowledge Management, 14*(2), 175–189. doi:10.1108/13673271011032337

Malhotra, Y. (2005). Integrating knowledge management technologies in organisational business processes: Getting real enterprises to deliver real business performance. *Journal of Knowledge Management*, *9*(1), 7–28. doi:10.1108/13673270510582938

Maqsood, T., Walker, D. H. T., & Finegan, A. D. (2007). Facilitating knowledge pull to deliver innovation through knowledge management: A case study. *Engineering, Construction, and Architectural Management*, *14*(1), 94–109. doi:10.1108/09699980710717007

Migdadi, M. (2009). Knowledge management enablers and outcomes in the small-and-medium sized enterprises. *Industrial Management & Data Systems*, *109*(6), 840–858. doi:10.1108/02635570910968072

Mohamed, M., Stankosky, M., & Murray, A. (2004). Applying knowledge management principles to enhance cross-functional team performance. *Journal of Knowledge Management*, *8*(3), 127–142. doi:10.1108/13673270410541097

Nathai-Balkissoon, M., & Pun, K. F. (2011). A holistic model and research agenda for organisational learning in manufacturing. *World Journal of Entrepreneurship, Management and Sustainable Development*, *7*(1), 27–38. doi:10.1108/20425961201000028

Neely, A. (2005). Performance measurement system and relationships with performance results: Developments in the last decade and a research agenda for the next. *International Journal of Operations & Production Management*, *25*(12), 1264–1277. doi:10.1108/01443570510633648

Neely, A., Adams, C., & Crowe, P. (2001). The performance prism in practice. *Measuring Business Excellence*, *5*(2), 6–13. doi:10.1108/13683040110385142

Neely, A., & Jarrar, Y. (2004). Extracting value from data: The performance planning value chain. *Business Process Management Journal*, *10*(5), 506–509. doi:10.1108/14637150410559180

Neely, A. D., Adams, C., & Kennerley, M. P. (2002). *Performance Prism: The Scorecard for Measuring and Managing Stakeholder Relationships*. London: Prentice Hall.

Neely, A. D., Gregory, M. J., & Platts, W. K. (2005). Performance measurement system design: A literature review and research agenda. *International Journal of Operations & Production Management*, *25*(12), 1228–1263.

NIST. (2017). *Malcolm Baldrige National Quality Award.* National Institute of Standards and Technology, US Department of Commerce. Retrieved September 1, 2017, from http://www.nist.gov/

Ogiwara, N. (2009). *Knowledge Management: Case Studies for Small and Medium Enterprises.* Asian Productivity Organisation.

Paulzen, O., Doumi, M., Perc, P., & Cereijo-Roibas, A. (2002) A maturity model for quality improvement in knowledge management. In: *ACIS 2002 Proceedings* (Paper 5). Retrieved March 1, 2013, from http://aisel.aisnet.org/acis2002/5

Pillania, R. K. (2009). Demystifying knowledge management. *Business Strategy Series, 10*(2), 96–99. doi:10.1108/17515630910942223

Pun, K. F. (2002). Development of an integrated TQM and performance measurement system for self-assessment: A method. *Total Quality Management, 13*(6), 759–777. doi:10.1080/0954412022000010127

Pun, K. F. (2003). *A Study of an Integrated Approach for Strategy Formulation and Performance Measurement in Manufacturing Enterprises* (Unpublished doctoral dissertation). Middlesex University School of Computing Science, London, UK.

Pun, K. F., & Chin, K. S. (2005). On-line assessment of new product development performance: An approach. *Total Quality Management & Business Excellence, 16*(2), 157–169. doi:10.1080/14783360500054244

Pun, K. F., Chin, K. S., & Yiu, M. Y. R. (2010). An AHP approach to assess new product development performance: An exploratory study. *International Journal of Management Science and Engineering Management, 5*(3), 210–218.

Pun, K. F., & Lau, C. W. (2003). Integration of total quality management and performance measurements in government departments: An empirical study in Hong Kong. *International Journal of Business Performance Management, 5*(4), 316–335. doi:10.1504/IJBPM.2003.003815

Pun, K. F., & White, A. S. (2005). A performance measurement paradigm for integrating strategy formulation: A review of systems and frameworks. *International Journal of Management Reviews, 7*(1), 25–46. doi:10.1111/j.1468-2370.2005.00106.x

Pun, K. F., & Yiu, M. Y. R. (2017). Management of knowledge and ignorance in the context of organisational learning: A research agenda. *West Indian Journal of Engineering, 40*(1), 24–30.

Ratnatunga, J., Gray, N., & Balachandran, K. R. (2004). CEVITA™: The valuation and reporting of strategic capabilities. *Management Accounting Research, 15*(1), 77–105. doi:10.1016/j.mar.2003.12.005

Sasson, J. R., & Douglas, I. (2006). A conceptual integration of performance analysis, knowledge management, and technology: From concept to prototype. *Journal of Knowledge Management, 10*(6), 81–99. doi:10.1108/13673270610709233

Serenko, A., & Bontis, N. (2009). Global ranking of knowledge management and intellectual capital academic journals. *Journal of Knowledge Management, 13*(1), 4–15. doi:10.1108/13673270910931125

Simons, R. (2000). *Performance Management and Control Systems for Implementing Strategy*. Englewood Cliffs, NJ: Prentice Hall.

St-Pierre, J., & Delisle, S. (2006). An expert diagnosis system for the benchmarking of SMEs' performance. *Benchmarking: An International Journal, 13*(1/2), 106–119. doi:10.1108/14635770610644619

Standards Australia. (2005). *Knowledge Management - A Guide AS 5037-2005*. Sydney: Standards Australia.

Sujatha, R. (2007). Critical success factors for implementation of knowledge management. *Icfai Journal of Knowledge Management, 5*(1), 61–72.

Talisayon, S. D. (2008). *Knowledge Management in Asia: Experience and Lessons*. Asian Productivity Organisation.

Taticchi, P., & Balachandran, K. R. (2008). Forward performance measurement and management integrated frameworks. *International Journal of Accounting and Information Management, 16*(2), 140–154. doi:10.1108/18347640810913807

Taticchi, P., Tonelli, F., & Cagnazzo, L. (2010). Performance measurement and management: A literature review and a research agenda. *Measuring Business Excellence, 14*(1), 4–18. doi:10.1108/13683041011027418

Turner, T., Creighton, S., Nudurupati, S., & Bititci, U. (2004). Implementing a Web enabled performance measurement system. *Measuring Business Excellence, 8*(2), 40–54. doi:10.1108/13683040410539427

Wang, C. L., & Ahmed, P. K. (2005). The knowledge value chain: A pragmatic knowledge implementation network. Handbook of Business Strategy, 321-326.

Wei, C. C., Choy, C. S., & Yew, W. K. (2009). Is the Malaysian telecommunication industry ready for knowledge management implementation? *Journal of Knowledge Management*, *13*(1), 69–87. doi:10.1108/13673270910931170

Wong, K. Y., & Aspinwall, E. (2005). An empirical study of the important factors for knowledge-management adoption in the SME sector. *Journal of Knowledge Management*, *9*(3), 64–82. doi:10.1108/13673270510602773

Wongrassamee, S., Simmons, J. E. L., & Gardiner, P. D. (2003). Performance measurement tools: The Balanced Scorecard and the EFQM Excellence Model. *Measuring Business Excellence*, *7*(1), 14–29. doi:10.1108/13683040310466690

Yiu, M. Y. R. (2012). *An Investigation into an Integrated Knowledge Management Approach for Industrial Organisations* (Unpublished MPhil dissertation). Faculty of Engineering, The University of the West Indies, St Augustine.

Yiu, M. Y. R., & Pun, K. F. (2014). Measures of knowledge management performance in organisations: An exploratory study based on an integrated model. *The Learning Organization*, *21*(5/6), 310–332.

Yiu, M. Y. R., & Sankat, C. K. (2008). A self-assessment model for evaluating knowledge management performance. *The Journal of the Associations of Professional Engineers of Trinidad and Tobago*, *37*(1), 39–47.

Yiu, M. Y. R., & Sankat, C. K. 2012, Measuring knowledge management performance in industrial enterprises: an integrated model. In *Proceedings of the ANQ Congress 2012*. Hong Kong University of Science and Technology.

Yiu, M. Y. R., Sankat, C. K., & Pun, K. F. (2013). In search of the knowledge management practices in organisations: A review. *West Indian Journal of Engineering*, *35*(2), 103–116.

Yu, S., Kim, Y., & Kim, M. (2004). Linking organisational knowledge management drivers to knowledge management performance: An exploratory study. In *Proceedings of the 37th Hawaii International Conference on System Sciences*. Piscataway NJ: IEEE Computer Society.

Chapter 10
Integration Between Knowledge Management and Total Quality Management in Jordanian Universities:
Empirical Study

Issa Mahmoud Shehabat
Yarmouk University, Jordan

Mohammad Berrish
Shaqra University, Saudi Arabia

ABSTRACT

Resulting from the development of management theory in the past few years, knowledge management has emerged, which is identifying information of value and how to use it well, as well as understanding cognitive assets of an organization and how to exploit them. Knowledge management is important and necessary for the survival of an organization and its superiority. It can be integrated into the philosophy of total quality management to play a role in the development of the basis of performance of modern organizations. The total quality management philosophy seeks to achieve customer satisfaction through the commitment of the leadership of an organization and its workers by a process of continuous improvement of the quality of performance in various aspects of an organization. This chapter aims to identify the integration between knowledge management and total quality management, when applied to public Jordanian universities as a sample for the study. The chapter found the presence of integrity and a strong correlation between knowledge management and total quality management.

DOI: 10.4018/978-1-5225-5655-8.ch010

INTRODUCTION

This paper is related to knowledge management (KM) and total quality management (TQM) in higher educational institutions (HEIs) in Jordan. These two issues have attracted the attention of many researchers and writers in higher education (HE). The significance of these issues is that they are both associated with the effectiveness of the academic process. Therefore, the growing interest worldwide among organisations in KM has increased because it is related to the quality of education and learning, as an alternative managerial action (Harvey, 2010). This is due to the competitiveness among higher educational institutions in their fields of interest. When TQM became more defined, it was successfully applied by many HE Institutions. However, as KM was more defined, it was extensively applied in developed countries, and good results were obtained. At the same time, some organisations still lack the proper use and benefits of KM in achieving distinction through the use of TQM. This is especially evident from the inefficient integration of KM and TQM.

The quality of education is vital for promoting society and contributes to the enduring wealth and security of a country (Babbar, 1995). Therefore, this research paper focuses on the integration of KM and TQM in HE in Jordan. The purpose of this research is to learn about KM and TQM and to identify their role in HE excellence in Jordan.

The last few decades have witnessed a widespread acceptance of KM and TQM as a means of acquiring and maintaining competitiveness in the global marketplace. Globalisation issues have triggered the move to improve performance, and HE institutions have been quick to adopt quality and knowledge models in order to survive in the increasingly global market. Indeed, worldwide, HE institutions have made steady progress in adopting KM and TQM models and institutional self-assessment approaches. The Malcolm Baldrige National Quality Award (MBNQA) in the United States and the European Foundation for Quality Management Excellence Model (EFQM) in the United Kingdom have been adapted for the HE sector, and institutions within the sector are recognising their benefits. The HE sector in Jordan has tried to adapt the model as well and is as susceptible to globalisation as any other country.

The study aim is to establish a framework that represents the integration of KM and TQM in HE. Therefore, the research was set up according to the following objectives:

- To review relevant literature on the concepts of KM and TQM and their significance for educational organisations and to contribute to the conceptual framing of them.
- To contextualise a framework for the integration of KM and TQM.

- To refine the framework scientific methods and techniques to be followed in order to take advantage of the integration of KM and TQM in enhancing the competitive advantage.

LITERATURE REVIEW

Universities in Jordan

Higher education in Jordon has expanded dramatically since 1951. It started with a one-year postsecondary teacher training college. Later, in 1962 the first public university was established, the University of Jordan. Fourteen years later the University of Yarmouk was established, which was followed by the establishment of six universities (Abu-El-Haija, 2014). Table 1 demonstrates the progress of the establishment of public and private universities in Jordan since 1962. Nine universities or approximately 43% were established within three years (1990-1992) since 1962.

The progress of the establishment of private universities was due the government policy to ease the pressure in providing HE to the increased number of students who wanted to continue their HE. As the Jordanian recourses are limited, the government was struggling to subsidise the increased number of enrolled students. Therefore, many private universities were set up to ease the pressure on public universities. The Ministry of Higher Education and Scientific Research (MoHESR) has emphasised

Table 1. Progress of the establishment of public and private universities by year

Year	Number of Public Universities	Number of Private Universities
1962	1	0
1976	1	0
1981	1	0
1986	1	0
1990	0	2
1991	0	5
1992	1	2
1993	1	1
1994	0	1
1995	0	1
1997	1	1
Total	7	13

quality of education as one of the important public sectors that can benefit society. Therefore, the establishment of the Higher Education Council in 1982 was aimed to maintain the quality and efficiency of education in Jordan.

The establishment of the Jordanian Council of Higher Education was in response to the need for the regulation and planning of HE policies and coordination among Jordanian public policies at the post-secondary level and of legislating up-to-date laws on HE.

Background of the Study

The last few decades have witnessed a widespread acceptance of KM and TQM as a means of acquiring and maintaining competitiveness in the global marketplace. Globalisation issues have triggered the move to improve performance, and HE institutions have been quick to adopt quality and knowledge models in order to survive in the increasingly global market. Indeed, worldwide, HE institutions have made steady progress in adopting KM and TQM models and institutional self-assessment approaches. The Malcolm Baldrige National Quality Award (MBNQA) in the United States and the European Foundation for Quality Management Excellence Model (EFQM) in the United Kingdom have been adapted for the HE sector, and institutions within the sector are recognising their benefits. The HE sector in Jordan has tried to adapt the model as well and is as susceptible to globalisation as any other country.

Types of Knowledge Management Technology

The term *knowledge management technology* can refer to the most significant developments in information and communications technology, which effectively contributed to the emergence and development of KM, which has made technology an important element and major component of KM. These technologies are detailed below.

Knowledge Work System (KWS) is a common set of systems for generating new information and knowledge. A KWS provides fast, low cost tools to gain and manage knowledge in a better way in various organisations and encompasses computer-aided design (CAD), virtual reality systems, and investment workstations (Laudon & Laudon, 2001, pp. 364-365).

- **Computer-Aided Design (CAD):** Computer and complex design software are being used to design products, where they can draw and test designs faster and less expensively. It also provides the ability to deliver more designs and better quality. Many organisations use such systems.

- **Virtual Reality Systems:** Virtual reality systems go beyond CAD systems through using computer simulation in which products are tested more realistically through the computer. Thus, the user feels as though the simulation is actually in a realistic world. Virtual reality systems provide many advantages in education, medical science, and business administration and provide three-dimensional images to illustrate products and their uses in an interactive manner; such software is available on the Internet (Laudon & Laudon, 2001, p. 365).
- **Office Automation Systems:** These system are used to automate administrative tasks into an integrated whole, these systems include e-mail, information retrieval, word processing, and teleconferencing.

Knowledge Team

In the early years of KM, many organisations lacked in the development of significant conformation of technological solutions before understanding the human factors of KM, but these organisations realised that when individual capabilities and technology are brought together effectively, KM may provide real value to an organisation (Baserda, 2006, p. 72). Knowledge management is partly a reaction and sequel to the patterns of the last decade, as downsizing and rearranging of business processes has led to the withdrawal of a lot of talented employees in organisations, after which the organisations realised the transfer of technology-based advantages to competitive organisations meant the only remaining advantage was their work force (Baserda, 2006, p. 72). Theorists of KM realised the importance of individuals and their value to the success of an organisation; therefore, they are now trying harder to reduce the number of individuals in the organisations and to increase the value of those remaining (Zorn, 2001, p. 5).

Knowledge workers are different from data makers, who are responsible for the processing that is conducted by secretaries, salespersons, and record holders, and are capable of generating information and knowledge. They are more informed and cultured than data makers, and they may be organisational consultants. Other knowledge makers include researchers, designers, writers, and judges who can generate knowledge (Laudon and Lauson, 2001, p. 359). Kubaisi (2002, p. 6) believes that knowledge makers are generally university graduates, and are often members of a recognised profession, such as engineers, scientists, financial analysts, marketing partners, and production planners. They are responsible for generation, development, and integration of new knowledge into existing knowledge and should have the ability to judge. They have a thorough understanding and independence to act. Laudon and Lauson (2001, p. 363) indicated that knowledge makers are essential for an organisation and managers. They can maintain a currency of knowledge in the organisation and communicate its development to the external environment in the fields of technology, science, social thought, and the arts. They also can serve as advisors in the field of internal knowledge and act as agents of change (i.e., perform and evaluate the process of organisational change).

Knowledge Managers

The introduction of the post of Chief Knowledge Officer (CKO), termed by Rastogi (2000, p. 41) as Chief Knowledge Manager (CKM), is the first step in helping organisations manage their own knowledge. The manager should have the basic task of the collection of various aspects of knowledge, especially with regards to personnel, the classification and transfer of the knowledge base to derive results, standards, and formulas necessary to improve work (Kuhn, 2001, p. 1). Organisations seeking to create the post of CKO must provide a measure of technical and regulatory infrastructures and must focus on critical initiatives to ensure the achievement of KM. There are three roles that can be performed by a CKO (Herschel, 2000, p. 37).

- Responsible for the development and execution of KM programs.
- Development of the strategy that guides the organisation on how to maintain and process their intellectual assets.
- Adoption of an organisational culture that focuses on permanent learning and growth. Turban et al. (2001, p. 40) indicated that the role of CKO represented the briefing of knowledge using the method of information technology and has three responsibilities, including generating infrastructure for KM, building a knowledge culture, and dissemination of the knowledge culture.

Customer KM (CKM) represents the management of an external structure (capital of the customer), which reflects the knowledge of external customers of an organisation. Most organisations that are geared toward the customer have managers that know their customers and mostly succeed in KM "about" the customer, but few could appropriately manage their knowledge of the customers (Baserda, 2006, p. 76). In this respect KM can be considered a dynamic combination of specific processes, which include the identification of knowledge and its acquisition. Knowledge gained from customers must be learnt, which aids in the application of sharing and the exploitation of knowledge application and knowledge technology.

The Concept of Knowledge

Different types of contemporary organisations are facing a surge of transitions and accelerating changes sweeping today's world, on top of informatics and technical evolution. This evolution relies on advanced scientific knowledge and the best use of information flow resulting from big developments in computer technology and world communication (Internet). As a result of those transitions, knowledge has become the most significant strategic source; moreover, it has become the strongest, most influential, and dominating factor in an organisation's success or failure (Schwandt & Marqurdt, 2000).

By looking at the subject in literature, it has been demonstrated that trend researchers classified the concept of knowledge as following two main streams: looking at a partial perspective of knowledge constructed only through implicit acquired knowledge of information and the causal links that assist in making sense of this information (Sarvary, 1999, p. 96). As described by Daft (2006, p. 258), "the basis for the sense to measure the personal experience of intuition (intuition) or special view." Knowledge, according to this definition, is the fundamental basis of the experience of an individual in judging things. Hislop et al. (2000, p. 1) add that the status of knowledge is defined as "a continuous process for the restructuring of ideas during exercise or ongoing experience." While to Drucker (1999) knowledge is "the ability to translate information into the performance to achieve a specific task, and this ability can only be done in humans with brains and intellectual skills"; this definition refers to performance as the director of knowledge to achieve specific targets, which was not mentioned in any of the above definitions.

Knowledge Management (KM)

Emergence of Knowledge Management

The field of KM is relatively new, particularly at the applied level. It has been recognised only in the past few years, and afterwards, some standards, as well as a growing awareness of the benefits, have been realised by its application. At the end of the twentieth century, attention was drawn to this subject because at the time the components involved in the management of knowledge needed to re-adjust the value that would be utilised in the market to meet the needs of the customer in the face of rapid changes taking place. The need to manage knowledge emerged because of the increasing growth and necessity to diagnose the flexibility of organisational structures, which could not cope with rapid change in the market. Perhaps the increasing use of information technology and communications that led to the evolution of information available, which impressed users, is one of the reasons that stimulated the emergence of KM. It was not a challenge limited to finding information on an order or a specific project, but how to search through the information available to find the most useful information for a particular purpose. Another method is the ability of the growing information and communications technology that has assisted additional processing and retrieval methods in obtaining the information on storage, which is necessary to the user.

Don Marchand first used the term *knowledge management* (KM) at the beginning of the 1980s as the final stage of hypotheses concerning the development of information systems but did not refer to it independently as a process (Kubaisi, 2002, p. 57). Then, a series of early pioneers of KM were credited with putting KM on the map. While featured organisations like Beckman Labs and the International Bank, by their creation of jobs in the field of KM, led Japanese companies to play an important role in the emergence of KM and its applications.

The emergence of KM and its development were the collections of outcomes of a set of elements, assets, and the roots of the relationship. The most important include the following (Abu Fara, 2004, pp. 9-10):

- Philosophical theoretical thinking.
- Consideration of realism in the requirements of experience in the labour market.
- The views of educators.
- The views of entrepreneurs and business leaders.
- The role of economic forces in the formation of knowledge.
- The efforts of business organisations during the twentieth century to increase effectiveness.

In this context, the accumulation of knowledge and the rapid changes and developments that have been seen by the world during the end of the last century contributed greatly to the creation of the term KM. Then, a series of writings and research on the development of this term determined the concept, models, and development as well as the development of applied standards. Many organisations attempted to apply KM in their work. Some of which succeeded and achieved good results in performance, while others failed due to the vagueness of the term, which was in the process of evolution.

The Concept of KM

The concept of KM is considered a modern concept in management science. It has gained increased interest in the last two decades that has led to the emergence of many definitions for the concept, which vary depending on the specialities of the researchers and various perspectives. To understand and define KM, there are three basic portals, as follows (Malhotra, 2000, p. 83):

The **Informational Portal** *includes data processing and the management of information flow activities, database development, and documenting business activities in an organisation.*

The **Technical Portal** *is concerned with building and developing KM systems based on information technology systems, such as data mining, data warehousing, expert systems, analytical processing systems, instant information systems based on the Web, and others.*

The **Cultural Portal** *is concerned with behavioural or intellectual dimensions of KM through handling the fields of collective learning, lifelong learning, and building learning-oriented organisations.*

In all these portals, KM seeks to provide administrative solutions through investing in knowledge resources, building memory for knowledge, and focusing on knowledge exchange and participation through a systematic methodological portal. Therefore, different definitions appear in the literature. As such, the researcher was concerned with providing a range of definitions that serve the purpose of the study. Thus, Chou (2005, p. 26) maintains that KM is the process whereby extraction and investment of intellectual capital of an organisation, in order to reach decisions efficiently, effectively, and innovatively, to give an organisation a competitive advantage and to

Figure 1.

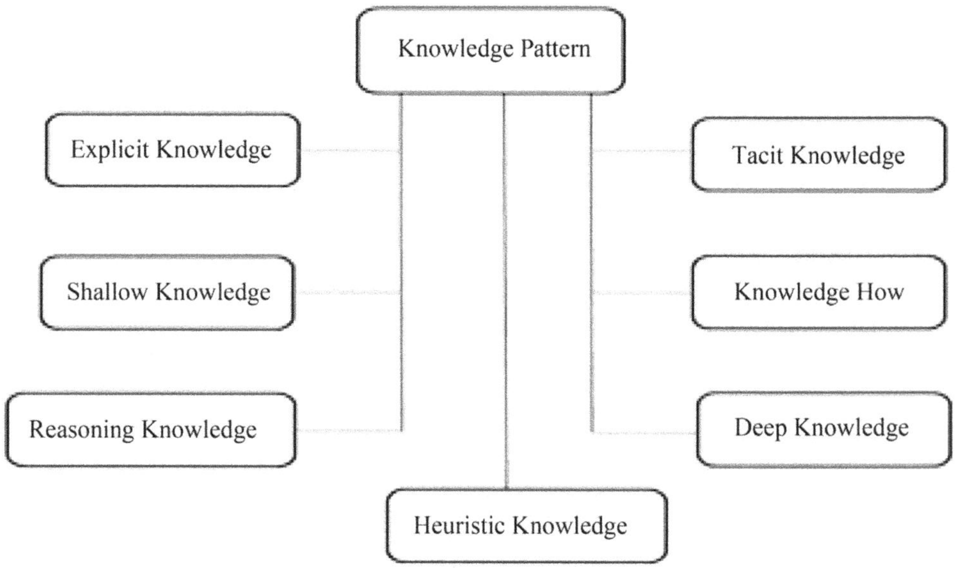

obtain the loyalty and commitment of customers. Delong (2004, p. 6) sees KM as an administrative system activity based on containment, collection, and formulation of all matters related to the important critical activities of an institution in order to enhance performance and ensure the continuity of the evolution of an institution in the face of the surrounding variables. In other words, it is the process aimed at institutional coordination and integration of data processing, information technology, human resources utilised, and the factors surrounding the institution (Delong, 2004, p. 6). Alokalpi (2008, p. 26) concluded a definition for KM as an administration that is working to identify the complete knowledge (in their minds) individuals have (whether employees, consultants, or beneficiaries), or to collect and create knowledge that appears in records and documents and organises it in a manner that facilitates its use, or the participation of employees of the institution in order to achieve higher performance and work success with the best methods at the lowest possible cost. In addition, Al-Zamel (2003) & Shehabat et.al. (2008) defined KM as a set of activities and processes that help the organisation generate knowledge, access it, select it, use it, organise it, and disseminate it, and work on transforming the knowledge (along with what it includes in terms of data, information, experiences, directives, and capabilities) to products (goods or services) and using KM outputs in decision-making, solving problems, drawing learning processes, and building an integrated system for strategic planning (Shehabat & Berrish 2013).

Based on the above, the researcher defines KM as a continuous dynamic process that includes a set of activities and practices designed to identify, find, develop, distribute, use, conserve, and facilitate the retrieval of knowledge, resulting in increased performance, cost reduction, and capacity improvement related to the process of adapting to the requirements of rapid change in the organisation's surrounding environment.

Models of KM

Many researchers introduced several models for KM in an attempt to understand and guide the efforts of KM activities in institutions. The goal is to direct organisations to build knowledge strategies to help achieve goals and solve problems, even in line with the changes of the times.

The SECI model is the model is the most famous KM Model developed by Nonaka and Takeuchi (1995), and it was abbreviated as SECI to shorten the four processes of KM. The full terms for this abbreviation represent respectively the following socialisation, externalisation, combination, and internalisation, which are illustrated in Figure 2.

Figure 2. The SECI model as developed by Nonaka and Takeuchi (1995)

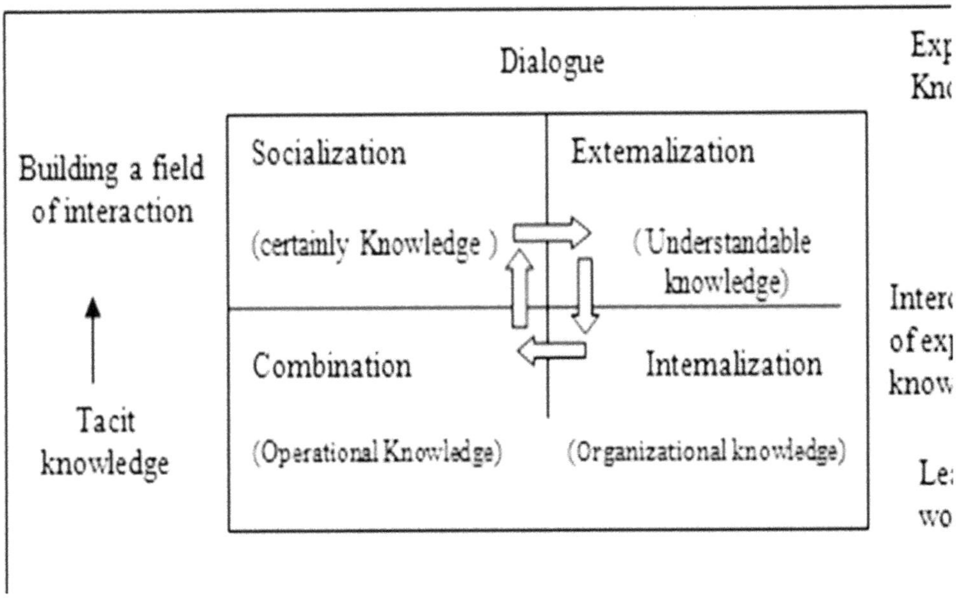

Justifications of the Application of KM in Educational Institutions

The regulatory environment of educational institutions, such as universities and colleges, is one of most suitable environments for the application of the concept of KM, but it is may be *the* most suitable environment to apply this concept compared to other organisations, due to their role in the community. As universities and colleges of various kinds are responsible for the preparation and the creation of a qualified and trained workforce, which is a vital element for all comprehensive community development, thus necessitating the adoption of the concepts, methods, and practices of modern management that contribute to raising levels of performance (teaching, research, innovation) that leads to upgrading the quality of its outputs (Abu Khudair, 2009, p. 14). A study of Kidwell et al. (2000) has shown that knowledge creation, innovation, discovery, dissemination, and circulation are the main reasons for the existence and establishment of universities and colleges, and they emphasise the importance of the inevitable adoption of KM as the entrance to the improvement and development that will contribute effectively in achieving its goals and mission. Mikulecka and Mikulecky (2005) have defined several justifications for considering the environment of universities and colleges and the role assigned to them as the most appropriate environment for the application of the principles and methods of KM, which are as follows:

- Universities and colleges usually own infrastructure and modern information.
- Knowledge sharing with others is normal between faculty members, lecturers, and students in general.
- Students seek access to knowledge sources that are accessible through joining universities and colleges.
- These institutions usually provide an organisational climate characterised by confidence; thus, no member would hesitate in the publication and distribution of knowledge.
- In light of the great similarity between universities and colleges and modern business organisations and in terms of the orientation of these institutions toward providing many activities and educational services, research, and advisory meetings in return for money, any style or method that may give the university a competitive advantage and the ability of better performance may be of interest to those institutions, and KM is one of the modern management techniques leading in this direction.

Jordanian universities are one of the developing country's educational institutions that are not less popular than international universities in terms of the desire to attend, and perhaps the most important justification for their popularity and increasing demand is because they contain comprehensive, flexible courses that meet the needs of individuals and society and with the needs of the development plans of human resources and the labour market (Amer, 2008). Jordanian universities provide a greater opportunity for those wishing to join the labour market and those who want to continue their advanced higher university education (Amer, 2008). The programme aims to provide students with the skills needed by the labour market, in addition to a broad base of information with multi-purpose use, such as basic science, language, communication skills, and the use of computers. Most Jordanian universities give a final certificate that qualifies the student for a specific career and emphasises proficiency for specialisation and familiarity with skills that can be refined by appropriate expertise from the corresponding profession (Amer, 2008).

Based on many researchers (Dacey, 2007; Mertins et al., 2001; Badr, 2010; Awdah, 2010; Alavi, 1997; Kiessling et al., 2009; al-Qubaisi, 2004; Alziyadat, 20008 Coakes, 2003; Alwani, 2006; Marquardt 2002; Alian, 2008; and others), the researcher considers knowledge identification and knowledge discovery as a single dimension referred to as knowledge identification. Other main processes of KM, which are storage, sharing, and application, make up the remainder of the processes (dimensions) of KM and, along with technology and KM teams, are considered in this study. A discussion of the three dimensions of KM follows in subsequent subsections

The Concept of TQM

Literature classified TQM from different perspectives, including as a system, approach, philosophy, culture, and strategy according to its components.

Definition of TQM

It is a universal fact that quality is an attribute that cannot be measured in absolute terms, as it is not quantifiable, and there are numerous meanings to different schools of thought (Ediger, 2010). The issue is still the same; when we focus on TQM, a common concept in management style emerges, which was put forward in 1961 as TQC or Total Quality Control by Feigenbaum (Black & Porter, 1996). Many considered TQM to be one of the most controversial topics within the management field; thus, this concept and its importance have been interpreted by many schools of thought differently over a period of time (Lee et al., 2010).

Regardless of all the differences, TQM is regarded by all schools of thought as a concept that is significant in constantly improving upon production and processes to improve and exceed customer satisfaction and expectations (Zakuan et al., 2010).

Total quality management ensures that every employee of the organisation is involved in the process of developing a complete picture of the effectiveness of the business (Fotopoulos & Psomas, 2010). This ensures the end customer's needs and satisfaction are met by the organisation, which may be internal or external to the business (Tanninen et al., 2010).

Thus, in the short and long term, TQM acts as a direction for businesses, providing advantages to all stakeholders (Perera & Kuruppuarachchi, 2010). By directing the energies and abilities of those on board, TQM tends to continuously meet customer needs and requirements at a feasible price (Black & Porter, 1996).

It has been certified many times that those organisations that adhere to TQM in their management processes tend to benefit from augmented performance and benefits (Lee et al., 2010). It is believed to enhance and improve the management of the organisation to its optimal best, ensuring more productivity (Fotopoulos & Psomas, 2010).

The ultimate goal when focusing on continuous improvement of the overall business with the application of TQM is to acquire a competitive advantage for the organisation (Tanninen et al., 2010). It is also significant to understand the importance of this advantage when TQM tends to incorporate all aspects of the organisation and take every factor into account, while directing energy in one direction to augment the effectiveness, competitiveness, and efficiency of the business for the sake of stakeholders (Zakuan et al., 2010).

Quality in Education

Haworth and Conrad (1997) wrote that in the context of the ever-changing dynamics of the norms that we adhere to, it is only consistent that the definition of the quality of education also changes with time. Thus, it can be characterised as what society deems necessary to standardise as the minimum level of education within a certain factor, the ethics, and monetary pledge to achieve it (Davok, 2007). Therefore, it can be put forward as the accomplishment of the implied and explicit objectives of society that it expects from the educational system, while assuring strategic requirements (Tikly & Barrett, 2011). All of the process, including its elements, inputs, and outputs, while justifying the expectations of the society, determines the direction of education, and the character of each of the mentioned elements determines the quality of education (Gidley et al., 2010).

The ultimate objective for a student within an educational institution is to successfully attain a certain noteworthy educational accolade that would enable the student to be a valuable contributor to society (Haworth & Conrad, 1997). Therefore, the extent to which any educational institution is able to maintain the same educational quality which was also put forward as being the definition of the quality of education by yet another school of thought (Jones, 2011). While considering the fact that (like health sector) education is also a service sector, we can define the quality of education as the extent to which the educational needs and expectations of the individual in general, and society as whole, are met (Davok, 2007).

TQM in Higher Education

To date, TQM has been globally adopted by organisations hailing from altogether different sectors successfully and has born fruits for management and stakeholders (Tanninen et al., 2011). However, it is a different story when it comes to HE. The HE sector has been bombarded with a number of hurdles while implementing TQM; some of these included performance measures, society and organisational culture issues, and customer identification and leadership (Ali & Shastri, 2010).

When it concerns the HE sector, there is a growing uncertainty over the adaptation of TQM practices (Pratasavitskaya & Stensaker, 2010). A school of thought believed that, although there were issues with the system, due to the slow progress of educational decline, they seldom were taken to task (Harvey & Williams, 2010). Instead, these issues become part of the systems and management of the institutions that allowed those to become part of the system, rather than treat them as an issue to be addressed and modified (Houston, 2007). The staff members also consider TQM to be a peril to the independence they have enjoyed within the HE sector since it presents added responsibilities, such as teamwork and customer satisfaction (Lomas, 2004).

One of the key issues that has been highlighted by management of institutions in the HE sector is that TQM has its roots within the manufacturing sector (Pratasavitskaya & Stensaker, 2010). As TQM developed from the manufacturing sector, its utility and implementation in the service sector seems pointless to them and poses a grave threat (Harvey & Williams, 2010). Therefore, in this situation, without any precedent, it is hard for supporters of TQM to persuade institutions to adapt to TQM practices and implement this within the HE sector (Houston, 2007).

It is a fact that adhering to TQM practices, especially in different scenarios, sectors, and different situations may pose a number of hurdles, but even then it is not impossible to adapt to it (Ali & Shastri, 2010). A school of thought is adamant that educational institutions may even be directed to continuous improvement with the application of TQM (Lomas, 2004). Although they may face a few issues while adhering to TQM, even then it is likely to boost the quality of education within these institutions (Harvey & Williams, 2010).

With reference to the HE sector, other schools of thought, however, refrain from suggesting the associated benefits of adapting to TQM (Houston, 2007). Instead, they propose doubts and uncertainty in adapting. This is because education belongs to a service sector that has no tangible end product, rather a satisfied customer, which would vary by individual (Pratasavitskaya & Stensaker, 2010). They suggest alterations to TQM in a manner that it is on the same lines that would be deemed feasible for the HE sector (Harvey & Williams, 2010).

As discussed earlier, these distinct attributes do pose a range of boundaries and limitations especially when considering TQM from the educational institution and HE perspective (Lomas, 2004). However, it is also evident that within the education industry there are very few quality performance indicators present, if any (Houston, 2007). Therefore, rather than focusing on the education itself, TQM should be used as an indicator of the service that educational institutions provide (Ali & Shastri, 2010). This is even more suggestive since the customers in the HE sector are relatively unknown compared to the manufacturing sector where the end customer is identified and the end product is tangible (Houston, 2007).

However, there are a number of institution within the HE sector that (in the span of last decade or two) began implementing facets of (if not the complete) TQM processes within their models (Lomas, 2004). The adherence was prompted in the development of the institutional cultures that were more prone to TQM principles and allowed the institutions to work more efficiently and effectively (Ali & Shastri, 2010).

However, the lack of quantifiable, tangibility of the end product and the lack of homogeneity for the customer pose a grave threat to TQM practice and somewhat segregates the manufacturing sector from the service sector, which encompasses the HE sector (Harvey & Williams, 2010).

Since TQM paves the way for quantifying and is suggestive of empirically calculating quality, it causes a point of concern for those who are employed by the HE sector (Houston, 2007). The quality management philosophy is beyond their comfort level (Pratasavitskaya & Stensaker, 2010). This is primarily because not only is education an on-going, long-term process, but also it is rather beyond the education sector to quantify and analyse the progress on the basis of goal achievement (Lomas, 2004). The education sector is more comfortable with dispersing knowledge, while enabling the students to understand and comprehend, imparting wisdom to tackle issues, and developing the characters of the students to ensure that they become contributing member of society (Ali & Shastri, 2010). From their perspective, it is unfair to be subjected to analysis and measurement on specific goal achievements as a part of a TQM system for the education sector (Harvey & Williams, 2010).

Due to this culture in place within the education sector, it was suggested that a modified version of TQM can be implemented within the HE sector (Houston, 2007). An excellence model was put forward with a suggestion that strong leaders were required to apply and ensure its implementation within the institution while modifying the institutional culture (Lomas, 2004). There was a mutual agreement by different schools of thought that application of TQM within the HE sector was not an impossibility. Instead, the management of the institution should focus on determining the modifications that are required within TQM to make it more adaptable from the HE perspective (Ali & Shastri, 2010).

There was a suggestion to modify the terms used in the context of TQM while applying it in the HE setting (Pratasavitskaya & Stensaker, 2010). However, it is significant to understand that each and every element of HE cannot be brought under the folds of TQM (Harvey & Williams, 2010). This notion is truer in the case of education, academics, teaching, and research where the tangibility of the effort poses a obstacle (Houston, 2007). Therefore, TQM is more applicable in the service sector if the independent nature of the professionals is taken into account or modifications are made in TQM to adhere to their liking within HE sector (Lomas, 2004).

Regardless of the fact that the manufacturing sector stimulated the creation of TQM philosophy, there is a belief that the approach would also help the education service sector (Ali & Shastri, 2010). The HE sector and the institutions within need to identify the issues and tackle them with modifications that are more suited to their industry (Harvey & Williams, 2010). The issues faced by HE pertaining to TQM should be addressed and the barriers identified to ensure that TQM is implemented within institutions, and the education sector is able to make full use of the philosophy while ensuring organisational successes (Houston, 2007).

In the 1980s, TQM started gaining acceptance in US colleges and universities (Harvey & Williams, 2010). Although TQM made its way into the educational sector at that time, it was not able to cement its position within HE culture comprehensively as compared to other industries in the country (Lomas, 2004). This was because there were a number of impediments that were almost contrary to the culture and traditional characteristics of institutions within the HE fold in the US (Houston, 2007). Within HE sector, there were also a number of variations within the culture being adopted by different institutions that are discussed here (Ali & Shastri, 2010).

Universities that were relatively older were found to be engaging in longer-range plans. There was status quo in the top management, and they would not readily agree to any change or introduction of any new ways, culture, or any measures that would gauge their performance (Pratasavitskaya & Stensaker, 2010). They would resist any such implementations, as they considered TQM a threat to their independence and method of working (Houston, 2007). However, the newer universities and colleges were more apprehensive of their image and status and were found to be very eager regarding various aspects of TQM implementation within the HE sector (Lomas, 2004). Since the 1990s there has been an increase in pressure from various sectors and businesses to enforce TQM practices within the HE sectors and all institutions to adhere to TQM philosophy (Ali & Shastri, 2010).

As the global application and success of TQM started to become evident, the HE sector also started to implement the same within its institutions in the 1980s (Harvey & Williams, 2010). The importance of TQM was soon realised, as it started playing a major role in national developmental programs and global economies (Pratasavitskaya & Stensaker, 2010).

With the ever-changing dynamics of the global stage, there was a need to revise the curriculum to maintain it with the changes (Houston, 2007). This would be applicable within an institution in the field of HE that would endorse consistent quality and performance improvement within the educational system (Lomas, 2004). Therefore curriculum, administrative units, non-academic units, and the core learning process were deemed to be four areas where the application of TQM could obtain better results for educational institutions (Ali & Shastri, 2010).

The issues that cropped up for the institutions in the HE sector during the application and adaptation to TQM practices were very common and suggestive ones (Harvey & Williams, 2010). These included issues, such as the variation and character of the end customer of their services, the traditional management style of the educational institute, and the research and academic nature of the work done within the HE sector (Houston, 2007). The compartmentalisation, conformance to traditional style, lack of competition, and adherence to the status quo were identified as the four main obstacles to the continuous improvement cycle of TQM (Lomas, 2004). There was a sense among the employees that they had already worked enough to attain whatever excellence they may have already achieved (Pratasavitskaya & Stensaker, 2010). Since TQM was considered to be the brain child of the manufacturing industry, it was hard to assure those within the HE sector that it would have the same impact on the improvement in the education sector, which proved to be a major impediment as well (Harvey & Williams, 2010).

There was an understanding that instead of the academic itself, TQM was beneficial for employee and process performance within the HE sector, which in the longer run on a continuous basis would be advantageous for teachers, students, and all stakeholders with the education sector (Houston, 2007). It would help in addressing the issues and problems of students, teachers, and other stakeholders, while assisting them in meeting the needs and expectations of those concerned (Lomas, 2004). Continuous improvement would analyse the effectiveness of the academic programs while, identifying areas of improvement and working on how to resolve these (Harvey & Williams, 2010). An important aspect for continuous improvement in HE is to allocate resources to tackle the same areas and issues on priority (Ali & Shastri, 2010).

It was important for management to fully understand the importance of TQM and the benefit of the application of continuous improvement (Pratasavitskaya & Stensaker, 2010). Management posed serious impediments to its application due to their misinformation and lack of relevant awareness (Houston, 2007). Even in the case of a willingness to continue with TQM, management failed to implement basic guidelines like teamwork, skill provision, agreed upon goals, and resource allocation (Lomas, 2004).

Literature indicated that the evidence in favour of the use of TQM in the educational context relates primarily to administrative tasks (Koch & Fisher, 1998; Koch, 2003; Venkatraman, 2007). The non-academic focus of TQM dramatically reduces its significance and impact in the world of higher education (Ahmed, 2008; Koch, 2003; Venkatraman, 2007). Hence, research is needed to test the position that TQM initiatives do, indeed, have a relationship with academic related improvements. In light of this, the present study will focus on the educational process as a core academic activity in HEIs. Moreover, the researcher aims to contribute in this direction by examining the TQM effect on performance through KM to further enhance the fitness of TQM in the higher-education context.

From the principles of TQM and other quality management theories, it is obvious that these are all highly structured towards one philosophy, which is continuous improvement in order to achieve a common goal—customer satisfaction (Adair, 2004). Therefore, it is necessary to identify the customers (stakeholders) of higher education. Kanji and Tambi (1999) classified the customers of higher education from two perspectives, primary and secondary customers, based on their position (i.e., internal or external) and the interactions the institution has with them. While the academic educator (as an employee) is defined as the primary internal customer, the students (as educational partners) are secondary internal customers.

Similarly, the student is also the primary external customer, and the government, business companies, and parents are secondary external customers.

Therefore, quality in higher education has become a public issue in Arab countries. Since Jordan is one of these Arab countries, the case is not different in the institutions of higher education in Jordan.

Then, quality in higher education became one of the public issues in Arab countries. Since Jordan is one of the Arab countries, this case is no different in the institutions of higher education in Jordan.

KM Integration With Total Quality Management

The integration of KM with TQM can enhance and develop the organisation's performance as well as achieve its goals of excellence and leadership. These two issues are closely related to each other (Table 2). shows the similarity and differences between KM and TQM.

METHODOLOGY

Background of the Study

According to Borhan and Ziarati (2002), a strong relationship exists between a country's competitiveness and the quality of HE. Hence, countries need to assure themselves that quality is certain in their HE institutions. In Jordan, where the number of HE institutions has expanded substantially, this has become an issue. The growth in Jordan's HE sector has begun with a large investment in state universities, which correspondingly attracted a large number of students. With an increase in the student market and the resulting pressure on the public budget to provide continually more HE locations, Jordan's HE policy allowed the local public administration and private sector to establish universities, colleges, and technical HE institutions. Now, with the increasing number of private universities, the need to establish and maintain a competitive advantage is becoming increasingly significant. In this respect, the University of Jordan has adopted TQM and information technology systems (Abu-El-Haija, 2014). However, research has reported poor implementation for TQM principles in Jordanian colleges of physical education (Altahayneh, 2014). In addition, in view of the increased demand on Jordanian universities, the King of Jordan and his prime minister encourage adopting a clear plan for improving quality in HE that enables universities to compete with the global market (Schwab, 2012).

Knowledge Management and TQM play vital roles in the modern management progress where KM contributes to innovation by creating, storing, transferring, and

Table 2. Similarities and Differences: KM and TQM

Similarities	
KM	**TQM**
• Continuous improvement and learning from others • Valuing employees/intellectual capital • People/competence development • Empowerment/ involvement • Teambuilding/collaboration • Acquiring knowledge of competitors, customers, suppliers and partners • Facilitating/improving access to knowledge • Improving quality and efficiency of decision-making	• Continuous improvement and measuring to achieve customer satisfaction • Valuing employees/intellectual capital • Employee training/education/development • Empowerment/involvement • Teambuilding/collaboration • Acquiring customer and market knowledge • Selection and use of information and data • Open channel of communication • Improving quality and efficiency of decision-making
Differences (Focus/Strategies)	
KM	**TQM**
• Embedding knowledge in staff, customer, products, process, and services • Regarding knowledge as the source of competitive advantage • Achieving greater productivity through the use of knowledge • Creating/disseminating new knowledge and embedding it in new technologies and products • Searching for new source of information • Adapting knowledge to market needs	• Better use of resources, to achieve effectiveness and efficiency • Striving for excellence through benchmarking, etc. • Consistent and precise performance to high standards in all areas of the organisation • Effective leadership and team commitment throughout • Customer focus • Results focus • Measurement of quality using data and tools • Management by facts and processes • Mutual respect, mutual trust, and mutual benefit of all stakeholders

applying knowledge. The implementation of TQM practices is considered one of the important factors that influenced innovation in a positive way (Honarpour et al., 2012). Despite the vast growth in the implementation of TQM in HE institutions worldwide, there is no universal model for adopting TQM within a higher education context (Sarvan & Anafarta, 2005, p. 12).

The HE institutions in Jordan have witnessed an increase in demand for enrolment in educational programmes in universities and colleges. The low level of the quality of education, which is has been nationally confirmed by King Abdullah II Ibn Al Hussein for the Prime Minister on Higher Education on February 7, 2007, during his visit to the University of Jordan on April 26, 2007. Between them, it is considered important to have a clear plan for HE quality improvement as well as improved quality for the coming years to ensure the right of young people at universities to have a high-quality education to enable both students and universities to compete in the global market.

The graduates of the Jordanian educational system in all disciplines are not sufficiently knowledgeable or trained to be productive, and graduates usually need extensive retraining. As such, the quality of HE in Jordan is illustrated in Table 3.

The global rank of the quality of HE was the highest in the availability of scientists and engineers (11 out of 144). However, the rank for university/industry research collaboration in research and development, extent staff training, and quality of scientific research institutions were, respectively, 93, 88, and 77 out of 144 (Schwab, 2012). Furthermore, the quality of the educational system was ranked 31 out of 144. The overall assessment of HE in Jordan is more likely to indicate that the quality of the educational process supports business in Jordan.

The primary motive for conducting this study is the lack of empirical research on the implementation of KM and TQM within Jordanian HE institutions. While there has been much attention devoted to the integration of KM and TQM, the vast majority of literature has been focused on its association with economies (McAdam & McCreed, 1999; Thiagarajan et al., 2001; Aboyassin et al., 2011). The growing awareness of the importance of KM and TQM in developing the economy has attracted the interest of many researchers. The study encourages empirical research to study TQM for improving the quality system of an organisation (Soltani & Lai, 2007). They found that the major discrepancy between the rhetoric of these systems and the reality of their practice is where organisations tend to develop a more strategic approach to managing soft aspects of quality management. However, literature indicated that the number of organisations that implemented TQM successfully is comparatively small (Gibbs, 2009). In particular, the number

Table 3. Higher education and training quality in Jordan

Indicators	Country Ranking/144
Quality of the educational system	31
University/industry research collaboration in research and development	93
Quality of math and science education	28
Quality of scientific research institutions	77
Quality of primary education	43
Availability of scientists and engineers	11
Quality of management schools	55
Internet access in schools	44
Availability of specialised research and training services	47
Extent staff training	88

(Schwab, 2012)

of studies on quality in HE is relatively few (Lagrosen et al., 2004). Furthermore, KM processes including diagnosing, acquiring, generating, sharing, storing, and applying have been found to influence TQM (Aboyassin et al., 2011). Therefore, KM could be considered a key tool to higher educational long-term success. Thus, change initiatives in organisations including educational institutions can improve shareholder value (i.e., economic value theory) and develop organisational capabilities (i.e., organisational capability theory) (Soltani et al., 2007). Like any organisation, educational institutions are confronted by competition due to the presence of many universities and colleges in Jordan and other countries in the region. In addition, globalisation, the advancements in communications and information technologies, economic recession, and a simultaneous search for excellence are also encouraging competition; therefore, the desire to take up change initiatives has interested the majority of leading organisations (Soltani et al., 2007). The current study will give the opportunity to explore the correlation between public and private universities in Jordan in terms of KM and TQM.

The Research Problem

Many organisations seek to keep abreast of modern administrative developments and implement them in their various activities—not only to thrive and survive but also to achieve and maintain high levels of leadership. In addition, due to the development of Quality Management (QM), many educational organisations moved to implement this concept in many countries. In this respect, the development of KM contributes to better achievement in the global competitive market. There is a close relationship between KM and TQM in achieving a competitive edge. Therefore, the emergence of KM has helped many educational institutions achieve a high educational standard and increase their rank. Therefore, research stands as a vital tool for investigating what is required to develop the effectiveness of the educational process in terms of QM and TQM. Initially, research in QM and TQM has attracted the interest of many universities in Jordan. Nonetheless, it is still necessary to continue the research to get insight into the real nature of what is required to achieve high standards using KM and implementing TQM. Therefore, it is possible to summarise the main issues that are related to KM and TQM.

1. How deficient is the lack of personal knowledge in KM and TQM in Jordanian universities?
2. What benefits are gained from KM in the development of quality in Jordanian universities?
3. How clear is the vision with regard to integration of KM and TQM?

Yet, the main question is *what distinguishes the current study from previous studies?* By reviewing past studies, it appears that some have examined the attitudes of managers towards the application of KM and other studies have examined the performance of KM, creativity, and decision-making as well as discovering the competitive advantage. The current study is distinguished from past studies and research by focusing on a vital aspect, which is the integration of KM and TQM in institutions of HE in Jordan and its role in achieving TQM in these institutions, where most efforts and research have been directed towards KM. In addition, the researchers have benefited from previous studies, and the benefit is reflected in two points. The first is the formulation of the theoretical framework for the study, and the second is the formulation of hypotheses.

Research Conceptual Framework

A research conceptual framework is a useful set-up tool for organising social research (Rodman, 1980). According to Labaree (2013), the theoretical assumption connects the researcher to existing knowledge and gives a basis for hypotheses and research methods. In other words, the concept of the research framework can be extended to involve the integration of concepts and methods. In this respect, the description of background information and research driving KM and TQM are associated with literature and the design of the current research. Literature helps develop research focus and its approach for deriving the required data for analysis and discussion (Labaree, 2013).

Research Framework

Figure 3 shows the research framework in the sequence in which the current research was developed. This starts from an idea, which is supported theoretically by the literature before setting up the main research elements, including objectives, hypotheses, and methodology.

The details of the current study encompass variables related to KM, TQM, and EP. The concept related to these different variables is governed by two main hypotheses (see the next section). The sequence for the development of the current framework is illustrated in Figure 4. The components of the three variables in the current study follow.

Theoretical Framework

This section proposes a theoretical framework along with a set of hypotheses. The theoretical framework demonstrates the relationship between the variables in this

Figure 3. Research framework through current sequence of research development

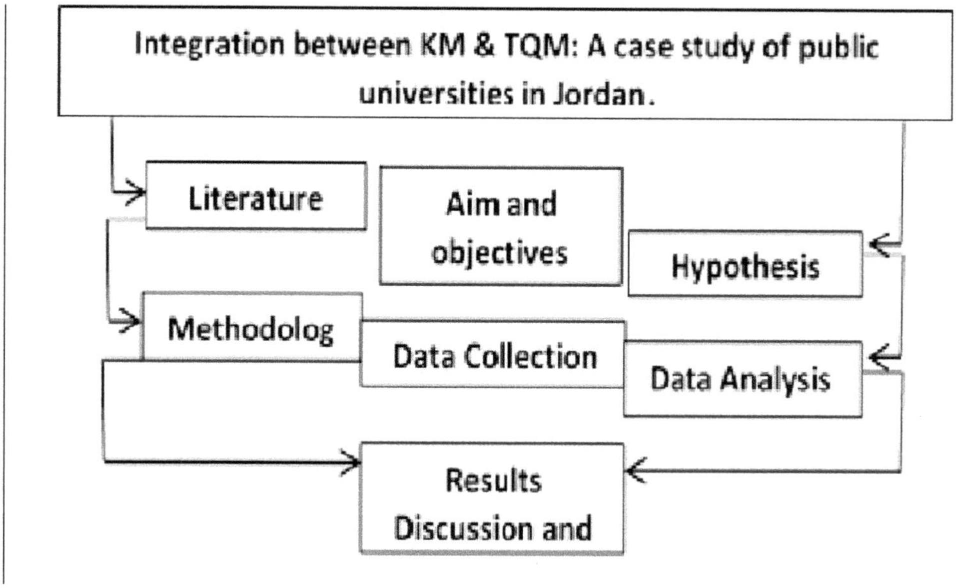

study (Cavana, Delahaye, & Sekaran, 2001; Sekaran & Bougie, 2010). Based on the theoretical foundations, which are reviewed in the literature, a framework has been developed to investigate the relationship and integration among the study's variables: TQM, KM, and its impact on EP for the Jordanian HE context. Figure 3 illustrates these relationships.

Research Variables

The first main independent variable is KM, which is subdivided into the following variables:

1. KM processes.
2. KM technology.
3. A team of knowledge.

The second main independent variable is the Total Quality Management, which is subdivided into the following variables:

1. Effective leadership.
2. Continuous improvement.
3. Employee involvement

Figure 4. Sequence for the development of the current framework and variables for KM and TQM

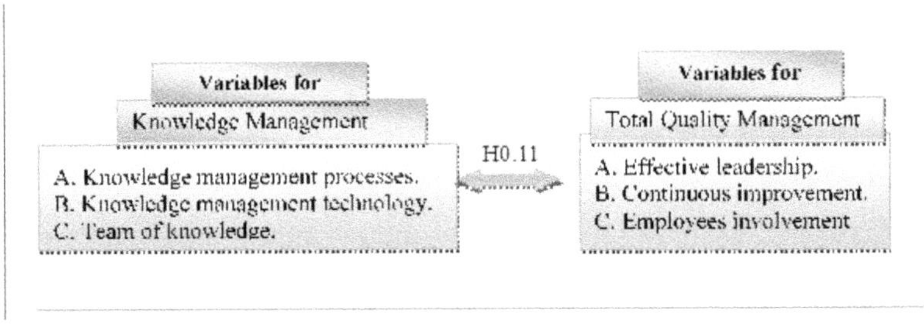

Research Hypotheses

The current study considers one main hypothesis. This hypothesis is subdivided into different hypotheses. Generally, the hypothesis is an important element of research for scientific studies, as it is one way to assist the researcher to focus on the main subject. It is also an alternative assumption made in order to draw out and test its logical or empirical consequences (Kenya, 2013). Here, the focus is on KM and TQM. These two contemporary terms are vital for the educational management process and its outcomes. Therefore, the hypothesis is one of the fundamental tools for research. Merriam (2009) and others (Bogdan & Biklen, 1992) maintained that the hypothesis serves to guide the researcher in seeking answers to tentatively adopted generalisations (Punch, 2005).

Two hypotheses to be justified are proposed in this study. The structure and their connection with the main variables of the current study are depicted in Figure 5.

The Hypothesis

- H0.1 There is no significant relationship between KM and TQM. This hypothesis is subdivided into the following hypotheses.
    - H0.1.1 There is no significant relationship between the processes of KM and effective leadership.
    - H0.1.2 There is no significant relationship between the processes of KM and continuous improvement.
    - H0.1.3 There is no significant relationship between KM processes and employee involvement.
    - H0.1.4 There is no significant relationship between the technology of KM and effective leadership.
    - H0.1.5 There is no significant relationship between the technology of KM and continuous improvement.

- H0.1.6 There is no significant relationship between the technology of KM and employee involvement.
- H0.1.7 There is no significant relationship between team knowledge and effective leadership.
- H0.1.8 There is no significant relationship between team knowledge and continuous improvement.
- H0.1.9 There is no significant relationship between the team and employee involvement.

The above are hypotheses effecting integration between KM and TQM in educational performance.

RESEARCH RESULTS

The Perception Towards KM

The Perception Towards KM Process

Tables 4 to 7 show the means and rank order for the first 20 statements in the questionnaire that were related to the KM process. Table 4 indicates that the average of the collective Likert rating scale ranged between 3.93 and 3.85 with a standard deviation ranging between 0.98-1.129.

Figure 5. The structure and connection of hypotheses with main variables of the study

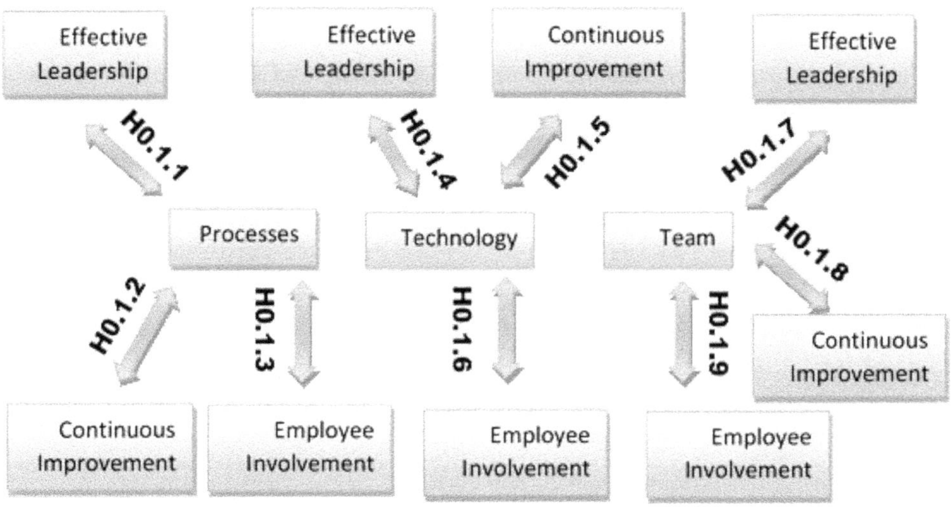

These results indicate that the perception of respondents at public universities rated knowledge creation-related statements close to ‘agree’ on the Likert rating scale (i.e., ‘agree’ is equivalent to four)

Similar results were obtained for storage of knowledge (Table 5), knowledge distribution (Table 6), and knowledge application (Table 7) statements. The range of the mean and standard deviation (SD) were, respectively, 3.91 to 3.87 (SD 1.03 to 1.088), 3.98 to 3.78 (SD 0.908 to 1.135) and 3.96 to 3.72 (SD 0.876 to 0.926).

Table 7 illustrates the overall mean for statements 1 to 20. The mean (3.879) perception of respondents at public universities agrees with the individual statement means (Tables 4 to 7), which also indicates that respondents agreed with the KM process at Jordanian public universities.

Table 4. Ranking order for statements related to knowledge creation: university administration depends on the following to create knowledge (N= 200).

No. of statement[1]	Statement	Rank order (Mean)	Std Deviation
2	Its external environment {customers (stakeholder), equipped, competitors, etc.}.	3.93	0.98
3	Contracting with experts from outside the university to work there.	3.93	1.032
1	Individual learning for workers.	3.86	1.044
4	Insight into the experiences of workers and the experiences of others.	3.85	1.129
5	The Internet, advanced computers’ software, and advisory research canters.	3.85	1.052

[1]Numbering appears as in the questionnaire.

Table 5. Rank order for statements related to storage of knowledge: University administration depends on the following to store knowledge: (N= 200).

No. of statement[1]	Statement	Rank order (Mean)	Std Deviation
7	Computer and other digital information media storage.	3.91	1.03
6	Records and written documents.	3.89	1.038
8	Activated incentives (financial and moral) to maintain individuals of high knowledge and talents.	3.87	1.059
9	A system to control the tangible knowledge assets.	3.87	1.088

[1]Numbering appears as in the questionnaire.

Table 6. Rank order for statements related to knowledge distribution: University administration depends on the following to distribute knowledge: (N= 200).

No. of statement[1]	Statement	Rank order (Mean)	Std Deviation
15	Metaphorical representation and imaging of knowledge.	3.98	0.908
11	Documents, publications, and internal information network.	3.96	0.968
10	Work Teams and regular meetings.	3.95	0.983
13	Relations and informal contacts.	3.92	0.948
14	The transfer of knowledge to others through lectures and services.	3.86	0.993
12	Training and openness in the exchange of thought and dialogue.	3.78	1.135

[1]Numbering appears as in the questionnaire.

Table 7. Rank order for statements related to knowledge application: to apply knowledge, the university administration should: (N= 200).

No. of statement[1]	Statement	Rank order (Mean)	Std Deviation
17	Grant faculty members freedom to apply their knowledge.	3.96	0.876
16	Make sure that faculty members realise the information they hold and feel able to find meaning for them.	3.95	0.955
19	Remove all Barriers that stand in the way of knowledge access for individuals, experts, and administrative units.	3.81	0.893
18	Hire knowledge managers who urge well application to it.	3.77	0.889
20	Make rules to control of knowledge.	3.72	0.926

[1]Numbering appears as in the questionnaire.

The Perception Towards TQM

Effective Leadership

Table 8 shows the rank order of seven means for statements related to effective leadership, which is all close to the fourth option, ‘agree’. The range of the seven means is 4.03 to 3.85, with a standard deviation of 0.740 to 0.903, respectively.

Table 8. Rank order for statements related to a variable of effective leadership (TQM Processes) (N= 200).

No. of statement[1]	Statement	Rank order (Mean)	Std Deviation
41	The senior administration adopts a strategic plan to achieve quality in all areas of the college.	4.03	0.740
42	The university adopts the principle of quality as the responsibility of all sections and parties involved in the college.	4.02	0.730
43	The university administration puts forward plans for continuous improvement and encouragement of employees to participate.	3.95	0.828
45	The university has competent research and development units to determine the needs and desires of the marketplace and exceed these requirements.	3.95	0.794
47	Higher administration makes radical changes to make quality a part of the organisation's systems and operations.	3.97	0.835
44	Communication channels to the marketplace should be available at the university to investigate grievances, opinions, and suggestions about the value-quality of graduates.	3.93	0.826
46	Higher administration decisions are made on the basis of facts and data or based on the recommendations of the staff and students.	3.85	0.903

[1]Numbering appears as in the questionnaire.

Continuous Improvements

Table 9 displays the means of respondent perceptions for five statements related to continuous improvement. The results indicated that respondents to these statements ranked these similarly to the previous perception level, or close to the fourth Likert rating option 'agree'. The range for the means of these five statements is between 3.95 and 3.82 with standard deviations of 0.794 to 0.890, respectively (Table 9). Computing these statements indicated that the overall mean of these five statements was 3.857.

Employee Involvement

Table 11 shows the means and standard deviations of eight statements related to the perceptions of the responses of Jordanian public universities. The means ranged between 3.96 and 3.82, which were arranged in a descending order. These results indicate that the perceptions towards these eight statements were again close to the fourth option on the Likert rating scale, or 'good' option. Further statistical analysis indicated that the overall mean of these eight statements was 3.914 (Table 11).

Table 9. Rank order for statements related to a variable of continuous improvement (N= 200).

No. of statement[1]	Statement	Rank order (Mean)	Std Deviation
48	There are teams in the university equipped with the necessary supplies with a task to improve quality.	3.95	0.794
49	There is a specific and clear-cut program to avoid mistakes.	3.86	0.786
50	There are statistical methods and tools at the university to test and measure the quality of graduates.	3.83	0.837
51	Devise ways and means continuously to help improve the quality of the educational performance of the university and the quality of its graduates	3.82	0.869
52	Workers receive training to be aware of the opportunities and improvement targets.	3.82	0.890

[1]Numbering appears as in the questionnaire.

Table 10 Statistics of the overall ranks for continuous improvement (statements 48-52, Tables 5.13).

	Mean	Standard Deviation
Statements 48-52	3.857	0.499

Table 11 Rank order for statements related to a variable of staff involved: (N= 200).

No. of statement[1]	Statement	Rank order (Mean)	Std Deviation
55	Faculty members and staff partake in solving the main problems at college.	3.96	0.795
60	There are efficient programs that help faculty members get acquainted with the objectives and mechanisms of the university functions.	3.96	0.876
56	The university has an efficient system to motivate individuals (financially and morally) to participate.	3.95	0.822
58	Administration works on removing fears among members of the faculty and staff while doing their job.	3.95	0.800
54	Faculty and staff are entitled to resolve major problems related to graduates.	3.94	0.31
53	Faculty members partake in making strategic decisions regarding quality.	3.90	0.823
59	Facilitate communication and strengthen links between the faculty and staff and senior management.	3.85	0.955
57	The university system rewards persons showing creativity.	3.82	0.935

[1]Numbering appears as in the questionnaire.

Correlation Coefficient of KM Process with Management-related Variables in Public Universities

The results in the previous section indicated that respondents agreed on the 83 statements, as the overall mean was 3.832, which is closer to the fourth option on the Likert rating scale. This section focuses on the analysis of the same statements above to find the degree of associations between different variables. Table 12 demonstrates the correlation between the KM process and effective leadership, continuous improvement, or employee involvement. The results of correlation revealed a positive, weak association between the KM process and TQM effective leadership (r= 0.368), continuous improvement (r= 0.284) or employee involvement (r= 0.368) at $p<0.000$ or $p<0.001$. These results indicated that the null hypotheses H0 1.1, H0 1.2, and H0 1.3 (see Chapter 1) are rejected, which means the alternative hypotheses are accepted.

Table 13 also displays a positive, but weak association between KM technology and TQM effective leadership (r= 0.383), employee involvement (r= 0.341), and continuous improvement (r= 0.398) in public universities at $p<0.000$ or $p<0.001$. These results indicate that the null hypotheses H0 1.4, H0 1.5, and H0 1.6 are rejected, which means the alternative hypotheses are accepted.

In addition, similar positive, but weak associations were found between team of knowledge and effective leadership (r= 0.390), continuous improvement (r= 0.301), and employee involvement (r= 0.354) at $p<0.000$. These results also indicate that the null hypotheses H0 1.7, H0 1.8, and H0 1.9 (se are rejected, which means the alternative hypotheses are accepted.

Furthermore, the correlation coefficient between KM and TQM in public universities demonstrated a moderate and positive correlation coefficient value (0.574) at $p<0.000$ (Table 13).

Table 12. Correlation coefficient of KM processes with TQM effective leadership, continuous improvement, or employee involvement in public universities (N=200).

		Variables		
		Effective Leadership	**Continuous Improvement**	**Employee Involvement**
KM Processes	Correlation Coefficient Value (r)	0.368	0.284	0,368
	Significance	0.000	0,001	0.000
Testing relevant hypothesis		H0.1.1 is rejected	H0.1.2 is rejected	H0.1.3 is rejected

Table 13. The correlation coefficient of KM technology with TQM effective leadership, employee involvement, and continuous improvement in the public universities (N=200).

		Variables		
KM Technology		**Effective Leadership**	**Employee Involvement**	**Continuous Improvement**
	Correlation Coefficient Value (r)	0.383	0.341	0.398
	Significance	0.000	0.000	0.000
Testing relevant hypothesis		H0.1.4 is rejected	H0.1.5 is rejected	H0.1.6 is rejected

DISCUSSION

The data collected from the 200 respondents who responded to the online questionnaire were analysed using various statistical methods, including skewness, kurtosis, Kolmogorov-Smirnov, Shapiro-Wilk, mean, standard division, correlation, and other methods. The results of analyses indicated close similarities between the perceptions of respondents in public universities. Therefore, the discussion will focus on the overall perceptions of respondents and justify null hypothesis.

This section focuses on the discussion of the results in relation to different variables related to KM and TQM. Generally, the overall perception at public universities indicated that respondents agreed with the statements in the questionnaire.

The perceptions of respondents from public universities on knowledge creation, storage of knowledge, knowledge distribution, and knowledge were significant related to the KM process. These results indicated that respondents agreed on statements

Table 14. Correlation coefficient KM team of knowledge with TQM effective leadership, continuous improvement, and employee involvement in public universities (N=200).

		Variables		
Team of Knowledge		**Effective Leadership**	**Continuous Improvement**	**Employee Involvement**
	Correlation Coefficient Value (r)	0.390	0.301	0.354
	Significance	0.000	0.000	0.000
Testing relevant hypothesis		H0.1.7 is rejected	H0.1.8 is rejected	H0.1.9 is rejected

Table 15 The correlation coefficient of KM with TQM in public universities.

	Variables	
KM		TQM
	Correlation Coefficient Value (r)	0.574
	Significance	0.000
	Testing relevant hypothesis	H0.1.10 is rejected

related to KM. In this respect, the three main aspects that were involved in this study included the KM process, KM technology, and team knowledge. These three elements are important in respect to higher educational success. We may argue that these three elements of KM are interrelated in a dynamic process, which supports how people assimilate information and knowledge, exchange and combine it to create new information. The results obtained from the questionnaire in the current study revealed respondent satisfaction in the creation, storage, distribution, and application of knowledge that was obtained internally or externally. These results demonstrate a sign of good implementation in modelling KM at public universities in Jordan. In addition, both technology and knowledge teams play an essential role in managing higher educational institutions in Jordon. Per se, the current results indicated that both public and private universities have adopted modern information technologies in managing the educational process through systems-based knowledge. However, the interrelationships between the KM process, KM technology, and team knowledge individually displayed weak, positive correlations among variables ($r = 0.249$-0.468) with effective leadership, continuous improvement, and employee involvement. It is difficult to give a clear explanation of these results. Literature indicated that KM is a complex process without end; however, effective KM can be a goal for any organisation (Clark & Rollo, 2001; Crawford, 2003). In addition, multiple regression analysis revealed that KM processes at public universities impact effective leadership. The individual's knowledge creation, distribution, storage, and knowledge application factors affected leadership indicating that all four process factors (awareness and acceptance of knowledge processes and perceived usefulness and ease of use) have significant and positive effects on effective leadership. The perceived usefulness have a positive effect on the relationship between KM processes and the relationship with effective leadership.

The direct effect of each KM process factor on the relationship with continuous improvement was tested through simple regression, and all four possible factor effects were found to be significant and positive. Conscientiousness, agreeability, and extraversion have significantly contributed to the relationship with continuous improvement. These results were found to be consistent with prior results showing that these are valid predictors of the relationship between them (Molina et al., 2007).

These findings suggest that KM process factors are easier to correlate with continuous improvement than with others, and these are important and more easily understandable for predicting KM processes. These results were also consistent with earlier studies (Hsu & Shen, 2005; Lee & Asllani, 1997; McAdam & Leonard, 2001). Extraversion has a slightly high effect in public universities (r = 0.284, p <.001). These results corroborate the relationship between KM processes and continuous improvement.

In addition, the four factors (knowledge creation, storage of knowledge, knowledge distribution, and knowledge application) significantly contributed to the relationship between KM processes and employee involvement. The results of the regression analysis supported the hypothesised relationships. Among all four factors, perceived usefulness had a high effect on the relationship between them (r = 0.368 p<0.001) in public universities. These results are consistent with earlier studies' results demonstrating the relationship between KM processes and employee involvement (Kaufman, 1992; Silos, 1999; Wilson & Asay, 1999; Bhatt, 2000; Choi, 2000; Hall, 2001; Binney, 2001; Ryan & Prybutok, 2001; Hung et al. 2005; Moffett et al., 2003).

The results regarding KM technology factors as predictors reveal that issues related to technology in KM were found to be significant predictors of the relationship with effective leadership. Literature revealed that conscientiousness, agreeability and extraversion were found to be valid predictors of this correlation. (Abdul-Wahab, 2007; Hijazi, 2005). These findings suggest that it was possible to correlate technology factors with effective leadership. These factors are important and more easily understandable for predicting the enhancement of this correlation (Baserda, 2006). These factors showed slightly high correlation for public universities (r = 0.383, p <0.001).

These results corroborate the hypothesis that there is a relationship between KM technology and continuous improvement in TQM (Dataware Technologies, 1988; Baserda, 2006).

The correlation between KM technology and employee involvement is another key factor in linking KM and TQM (Ju et al., 2006). Therefore, the correlation between KM technology and employee involvement in TQM were tested. This subsection presents the discussion and interpretation of the results regarding the direct effect of KM technology factors on employee involvement through simple regression. These results revealed that all factors significantly contributed to the correlation between them and fully supported the hypothesised relationships. Among all factors, perceived usefulness has a slightly high effect on the relationship between KM technology and employee involvement (r =0.341p<0.001) at public universities.

While making comparisons among the effect sizes of KM factors on TQM, the results showed that KM factors have a greater impact on TQM. One possible implication of these results suggests that KM can be taken as a mediator of the relationship between TQM factors. This means the results supported the main hypothesised relationships and were consistent with earlier studies' results (Gloet & Terziovski, 2004; Hsu & Shen, 2005; Lee & Asllani, 1997; McAdam & Leonard, 2001).

Contribution Knowledge

This study was able to contribute much towards widening the boundaries of knowledge as well as providing more empirical evidence for practitioners as detailed here.

There are many concepts of management models from literature that have been scrutinised and have become general management terminologies. For example, TQM has survived and grown by the efforts of researchers and practitioners. As mentioned earlier, the link between TQM and KM has been recognised, yet the studies have thus far have been few and incomplete (Ju et al., 2006; Ooi, 2009).

In this study, the researchers use a theoretical and empirical approach to investigate the integration and correlation between TQM and KM using a quantitative research method.

The discoveries of this study have a theoretical and practical contribution. From the theoretical standpoint, the results acquired from this study were consistent with theories and earlier literature, which advocated these theories. The empirical evidence from this study contributes to the body of knowledge in the fields of TQM and KM as well. Moreover, to the best of our knowledge, this research paper is one of the first KM and TQM initiatives for universities in Jordan. Thus, it has adopted a case study strategy in conducting this study.

CONCLUSION

The results of this work indicated that the average perceptions of Jordanian public university respondents towards different issues related to KM and TQM were close to the fourth option on the Likert rating scale. This option indicates that respondents agreed with the 83 statements on the questionnaire. In addition, all correlation analyses demonstrated that the respondent's perceptions exhibited positive correlations at mostly high significance levels, which resulted in the rejection of the null hypotheses.

SUGGESTIONS AND RECOMMENDATIONS

In view of the results and discussions in the current paper, the researchers suggest universities should consider the following issues:

1. It is vital to consider each individual in the university the main element for supporting the dynamic pace of the development of intellectual capacity among the members of the university. This will likely improve the experiences of others and derive lessons from successes and failures in order to capture useful knowledge that will enable the university to achieve its objectives with respect to KM and TQM.
2. The university should support distinctive staff, those who have a high level of knowledge and experience to ensure that the university achieves a high level of KM and TQM.
3. The university should manage a high level of communication between workers to facilitate a better understanding of how to achieve better KM and TQM. It should also take advantage of experts and prominent research centres and use modern technologies to increase the university's ability to transfer knowledge to students and others via its services for a better reputation.
4. Encourage holding seminars, conferences, and workshops with the aim to increase understanding the nature of the current KM and TQM for better strategy implementation and development.
5. Establish comprehensive and precise data warehouses that can be used for developing an internal system to encourage better KM and TQM.
6. Invest in areas that improve communication and collaboration for developing skills and improve the intellectual perspectives of workers for better creativity.

REFERENCES

Aboyassin, N. A., Alnsour, M., & Alkloub, M. (2011). Achieving total quality management using knowledge management practices: A field study of the Jordanian insurance sector. *International Journal of Commerce and Management*, *21*(4), 394–409. doi:10.1108/10569211111189383

Abu-El-Haija, A. (2014). Higher education in Jordan. National TEMPUS Office.

Amer, N. (2008). *Mechanisms for the advancement of the Arab mission of technical colleges in the light of the experiences of some leading conferences of technical and vocational education in Palestine*. Academic Press.

Babbar, S. (1995). Applying total quality management to educational instruction. *International Journal of Public Sector Management*, *8*(7), 35–55. doi:10.1108/09513559510103175

Baserda, T. (2006). *Integration of Knowledge Management and Total Quality Management and Performance* (Doctoral dissertation). Damascus University.

Bogdan, R. C., & Biklen, S. K. (1982). *Qualitative research for education. In Educational Research: Contemporary Issues and Practical Research.* London: Continuum.

Borahan, N. G., & Ziarati, R. (2002). Developing quality criteria for application in the higher education sector in Turkey. *Total Quality Management*, *13*(7), 913–926. doi:10.1080/0954412022000017021

Daft, R. (2006). Organization theory and design. South-Western College Pub,.

Davok, D. F. (2007). Quality in education. *Avaliação: revista da avaliação da Educação Superior (Campinas)*, *12*(3), 505-513.

DeLong, D. W. (2004). *Lost knowledge: Confronting the threat of an ageing workforce*. Oxford University Press. doi:10.1093/acprof:oso/9780195170979.001.0001

Drucker, P. F. (1999). Knowledge-worker productivity: The biggest challenge. *California Management Review*, *41*(2), 79–94. doi:10.2307/41165987

Gibbs, P. (2009). Quality in work-based studies not lost, merely undiscovered. *Quality in Higher Education*, *15*(2), 167–176. doi:10.1080/13538320902995782

Gidley, J. M., Hampson, G. P., Wheeler, L., & Bereded-Samuel, E. (2010). From access to success: An integrated approach to quality higher education informed by social inclusion theory and practice. *Higher Education Policy*, *23*(1), 123–147. doi:10.1057/hep.2009.24

Gray, P. H., & Durcikova, A. (2005). The role of knowledge repositories in technical support environments: Speed versus learning in user performance. *Journal of Management Information Systems*, *22*(3), 159–190. doi:10.2753/MIS0742-1222220306

Harvey, L., & Williams, J. (2010). *Fifteen years of quality in higher education.* Part Two.

Haworth, J. G., & Conrad, C. F. (1997). Emblems of Quality in Higher Education. Developing and Sustaining High-Quality Programs. Allyn & Bacon.

Herschel, R. T., & Nemati, H. R. (2000). Chief knowledge officer: Critical success factors for knowledge management. *Information Strategy: The Executive's Journal, 16*(4), 37–45.

Hislop, D., Newell, S., Scarbrough, H., & Swan, J. (2000). Networks, knowledge and power: Decision making, politics and the process of innovation. *Technology Analysis and Strategic Management, 12*(3), 399–411. doi:10.1080/713698478

Honarpour, A., Jusoh, A., & Md Nor, K. (2012). Knowledge management, total quality management and innovation: A new look. *Journal of Technology Management & Innovation, 7*(3), 22–31. doi:10.4067/S0718-27242012000300003

Houston, D. (2007). TQM and higher education: A critical systems perspective on fitness for purpose. *Quality in Higher Education, 13*(1), 3–17. doi:10.1080/13538320701272672

Jones, J. K. (2011). *Re-framing education as a thirdspace: neonarratives of pedagogy, power and transformation* (Doctoral dissertation). University of Southern Queensland.

Ju, T. L., Lin, B., Lin, C., & Kuo, H. J. (2006). TQM critical factors and KM value chain activities. *Total Quality Management & Business Excellence, 17*(3), 373–393. doi:10.1080/14783360500451614

Kidwell, J. J., Vander Linde, K., & Johnson, S. L. (2000). Applying Corporate Knowledge Management Practices in Higher Education. *EDUCAUSE Quarterly, 23*(4), 28–33.

Kubaisi, S. A. (2002). *Knowledge management and its impact on organisational innovation: an exploratory study comparing a sample of mixed industrial sector companies* (Doctoral dissertation). Baghdad.

Kubaisi, S. A. (2004). *Knowledge management and development organisations*. The Office of Universities, the Arab Republic of Egypt.

Kuhn, J. (2000). *Linking Continuous Improvement and Knowledge Management*. Retrieved from www.continuous-innovation.net/publications/papers-3rd/kuhn-ci.2000.pdf

Kulkarni, U. R., Ravindran, S., & Freeze, R. (2007). A knowledge management success model: Theoretical development and empirical validation. *Journal of Management Information Systems, 23*(3), 309–347. doi:10.2753/MIS0742-1222230311

Lagrosen, S., Seyyed-Hashemi, R., & Leitner, M. (2004). Examination of the dimensions of quality in higher education. *Quality Assurance in Education, 12*(2), 61–69. doi:10.1108/09684880410536431

Lomas, L. (2004). Embedding quality: The challenges for higher education. *Quality Assurance in Education, 12*(4), 157–165. doi:10.1108/09684880410561604

McAdam, R., & McCreedy, S. (1999). A critical review of knowledge management models. *The Learning Organization, 6*(3), 91–101. doi:10.1108/09696479910270416

Pratasavitskaya, H., & Stensaker, B. R. (2010). Quality management in higher education: Towards a better understanding of an emerging field. *Quality in Higher Education, 16*(1), 37–50. doi:10.1080/13538321003679465

Punch, K. F. (2013). Introduction to social research: Quantitative and qualitative approaches. *Sage (Atlanta, Ga.)*.

Rastogi, P. N. (2000). Knowledge management and intellectual capital–the new virtuous reality of competitiveness. *Human Systems Management, 19*(1), 39–48.

Rodman, H. (1980). Are Conceptual Frameworks Necessary for Theory Building? The Case of Family Sociology*. *The Sociological Quarterly, 21*(3), 429–441. doi:10.1111/j.1533-8525.1980.tb00623.x

Sarvan, F. (2005). *Fundamentals of Quality Management in the University Context: The Example of Akdeniz University/Sarvan F., Anafarta N.* In The 49th EOQ Congress, Antalya, Turkey.

Sarvary, M. (1999). Knowledge management and competition in the consulting industry. *California Management Review, 41*(2), 95–107. doi:10.2307/41165988

Schwab, K. (2012). *The global competitiveness report 2012– 2013*. Geneva: World Economic Forum. Available from http://www3. weforum. org/docs/WEF_GlobalCompetitivenessReport_2012-13. pdf

Schwandt, D. R., & Marquardt, M. (2000). *Organizational learning: From worldclass theories to global best practices*. Boca Raton, FL: CRC Press.

Shehabat, I., & Berrish, M. (2013). E-learning content enhanced by active knowledge management techniques. *International Council for Educational Media (ICEM), 2013 IEEE 63rd Annual Conference,* 1-8. doi:10.1109/CICEM.2013.6820185

Shehabat, I., & Mahdi, S. A. (2008). E-learning as a Knowledge Management Approach for Intellectual Capital Utilization. *Turkish Online Journal of Distance Education, 9*(1), 205–216.

Soltani, E., & Lai, P. C. (2007). Approaches to quality management in the UK: Survey evidence and implications. *Benchmarking: An International Journal, 14*(4), 429–454. doi:10.1108/14635770710761852

Soltani, E., Lai, P. C., & Mahmoudi, V. (2007). Managing change initiatives: Fantasy or Reality? The case of public sector organisations. *Total Quality Management & Business Excellence, 18*(1–2), 153–179. doi:10.1080/14783360601053319

Thiagaragan, T., Zairi, M., & Dale, B. G. (2001). A proposed model of TQM implementation based on an empirical study of Malaysian industry. *International Journal of Quality & Reliability Management, 18*(3), 289–306. doi:10.1108/02656710110383539

Tikly, L., & Barrett, A. M. (2011). Social justice, capabilities and the quality of education in low income countries. *Total Quality Management & Business Excellence, 17*(3), 303–323.

Zorn, T. (2001). *Knowledge Management, E-Organization and the Technology of Management Fashion*. Retrieved from www.mngt.waikato.ac.nz/ict/mgtfashion.pdf.may

Compilation of References

Aboyassin, N. A., Alnsour, M., & Alkloub, M. (2011). Achieving total quality management using knowledge management practices: A field study of the Jordanian insurance sector. *International Journal of Commerce and Management*, *21*(4), 394–409. doi:10.1108/10569211111189383

Abu-El-Haija, A. (2014). Higher education in Jordan. National TEMPUS Office.

Abu-Ghazaleh, T. (2012). *Abu-Ghazaleh announces the UAE as knowledge and connectivity hub for the Arab region at e-AGE 2012*. Academic Press.

Abzari, M., Barzaki, A. S., & Abbasi, R. (2011). Knowledge sharing behavior: Organizational reputation or losing organizational power: Case study. *International Journal of Business and Social Science*, *2*(17), 144–149.

Ackoff, R. L. (1981). *Creating the corporate future*. New York: John Wiley & Son.

Ackoff, R. L. (1989). From data to wisdom. *Journal of Applied Systems Analysis*, *16*, 3–9.

Ackoff, R. L., & Emery, F. E. (1972). *On purposeful systems*. Chicago: Aldine-Atherton.

Ahmad, N., & Daghfous, A. (2010). Knowledge sharing through inter-organizational knowledge networks: Challenges and opportunities in the United Arab Emirates. *European Business Review*, *22*(2), 153–174. doi:10.1108/09555341011023506

Ahmad, S. M. (2012). Impact of Organizational Culture on Performance Management Practices in Pakistan. *Business Intelligence Journal*, *5*(1), 50–55.

Ajzen, I. (2006). *Constructing a theory of planned behavior questionnaire*. Retrieved December 1, 2012, from http://people.umass.edu/aizen/pdf/tpb.measurement.pdf

Ajzen, I. (1985). From intentions to actions: A theory of planned behavior. In J. Kuhl & J. Beckman (Eds.), *Action control: From cognition to behavior* (pp. 11–39). Berlin, Germany: Springer-Verlag. doi:10.1007/978-3-642-69746-3_2

Ajzen, I. (1991). The theory of planned behavior. *Organizational Behavior and Human Decision Processes*, *50*(2), 179–211. doi:10.1016/0749-5978(91)90020-T

Ajzen, I. (2002). Perceived behavioral control, self-efficacy, locus of control, and the theory of planned behavior. *Journal of Applied Social Psychology, 32*(4), 665–683. doi:10.1111/j.1559-1816.2002.tb00236.x

Akhavan, P., Ghojavand, S., & Abdali, R. (2012). Knowledge sharing and its impact on knowledge creation. *Journal of Information & Knowledge Management, 11*(2), 1–12. doi:10.1142/S0219649212500128

Al-Adaileh, R., & Al-Atawi, M. (2011). Organizational culture impact on knowledge: Saudi telecom context. *Journal of Knowledge Management, 15*(2), 212–230. doi:10.1108/13673271111119664

Alajmi, B. M. (2011). *The intention to share: Professional's knowledge sharing behaviors in online communities* (Doctoral thesis). The State University of New Jersey.

Alajmi, B. M., McInerney, C. R., Orzano, A. J., Tallia, A. F., Meese, A., & Vamanu, I. (2008). Knowledge sharing processes and tools in U.S.A. primary health care: Analysis of four case studies. In *Proceedings of the International Conference on Knowledge Management 2008*. Hackensack, NJ: World Scientific doi:10.1142/9789812837578_0009

Al-Alawi, A. I., Al-Marzooqi, N. Y., & Mohammed, Y. F. (2007). Organizational culture and knowledge sharing: Critical success factors. *Journal of Knowledge Management, 11*(2), 22–42. doi:10.1108/13673270710738898

Alamiyan, M. (2002). *Organizational Behavior in Business Organizations* (in Arabic). Amman: Dar Wael for publication.

Alavi, M., & Leidner, D. E. (2001). Knowledge management and knowledge management systems: Conceptual foundations and research issues. *Management Information Systems Quarterly, 25*(1), 107–113. doi:10.2307/3250961

Albres, J. A. (2009). A practical approach to implementing knowledge management. *Journal of Knowledge Management Practice, 10*(1).

Al-Nahyan, N. b. M. (2012a). *Executive focus: Government commitment and investment in education, research initiatives, and improving the quality of education.* UAE Minister of Higher Education & Scientific Research HH Sheikh Nahyan bin Mubarak Al Nahyan.

Al-Nahyan, N. b. M. (2012b). *Education in the United Arab Emirates (UAE)*. The Prospect Group.

Al-Shammari, M. (2008). Toward a knowledge management strategic framework in the Arab region. *International Journal of Knowledge Management, 4*(3), 44–63. doi:10.4018/jkm.2008070104

Amer, N. (2008). *Mechanisms for the advancement of the Arab mission of technical colleges in the light of the experiences of some leading conferences of technical and vocational education in Palestine*. Academic Press.

Anantatmula, V. S., & Kanungo, S. (2010). Modeling enablers for successful KM implementation. *Journal of Knowledge Management, 14*(1), 100–113. doi:10.1108/13673271011015598

Anne, A., & Maaja, V. (2007). Relationships between Organizational Culture and Performance in Estonian Schools with Regard to Their Size and Location. *Baltic Journal of Economics*, *7*(1), 3–17. doi:10.1080/1406099X.2007.10840438

Arif, M., Egbu, C., Alom, O., & Khalfan, M. M. A. (2009). Measuring knowledge retention: A case study of a construction consultancy in the UAE. *Engineering, Construction, and Architectural Management*, *16*(1), 92–108. doi:10.1108/09699980910927912

Arif, M., Khalfan, M., Barnard, J. H., & Heller, N. A. (2012). Assessing Knowledge Retention in Construction Organisations: Cases from the UAE. *The Australasian Journal of Construction Economics and Building*, *12*(2), 55–71. doi:10.5130/ajceb.v12i2.2461

Ashby, W. R. (1956). *An introduction to cybernetics*. London: Chapman & Hall. doi:10.5962/bhl.title.5851

Babalhavaeji, F., & Kermani, Z. J. (2011). Knowledge sharing behaviour influences: A case of library and information science faculties in Iran. *Malaysian Journal of Library and Information Science*, *16*(1), 1–14.

Babbar, S. (1995). Applying total quality management to educational instruction. *International Journal of Public Sector Management*, *8*(7), 35–55. doi:10.1108/09513559510103175

Bach, C., & Belardo, S. (2003). Scientific and philosophical aspects of information and the relationships among data, information, and knowledge. *Proceedings of the 9th ACIS*, 2629-2639.

Badaracco, J. (2010). Effective knowledge sharing and its benefits. *The Point Newspaper Online*.

Balachandran, K. R., Shu, H. L., & Suresh, R. (2007, Winter). A framework for unused capacity: Theory and empirical analysis. *Journal of Applied Management Accounting Research*, 21-38.

Ball, R. (1979). The dialectical method: Its application to social theory. *Social Forces*, *57*(3), 785–798. doi:10.1093/sf/57.3.785

Balthazard, A. P., Cooke, A. R., & Potter, E. R. (2006). Dysfunctional Culture, Dysfunctional Organization: Capturing the Behavioral Norms that Form Organizational Culture and Drive Performance. *Journal of Managerial Psychology*, *21*(8), 709–732. doi:10.1108/02683940610713253

Bandura, A. (1997). Self-efficacy: The exercise of control. New York: W.H. Freeman. British Journal of Social Psychology, 44, 479-496.

Bandura, A. (1977). Self-efficacy: Toward a unifying theory of behavioral change. *Psychological Review*, *84*(2), 191–215. doi:10.1037/0033-295X.84.2.191 PMID:847061

Bandura, A. (1997). *Self-efficiency: The Exercise of Control*. New York: W. H. Freeman Publishers.

Barney, J. B. (1991). Firm Resources and Sustained Competitive Advantage. *Journal of Management*, *1*(17), 99–120. doi:10.1177/014920639101700108

Bartélémy, J.-H. (2012). Fifty key terms in the works of Gilbert Simondon. In Gilbert Simondon: Being and technology. Edinburgh, UK: Edinburgh University Press.

Baserda, T. (2006). *Integration of Knowledge Management and Total Quality Management and Performance* (Doctoral dissertation). Damascus University.

Batinic, B., Reips, U., & Bosnjak, M. (2002). *Online social sciences*. Seattle, WA: Hogrefe & Hubber Publisher.

Battour, M. (2011). *The impact of the Islamic attributes of destination on tourists' motivation, satisfaction and destination loyalty* (Unpublished doctoral dissertation). University of Malaya.

Bensaude-Vincent, B. (2007). Nanobots and nanotubes: Two alternative biomimetic paradigms of nanotechnology. In Genesis redux: Essays in the history and philosophy of artificial life. Chicago: University of Chicago Press.

Bereiter, C. (2002). *Education and Mind in the Knowledge Age*. Mahwah, NJ: Lawrence Erlbaum Associates.

Bergson, H. (1998). *Creative evolution*. Dover Publishers.

Bessick, J., & Naicker, V. (2013). Barriers to tacit knowledge retention: An understanding of the perceptions of the knowledge management of people inside and outside the organization. SA Journal of Information Management, 15(2).

Bigliardi, B., Dormio, A. I., & Galati, F. (2010). ICTs and knowledge management: An Italian case study of a construction company. *Measuring Business Excellence*, *14*(3), 16–29. doi:10.1108/13683041011074182

Bititci, U. S., Turner, T., & Begemann, C. (2000). Dynamics of performance measurement systems. *International Journal of Operations & Production Management*, *20*(6), 692–704. doi:10.1108/01443570010321676

Biygautane, M., & Al-Yahya, K. (2011). *Knowledge management in the UAE's public sector: The case of Dubai*. Paper presented at the Gulf Research Meeting Conference at the University of Cambridge.

Blankenship, L., & Brueck, T. (2008). Planning for knowledge retention now saves valuable organizational resources later. *Journal - American Water Works Association*, *100*(8), 57–61.

Bock, G. W., Zmud, R. W., Kim, Y. G., & Lee, J. N. (2005). Behavioral intention formation in knowledge sharing: Examining the roles of extrinsic motivators, social-psychological forces, and organizational climate. *Management Information Systems Quarterly*, *29*(1), 87–111.

Bogdan, R. C., & Biklen, S. K. (1982). *Qualitative research for education. In Educational Research: Contemporary Issues and Practical Research*. London: Continuum.

Boone, T., Ganeshan, R., & Hicks, R. L. (2008). Learning and Knowledge Depreciation in Professional Services. *Management Science*, *54*(7), 1231–1236. doi:10.1287/mnsc.1070.0849

Borahan, N. G., & Ziarati, R. (2002). Developing quality criteria for application in the higher education sector in Turkey. *Total Quality Management*, *13*(7), 913–926. doi:10.1080/0954412022000017021

Bordia, P., Irmer, B. E., Garden, M., Phair, K., & Abusah, D. (2004). *Knowledge sharing in response to a supportive work environment: evidence from an Australian engineering firm.* Paper presented at the iKMS International Conference on Knowledge Management, Singapore.

Bose, R. (2004). Knowledge management metrics. *International Journal of Technology Management*, *24*(4), 457–468.

Boucher, M.-P. (2012), Infra-psychic individualization: Transductive connections and the genesis of living techniques. In Gilbert Simondon: Being and technology. Edinburgh, UK: Edinburgh University Press.

Britton, G., & McCallion, H. (1994). An overview of the Singer/Churchman/Ackoff school of thought. *Systemic Practice and Action Research*, *5*(5), 487–521.

Buckland, M. (1991). Information as thing. *Journal of the American Society for Information Science*, *42*(5), 351–360. doi:10.1002/(SICI)1097-4571(199106)42:5<351::AID-ASI5>3.0.CO;2-3

Burgess, D. (2005). What motivates employees to transfer knowledge outside their work unit? *Journal of Business Communication*, *42*(4), 324–348. doi:10.1177/0021943605279485

Cabinet Office, Government of Japan. (2015). *Measures against the declining birthrate of the past.* Retrieved from http://www8.cao.go.jp/shoushi/shoushika/data/torikumi.html

Callaos, N., & Callaos, B. (2002). Toward a systemic notion of information: Practical consequences. *Informing Science*, *5*(1), 1–11. doi:10.28945/532

Cascio, W. F. (2006). *Managing Human Resources: Productivity, Quality of Life, Profits.* McGraw-Hill Irwin.

Chabot, P. (2005). The philosophical August 4th: Simondon as a reader of Bergson. *Angelaki*, *10*(3), 103–108. doi:10.1080/09697250500417290

Chalkiti, K., & Sigala, M. (2010). Staff turnover in the Greek tourism industry: A comparison between insular and peninsular regions. *International Journal of Contemporary Hospitality Management*, *22*(3), 335–359. doi:10.1108/09596111011035945

Chatzoudes, D., Chatzoglou, P., & Vraimaki, E. (2015). The central role of knowledge management in business operations: Developing a new conceptual framework. *Business Process Management Journal*, *21*(5), 1117–1139. doi:10.1108/BPMJ-10-2014-0099

Chaudhry, A. S., & Sivakamasundari, B. (2004). *Perceptions of teachers about knowledge sharing in schools.* Paper presented at the Information Resources Management Association International Conference 2004, New Orleans, LA.

Checkland, P., & Holwell, S. (1998). *Information, systems and information systems*. New York: John Wiley & Sons.

Cheng, E. C. K. (2012). Knowledge strategies for enhancing school learning capacity. *International Journal of Educational Management*, *26*(6), 557–592. doi:10.1108/09513541211251406

Chen, I., & Chen, N.S., & Kinshuk. (2009). Examining the factors influencing participants' knowledge sharing behavior in virtual learning communities. *Journal of Educational Technology & Society*, *12*(1), 134–148.

Chen, L., & Mohamed, S. (2010). The strategic importance of tacit knowledge management activities in construction. *Construction Innovation*, *10*(2), 138–163. doi:10.1108/14714171011037165

Chennamaneni, A. (2006). *Determinants of knowledge sharing behaviors: Developing and testing an integrated theoretical model* (Doctoral dissertation). The University of Texas at Arlington.

Chennell, A. F., Dransfield, S. B., Field, J. B., Fisher, N. I., Saunders, I. W., & Shaw, D. E. (2000). *OPM®: A system for organisational performance measurement. In Performance Measurement – Past, Present and Future*. Cambridge, UK: University of Cambridge.

Chikoore, L., & Ragsdell, G. (2013, March). Knowledge sharing in higher education: A study of students preparing assessed group work. *Journal of Knowledge Management Practice*, *14*(1).

Child, J. (1974). What Determines Organization Performance? The universals vs. the it-all-depends. *Organizational Dynamics*, *3*(1), 2–18. doi:10.1016/0090-2616(74)90001-1

Chin, K. S., & Pun, K. F. (2002). A proposed framework for implementing TQM in Chinese organisations. *International Journal of Quality & Reliability Management*, *19*(3), 272–294. doi:10.1108/02656710210415686

Chin, K. S., Pun, K. F., & Lau, H. C. W. (2003). Development of a knowledge-based self-assessment system for organisational performance. *Expert Systems with Applications*, *24*(4), 443–455. doi:10.1016/S0957-4174(02)00192-6

Chin, K. S., Yeung, I. K., & Pun, K. F. (2006). Development of a web-based system for assessing supplier quality performance. *International Journal of Quality & Reliability Management*, *23*(7), 743–765. doi:10.1108/02656710610679806

Chin, W. W. (2010). How to write up and report PLS analyses. In V. E. Vinzi, W. W. Chin, J. Henseler, & H. Wang (Eds.), *Handbook of partial least squares: Concepts, methods and applications* (pp. 655–690). Berlin, Germany: Springer-Verlag. doi:10.1007/978-3-540-32827-8_29

Chiou, J.-S. (1998). The effects of attitude, subjective norm, and perceived behavioral control on consumers' purchase intentions: The moderating effects of product knowledge and attention to social comparison information. *Proceedings of the National Science Council*, *9*(2), 298–308.

Chong, H. G. (2008). Measuring performance of small-and-medium sized enterprises: The grounded theory approach. *Journal of Business and Public Affairs*, *2*(1), 10–20.

Chow, C. W., Harrison, G. L., Mckinnon, J. L., & Wu, A. (2001). *Organizational culture: Association with affective commitment, job satisfaction, propensity to remain and information sharing in a Chinese cultural context.* San Diego University CIBER Working Paper Series.

Chugh, R. (2012). Knowledge sharing with enhanced learning and development opportunities. *IEEEXplore*, 100-104.

Chu, K. W., Wang, M., & Yuen, A. J. K. (2011). Implementing knowledge management in school environment: Teachers' perception. *Knowledge Management & E-Learning: An International Journal*, *3*(2), 139–152.

Churchman, C. (1971). *Design of inquiring systems*. New York: Basic Books.

Churchman, C., & Schainblatt, A. (1965). The researcher and the manager: A dialectic of implementation. *Management Science*, *11*(4), B69–B87. doi:10.1287/mnsc.11.4.B69

Cochran, D. S., Arinez, J. F., Duda, J. W., & Linck, J. (2001). A decomposition approach for manufacturing system design. *Journal of Manufacturing Systems*, *20*(6), 371–380. doi:10.1016/S0278-6125(01)80058-3

Cohen, J. (1988). *Statistical power analysis for the behavioral sciences*. Hillsdale, NJ: Lawrence Erlbaum Associates.

Collet, P., Lutton, E., Schoenauer, M., & Louchet, J. (2000). Take it EASEA. In Parallel Problem Solving from Nature PPSN VI (vol. 1971, pp. 891–901). Springer Berlin Heidelberg. doi:10.1007/3-540-45356-3_87

Combes, M. (2012). *Gilbert Simondon and the philosophy of the transindividual*. Cambridge, MA: The MIT Press.

Coole, D., & Frost, S. (Eds.). (2010). *New materialisms: Ontology, agency, and politics*. Durham, NC: Duke University Press. doi:10.1215/9780822392996

Cooper, R., & Kaplan, R. S. (1988, September). Measure costs right: Make the right decisions. *Harvard Business Review*, 10–20.

Cousins, K. C., & Robey, D. (2005). Human agency in a wireless world: Patterns of technology use in neomadic computational environments. *Information and Organization*, *15*(2), 151–180. doi:10.1016/j.infoandorg.2005.02.008

Creswell, J. (2009). *Research design: Qualitative, quantitative, and mixed methods approaches*. Thousand Oaks, CA: Sage Publications.

Cross, K. F., & Lynch, R. L. (1988). The SMART way to define and sustain success. *National Productivity Review*, *8*(1), 23–33. doi:10.1002/npr.4040080105

Cummings, J. N. (2004). Work groups, structural diversity and knowledge sharing in a global organization. *Management Science*, *50*(3), 352–364. doi:10.1287/mnsc.1030.0134

Daft, R. (2006). Organization theory and design. South-Western College Pub,.

Daft, R. L. (2000). *Organization Theory and Design* (7th ed.). South-Western College Publishing. Thomson Learning.

Daft, R., & Weick, K. (1984). Toward a model of organizations as interpretation systems. *Academy of Management Review*, *9*(2), 284–295.

Davenport, T. H., De Long, D. W., & Beers, M. C. (1998). Successful knowledge management projects. *Sloan Management Review*, *39*(2), 43–57.

Davok, D. F. (2007). Quality in education. *Avaliação: revista da avaliação da Educação Superior (Campinas)*, *12*(3), 505-513.

de Beistegui, M. (2005). From Merleau-Ponty's "reduction" to Simondon's "transduction". *Angelaki*, *10*(3), 109–122. doi:10.1080/09697250500417316

De Boever, A., Murray, A., Roffe, J., & Woodward, A. (Eds.). (2012). *Gilbert Simondon: Being and technology*. Edinburgh, UK: Edinburgh University Press.

de Gooijer, J. (2000). Designing a knowledge management performance framework. *Journal of Knowledge Management*, *4*(4), 303–310. doi:10.1108/13673270010379858

de Lima, E. P., da Costa, S. E. G., & Angelis, J. J. (2009). Strategic performance measurement systems: A discussion about their roles. *Measuring Business Excellence*, *13*(3), 39–48. doi:10.1108/13683040910984310

Deb, K., Pratap, A., Agarwal, S., & Meyarivan, T. (2002, April). A fast and elitist multiobjective genetic algorithm: NSGA-II. *Trans. Evol. Comp*, *6*(2), 182–197. doi:10.1109/4235.996017

Delong, D. W. (2004). *Lost Knowledge - Confronting the Threat of an Aging Workforce*. New York: Oxford University Press, Inc. doi:10.1093/acprof:oso/9780195170979.001.0001

Delong, D. W., & Davenport, T. (2003). Better practices for retaining organizational knowledge: Lessons from the leading edge. *Employment Relations Today*, *30*(3), 51–63. doi:10.1002/ert.10098

del-Rey-Chamorro, F. M., Roy, R., van Wegen, B., & Steele, A. (2003). A framework to create key performance indicators for knowledge management solutions. *Journal of Knowledge Management*, *7*(2), 46–62. doi:10.1108/13673270310477289

Demarest, M. (1997). Understanding knowledge management. *Long Range Planning*, *30*(3), 374–384. doi:10.1016/S0024-6301(97)90250-8

Demirkan, H., Spohrer, J. C., & Krishna, V. (Eds.). (2011). *The Science of Service Systems*. New York: Springer. doi:10.1007/978-1-4419-8270-4

Denison, D. R., & Mishra, A. K. (1995). Toward a Theory of Organizational Culture and Effectiveness. *Organization Science*, *6*(2), 204–223. doi:10.1287/orsc.6.2.204

DeWaal, A. (2002). The Power of world-class performance management: Use It! *Measuring Business Excellence*, *6*(3), 9–19. doi:10.1108/13683040210441940

Dokhtesmati, M., & Bousari, R. G. (2013). Knowledge sharing in Iranian academic institutions: Meta-analysis approach. *Procedia: Social and Behavioral Sciences*, *73*, 383–387. doi:10.1016/j.sbspro.2013.02.065

Drucker, P. (1999). *Management challenges for the 21st century*. New York: HarperBusiness.

Drucker, P. F. (1999). Knowledge-worker productivity: The biggest challenge. *California Management Review*, *41*(2), 79–94. doi:10.2307/41165987

Du Plessis, M. (2007). The role of knowledge management in innovation. *Journal of Knowledge Management*, *11*(4), 20–29. doi:10.1108/13673270710762684

Dunegan, K. J., & Hrivnak, M. W. (2003). Characteristics of mindless teaching evaluations and the moderating effects of image compatibility. *Journal of Management Education*, *27*(3), 280–303. doi:10.1177/1052562903027003002

Durmusoglu, S., Jacobs, M., Nayir, D., Khilji, S., & Wang, X. (2014). The quasi-moderating role of organizational culture in the relationship between rewards and knowledge shared and gained. *Journal of Knowledge Management*, *18*(1), 19–37. doi:10.1108/JKM-05-2013-0183

Durst, S., & Edvardsson, I. R. (2012). Knowledge Management in SMEs: A Literature Review. *Journal of Knowledge Management*, *16*(6), 637–649. doi:10.1108/13673271211246194

Dwivedi, Y. K., Venkitachalam, K., Sharif, A. M., Al-Karaghouli, W., & Weerakkody, V. (2011). Research Trends in Knowledge Management: Analyzing the Past and Predicting the Future. *Information Systems Management*, *28*(1), 43–56. doi:10.1080/10580530.2011.536112

Earl, M. (2001). Knowledge management strategies: Toward a taxonomy. *Journal of Management Information Systems*, *18*(1), 215–233. doi:10.1080/07421222.2001.11045670

EFQM. (2000). *The EFQM Excellence Model: Assessor Training Modules. MC Modules (Company) V1/E*. Brussels: European Foundation for Quality Management.

EFQM. (2017). *European Foundation for Quality Management*. Retrieved March 1, 2017, from http://www.EFQM.org/

Egger, U.K. (2013). *Cross Organisational Knowledge Sharing*. Academic Press.

Eisenhardt, K. (1989). Making fast strategic decisions in high velocity environments. *Academy of Management Journal*, *32*(3), 543–576. doi:10.2307/256434

Eisenhardt, K. M. (1989). Building Theories from Case Study Research. *Academy of Management Review*, *14*(4), 532–550.

Elogie, A. A., & Asemota, A. E. (2013). Knowledge sharing behaviour of academics in the University of Ibadan, Nigeria. *PNLA Quarterly*, *77*(4), 1–16.

Ensign, P. C., & Hebert, L. (2010, Winter). How reputation affects knowledge sharing among colleagues. *MIT Sloan Management Review*, *51*(2), 78–82.

Epstein, M. J., & Westbrook, R. A. (2001). Linking action to profits in strategic decision making. *MIT Sloan Management Review*, *42*(3), 39–49.

Erikson, E. H., Erikson, J. M., & Kivnick, H. Q. (1986). *Vital Involvement in Old Age: The Experience of Old Age in Our Time*. Penguin Books Canada Ltd.

Falkenauer, E. (1998). *Genetic Algorithms and Grouping Problems*. New York, NY: John Wiley & Sons, Inc.

Farashahi, M., Hafso, T., & Molz, R. (2005). Institutionalized norms of conducting research and social realities: A research synthesis of empirical works from 1983 to 2002. *International Journal of Management Reviews*, *7*(1), 1–24. doi:10.1111/j.1468-2370.2005.00104.x

Federal Authority for Government Human Resources (FAHR). (2017). *Guide of Knowledge Management in the Federal Government*. Retrieved from https://www.fahr.gov.ae/Portal/Userfiles/Assets/Documents/a3f395eb.pdf

Feldman, K. A. (1998). Identifying exemplary teachers and teaching: Evidence from student ratings. In K. A. Feldman & M. B. Paulsen (Eds.), *Teaching and Learning in the College Classroom* (2nd ed.; pp. 391–414). Needham Heights, MA: Simon and Schuster.

Fink, L. D. (2003). *Creating Significant Learning Experiences: An Integrated Approach to Designing College Courses, Revised and Updated*. San Francisco, CA: Jossey-Bass.

Fink, L. D. (2003). *Creating Significant Learning Experiences: An Integrated Approach to Designing College Courses*. San Francisco: Jossey-Bass.

Flapper, S. D., Fortuin, L., & Stoop, P. M. (1996). Towards consistent performance management systems. *International Journal of Operations & Production Management*, *6*(7), 27–36. doi:10.1108/01443579610119144

Fong, P. S. W., & Lee, H. F. (2009). Acquisition, reuse and sharing of knowledge in property management firms. *Facilities*, *27*(7), 291–314. doi:10.1108/02632770910956148

Forehand, G. A., & Von Haller, G. (1964). Environmental variation in studies of organizational behavior. *Psychological Bulletin*, *62*(6), 361–382. doi:10.1037/h0045960 PMID:14242590

Fornell, C. G., & Larcker, D. F. (1981). Evaluating structural equation models with unobservable variables and measurement error. *JMR, Journal of Marketing Research*, *18*(1), 39–50. doi:10.2307/3151312

Fornell, C., & Cha, J. (1994). Partial least squares. In R. P. Bagozzi (Ed.), *Advanced methods of marketing research* (pp. 52–78). Cambridge, UK: Blackwell.

Francis, J. J., Eccles, M. P., Johnston, M., Walker, A., Grimshaw, J., Foy, R., & Bonetti, D. et al. (2004). *Constructing questionnaires based on the theory of planned behavior: Manual for health services researchers*. University of Newcastle.

Frandsen, A.-C. (2009). From psoriasis to a number and back. *Information and Organization*, *19*(2), 103–128. doi:10.1016/j.infoandorg.2009.02.001

Franklin, S. (2005). Stem Cell R Us: Emergent life forms and the global biological. In Global assemblages: Technology, politics, and ethics as anthropological problems. Malden, MA: Blackwell.

Frigo, M. (2006). Knowledge retention: A guide for utilities. *Journal - American Water Works Association*, *98*(9), 81–84.

Fujimori, J. (2012). Practice and verification of the "Care-Will Model" toward autonomous social support implementation with a focus on male retirees. First Research Report of a Policy Sciences Project, Grant-in-Aid for Scientific Research, Health, Labor and Welfare Ministry, Japan.

Fujimori, J. (2014). Practice and verification of the "Care-Will Model" toward autonomous social support implementation with a focus on male retirees. Final Research Report of a Policy Sciences Project, Grant-in-Aid for Scientific Research, Health, Labor and Welfare Ministry, Japan.

Fullwood, R., Rowley, J., & Delbridge, R. (2013). Knowledge sharing amongst academics in UK universities. *Journal of Knowledge Management*, *17*(1), 123–136. doi:10.1108/13673271311300831

Galloway, A. R. (2004). *Protocol: How power controls after decentralization*. Cambridge, MA: The MIT Press.

Galloway, A. R. (2006). Language wants to be overlooked: On software and ideology. *Journal of Visual Culture*, *5*(3), 315–331. doi:10.1177/1470412906070519

Galloway, A. R., & Thacker, E. (2007). *The exploit: A theory of networks*. Minneapolis, MN: The University of Minnesota Press.

Gammelgaard, J., & Ritter, T. (2005). The knowledge retrieval matrix: Codification and personification as separate strategies. *Journal of Knowledge Management*, *9*(4), 133–143. doi:10.1108/13673270510610387

Garcia-Najera, A., & Bullinaria, J. A. (2011). An improved multi-objective evolutionary algorithm for the vehicle routing problem with time windows. *Computers & Operations Research*, *38*(1), 287–300. doi:10.1016/j.cor.2010.05.004

Garson, G. D. (2012). *Partial least squares: Regression and path modeling*. Statistical Associates Publishing.

Gary, C. (2009). *Performance Management - Integrating Strategy Execution, Methodologies, Risk, and Analytics*. John Wiley & Sons, Inc.

Gay, L., & Airasian, P. (2003). *Educational research: Competencies for analysis and applications*. Pearson Education.

Geisler, E. (2007). The metrics of knowledge: Mechanisms for preserving the value of managerial knowledge. *Business Horizons*, *50*(6), 467–477. doi:10.1016/j.bushor.2007.07.001

George, J. M., & Jones, G. R. (2000). The role of time in theory and theory building. *Journal of Management, 26*(4), 657–684. doi:10.1177/014920630002600404

Gephart, R. (2004). Qualitative research and the Academy of Management journal. *Academy of Management Journal, 47*(4), 454–462. doi:10.5465/AMJ.2004.14438580

Gharajedaghi, J. (2011). *Systems thinking – managing chaos and complexity: a platform for designing business architecture* (3rd ed.). Burlington, MA: Morgan Kaufmann.

Gibbs, P. (2009). Quality in work-based studies not lost, merely undiscovered. *Quality in Higher Education, 15*(2), 167–176. doi:10.1080/13538320902995782

Gidley, J. M., Hampson, G. P., Wheeler, L., & Bereded-Samuel, E. (2010). From access to success: An integrated approach to quality higher education informed by social inclusion theory and practice. *Higher Education Policy, 23*(1), 123–147. doi:10.1057/hep.2009.24

Ginevicius, R., & Vaitkunaite, V. (2006). Analysis of organizational culture dimensions impacting performance. *Journal of Business Economics and Management, 7*(4), 201–211.

Giraldez, R., Aguilar-Ruiz, J. S., & Riquelme, J. C. (2005, May). Knowledge-based fast evaluation for evolutionary learning. *Systems, Man, and Cybernetics, Part C: Applications and Reviews. IEEE Transactions on, 35*(2), 254–261.

Glaser, B., & Strauss, A. (1967). *The discovery of grounded theory: strategies for qualitative research*. Chicago: Aldine.

Golden, B., Raghavan, R., & Wasil, E. (Eds.). (2008). The Vehicle Routing Problem: Latest Advances and New Challenges. Springer. doi:10.1007/978-0-387-77778-8

Gorry, G. & Morton, M. (1989). A framework for management information systems. *Sloan Management Review*, 49–61.

Gourlay, S. (2006). Conceptualizing knowledge creation: A critique of Nonaka's theory. *Journal of Management Studies, 43*(7), 1415–1436. doi:10.1111/j.1467-6486.2006.00637.x

Grau, B.C., Horrocks, I., Motik, B., Parsia, B., Patel-Schneider, P., & Sattler, U. (2008). OWL 2: The next step for OWL. *Web Semantics: Science, Services and Agents on the World Wide Web, 6*(4), 309 – 322.

Gray, P. H., & Durcikova, A. (2005). The role of knowledge repositories in technical support environments: Speed versus learning in user performance. *Journal of Management Information Systems, 22*(3), 159–190. doi:10.2753/MIS0742-1222220306

Gregory, D. (Eds.). (2009). *Dictionary of Human Geography* (5th ed.). Oxford, UK: Wiley-Blackwell.

Guenzi, P., Georges, L., & Pardo, C. (2009). The impact of strategic account managers' behaviors on relational outcomes: An empirical study. *Industrial Marketing Management, 38*(3), 300–311. doi:10.1016/j.indmarman.2007.09.011

Gupta, B., Iyer, L. S., & Aronson, J. E. (2000). Knowledge management: Practices and challenges. *Industrial Management & Data Systems*, *100*(1), 17–21. doi:10.1108/02635570010273018

Gurbuz, T. (2008). *Knowledge management in education and research: An elective course at METU*. Paper presented at the 8th International Educational Technology Conference (IETC2008).

Haapalainen, P., & Makiranta, A. (2013, March). Acquiring and sharing knowledge in SMEs: A case study in the manufacturing industry. *Journal of Knowledge Management Practice*, *14*(1).

Haas, M. R., & Hansen, M. T. (2005). Different knowledge, different benefits: Toward a productivity perspective on knowledge sharing in organizations. *Strategic Management Journal*, *28*(11), 1133–1153. doi:10.1002/smj.631

Hair, J. F., Black, W. C., Babin, B. J., & Anderson, R. E. (2010). *Multivariate data analysis*. Pearson education.

Hair, J. F. Jr, Ringle, C. M., & Sarstedt, M. (2013). Editorial: Partial least squares structural equation modeling: Rigorous applications, better results and higher acceptance. *Long Range Planning*, *46*(1-2), 1–12. doi:10.1016/j.lrp.2013.01.001

Hair, J. F., Ringle, C. M., & Sarstedt, M. (2011, Spring). PLS-SEM: Indeed a silver bullet. *Journal of Marketing Theory and Practice*, *19*(2), 139–151. doi:10.2753/MTP1069-6679190202

Halavais, A. (2009). *Search engine society*. Cambridge, UK: Polity.

Handy, C. B. (1993). *Understanding Organizations*. New York: Oxford University Press.

Hansen, M. B. N. (2004). *New philosophy for new media*. Cambridge, MA: The MIT Press.

Hansmann, K. W., & Ringle, C. M. (2005). Enterprise-networks and strategic success- An empirical analysis. In T. Theurl & E. C. Meyer (Eds.), *Strategies for cooperation* (pp. 133–152). Shaker Verlag.

Harris, A. (2008). Leading innovation and change: Knowledge creation by schools for schools. *European Journal of Education*, *43*(2), 219–228. doi:10.1111/j.1465-3435.2008.00343.x

Harvey, L., & Williams, J. (2010). *Fifteen years of quality in higher education*. Part Two.

Hassanzadeh, M. (2012). *Factors that affect scientists' behavior to share scientific knowledge*. Paper presented at the 7th International Conference on Webometrics, Informetrics and Scientometrics (WIS) and 12th COLLNET Meeting, Istanbul, Turkey. doi:10.1142/9789812702081_0013

Hausman, C. (1993). *Charles S Peirce's Evolutionary Philosophy*. Cambridge, UK: Cambridge University Press. doi:10.1017/CBO9780511570773

Havighurst, R. J. (1969). *Adjustment to Retirement: A Cross-national Study*. Assen: Van Gorcum.

Haworth, J. G., & Conrad, C. F. (1997). Emblems of Quality in Higher Education. Developing and Sustaining High-Quality Programs. Allyn & Bacon.

Heffernan, M. M., & Flood, P.C. (2000). An Exploration of the Relationship between Managerial Competencies Organizational, Characteristic and Performance in an Irish organization. *Journal of European Industrial Training*, 128-136.

Heisig, P. (2009). Harmonisation of knowledge management: Comparing 160 KM frameworks around the globe. *Journal of Knowledge Management*, *13*(4), 4–31. doi:10.1108/13673270910971798

Henseler, J., Ringle, C. M., & Sinkovics, R. R. (2009). The use of partial least squares path modeling in international marketing. *Advances in International Marketing*, *20*, 277–319.

Henseler, J., & Sarstedt, M. (2013). Goodness-of-fit indices for partial least squares path modeling. *Computational Statistics*, *28*(2), 565–580. doi:10.1007/s00180-012-0317-1

Herschel, R. T., & Nemati, H. R. (2000). Chief knowledge officer: Critical success factors for knowledge management. *Information Strategy: The Executive's Journal*, *16*(4), 37–45.

Hew, K. F., & Hara, N. (2007). Empirical study of motivators and barriers of teacher online knowledge sharing. *Educational Technology Research and Development*, *55*(6), 573–595. doi:10.1007/s11423-007-9049-2

Hickman, L. (1994). The products of pragmatism. In G. Debrock & M. Hulswit (Eds.), *Living Doubt* (pp. 13–24). Kluwer Academic Publishers. doi:10.1007/978-94-015-8252-0_2

Hislop, D. (2013). Knowledge Management in Organizations: A Critical Introduction (3rd ed.). Oxford University Press.

Hislop, D., Newell, S., Scarbrough, H., & Swan, J. (2000). Networks, knowledge and power: Decision making, politics and the process of innovation. *Technology Analysis and Strategic Management*, *12*(3), 399–411. doi:10.1080/713698478

Ho, C.-T. (2009). The relationship between knowledge management enablers and performance. *Industrial Management & Data Systems*, *109*(1), 98–117. doi:10.1108/02635570910926618

Holland, J. H. (1975). *Adaptation in natural and artificial systems: An introductory analysis with applications to biology, control, and artificial intelligence*. University of Michigan Press.

Honarpour, A., Jusoh, A., & Md Nor, K. (2012). Knowledge management, total quality management and innovation: A new look. *Journal of Technology Management & Innovation*, *7*(3), 22–31. doi:10.4067/S0718-27242012000300003

Horwitz, F. M., Anderssen, K., Bezuidenhout, A., Cohen, S., Kirsten, F., Mosoeunyane, K., & Van Heerden, A. (2002). Due diligence neglected: Managing human resources and organizational culture in mergers and acquisitions. *ŃSA. Journal of Business and Management*, (33): 1–10.

House, E. R. (1978). Assumptions underlying evaluation models. *Educational Researcher*, *7*(3), 4–12. doi:10.3102/0013189X007003004

Houston, D. (2007). TQM and higher education: A critical systems perspective on fitness for purpose. *Quality in Higher Education*, *13*(1), 3–17. doi:10.1080/13538320701272672

Hsu, I. (2008). Knowledge sharing practices as facilitating factor for improving organizational performance through human capital: A preliminary study. *Expert Systems with Applications*, *35*(3), 1316–1326. doi:10.1016/j.eswa.2007.08.012

Hsu, M. H., Ju, T. L., Yen, C. H., & Chang, C. M. (2007). Knowledge sharing behavior in virtual communities: The relationship between trust, self-efficacy, and outcome expectations. *International Journal of Human-Computer Studies*, *65*(2), 153–169. doi:10.1016/j.ijhcs.2006.09.003

Hudson, M., Smart, A., & Bourne, M. (2001). Theory and practice in SME performance measurement systems. *International Journal of Operations & Production Management*, *21*(8), 1096–1115. doi:10.1108/EUM0000000005587

Hung, S. Y., Lai, H. M., & Chou, Y. C. (2010). *The determinants of knowledge sharing intention in professional virtual communities: An integrative model.* Paper presented at the 14th Pacific Asia Conference in Information Systems, Taipei, Taiwan.

Hyde, M., Wiggins, R. D., Higgs, P., & Blane, D. B. (2003). A measure of quality of life in early old age: The theory, development and properties of a needs satisfaction model (CASP-19). *Aging & Mental Health*, *7*(3), 186–194. doi:10.1080/1360786031000101157 PMID:12775399

Ida, R. I., Roshayati, H., & Fazli, I. (2012). *PLS application in journal of operations management: A review*. Paper presented at the Global Conference on Operations and Supply Chain Management, Bandung, Indonesia.

Ihaka, R., & Gentleman, R. C. (1996). R: a language for data analysis and graphics. *Journal of Computational and Graphical Statistics*, *5*, 299–314.

Ihlau, S., Duscha, H., & Gödecke, S. (2013). *Besonderheiten bei der Bewertung von KMU: Planungsplausibilisierung, Steuern, Kapitalisierung*. Wiesbaden: Springer Fachmedien. doi:10.1007/978-3-658-00945-8

Imai, T., Osada, H., & Nishimura, Y. (2009). Structure of concepts of purpose in life of retirees of 60 years of age or older: The difference of purpose in life concepts and subjective sense of well-being. [In Japanese]. *Japanese Journal of Gerontology*, *31*(3), 366–377.

Imai, T., Osada, H., & Nishimura, Y. (2012). Verification of the reliability and validity of the measure of purpose in life consciousness (Ikigai-9). [In Japanese]. *Japanese Journal of Public Health*, *59*(7), 433–439. PMID:22991767

Iqbal, J., & Mahmood, Y. (2012). Reviewing knowledge management literature. *Interdisciplinary Journal of Contemporary Research in Business*, *4*(6), 1005–1026.

ISO. (2015). *ISO 9001: Quality Management Systems - Requirements.* Geneva: International Organisation of Standardisation.

Jackson, P. (2010). Capturing, structuring and maintaining knowledge: A social software approach. *Industrial Management & Data Systems*, *110*(6), 908–929. doi:10.1108/02635571011055117

Jain, K. K., Sandhu, M. S., & Sidhu, G. K. (2007). Knowledge sharing among academic staff: A case study of business schools in Klang Valley, Malaysia. *JASA*, *2*, 23–28.

Janikow, C. Z. (1993). A knowledge-intensive genetic algorithm for supervised learning. *Machine Learning*, *13*(2-3), 189–228. doi:10.1007/BF00993043

Jayawardhena, C. (2010). The impact of service encounter quality in service evaluation: Evidence from a business-to-business context. *Journal of Business and Industrial Marketing*, *25*(5), 338–348. doi:10.1108/08858621011058106

Joia, L. A. (2002). Assessing unqualified in-service teacher training in Brazil using knowledge management theory: A case study. *Journal of Knowledge Management*, *6*(1), 74–86. doi:10.1108/13673270210417718

Jones, J. K. (2011). *Re-framing education as a thirdspace: neonarratives of pedagogy, power and transformation* (Doctoral dissertation). University of Southern Queensland.

Ju, T. L., Lin, B., Lin, C., & Kuo, H. J. (2006). TQM critical factors and KM value chain activities. *Total Quality Management & Business Excellence*, *17*(3), 373–393. doi:10.1080/14783360500451614

Kallinikos, J. (2006), Farewell to constructivism. Technology and context-embedded action. In The social study of information and communication technology: Innovations, actors, and contexts. Oxford, UK: Oxford University Press.

Kallinikos, J. (2005). The order of technology: Complexity and control in a connected world. *Information and Organization*, *15*(3), 185–202. doi:10.1016/j.infoandorg.2005.02.001

Kallinikos, J. (2006). *The consequences of information: Informational implications of technological change*. Cheltenham, UK: Edward Elgar. doi:10.4337/9781847204301

Kallinikos, J. (2009). On the computational rendition of reality: Artefacts and human agency. *Organization*, *16*(2), 183–202. doi:10.1177/1350508408100474

Kallinikos, J., Aaltonen, A., & Marton, A. (2013). The ontological ambivalence of digital objects. *Management Information Systems Quarterly*, *37*(2), 357–370. doi:10.25300/MISQ/2013/37.2.02

Kamarul Faizal Hashim. (2012). *Understanding the determinants of continuous knowledge sharing intention within business online communities* (Doctoral dissertation). Auckland University of Technology.

Kandula, S. R. (2006). *Performance Management*. New Delhi: Prentice Hall of India private limited.

Kangis, P., Williams, D., & Gordon, S. (2000). Organizational climate and corporate performance: An empirical investigation. *Management Decision*, *38*(8), 531–540. doi:10.1108/EUM0000000005371

Kanji, G. K. (2009). *An integrated Approach of Organisational Excellence*. Retrieved February 1, 2009, from http://www.gopal-kanji.com

Kanji, G. K., & Moura e Sá, P. (2002). Kanji's Business Scorecard. *Total Quality Management, 13*(1), 13–27. doi:10.1080/09544120120098537

Kankanhalli, A., Tan, B. C., & Wei, K. K. (2005). Contributing knowledge to electronic knowledge repositories: An empirical investigation. *Management Information Systems Quarterly, 29*(1), 113–143.

Kaplan, R. S., & Norton, D. P. (2000). *The Strategy-Focused Organisation: How Balanced Scorecard Companies Thrive in the New Business Environment*. Boston, MA: Harvard Business School Press.

Kay, K., & Shipman, C. (2014). *The Confidence Code: The Science and Art of Self-Assurance – What Women Should Know*. New York: HarperCollins.

Kennerley, M., & Neely, A. (2002). A framework of the factors affecting the evolution of performance measurement systems. *International Journal of Operations & Production Management, 22*(11), 1222–1245. doi:10.1108/01443570210450293

Khan, A. (2005). *Matching people with organizational culture*. Business Management Group.

Khan, Z., Bali, R. K., & Wickramasinghe, N. (2007). Developing a BPI framework and PAM for SMEs. *Industrial Management & Data Systems, 107*(3), 345–360. doi:10.1108/02635570710734262

Khatibian, N., Hasan gholoi pour, T., & Abedi Jafari, H. (2010). Measurement of knowledge management maturity level within organisations. *Business Strategy Series, 11*(1), 54–70. doi:10.1108/17515631011013113

Kidwell, J. J., Vander Linde, K., & Johnson, S. L. (2000). Applying Corporate Knowledge Management Practices in Higher Education. *EDUCAUSE Quarterly, 23*(4), 28–33.

Kilic, C., & Dursun, T. (2010). The Effect of Organizational Culture on Customer Orientation. *The Journal of American Academy of Business, Cambridge, 15*(2), 1–7.

Kirchler, D., & Wolfler Calvo, R. (2013). A Granular Tabu Search algorithm for the Dial-a-Ride Problem. *Transportation Research Part B: Methodological, 56*(October), 120–135. doi:10.1016/j.trb.2013.07.014

Knorr-Cetina, K. (1999). *Epistemic cultures: How the sciences make knowledge*. Cambridge, MA: Harvard University Press.

Knott, A. M. (2002). Exploration and exploitation as complements. In C. W. Choo & N. Botis (Eds.), *The strategic management of intellectual capital and organizational knowledge* (pp. 339–358). New York: Oxford University Press.

Kock, N. (2013). *Easily calculating the GoF fit index with WarpPLS outputs*. Retrieved from http://warppls.blogspot.com/2013/01/easily-calculating-gof-fit-index-with.html

Kosaka, M., & Shirahada, K. (Eds.). (2013). *Progressive Trends in Knowledge and System-Based Science for Service Innovation*. Hershey, PA: IGI Global.

Kotter, J. P. (2008). *Corporate culture and performance*. Simon and Schuster.

Kotter, P., & Heskett, L. (1992). *Corporate Culture and Performance*. New York: Free Press.

Kraft, P., Rise, J., Sutton, S., & Roysamb, E. (2005). Perceived difficulty in the theory of planned behaviour: Perceived behavioural control or affective attitude? *British Journal of Social Psychology*, *44*(3), 479–496. doi:10.1348/014466604X17533 PMID:16238850

Krejcie, R. V., & Morgan, D. W. (1970). Determining sample size for research activities. *Educational and Psychological Measurement*, *30*(3), 607–610. doi:10.1177/001316447003000308

Kuang, L. C., Davidson, E., & Yao, L. (2012). *A planned behavior-based investigation of knowledge sharing in construction industry*. Paper presented at the International Management and Engineering Conference.

Kubaisi, S. A. (2002). *Knowledge management and its impact on organisational innovation: an exploratory study comparing a sample of mixed industrial sector companies* (Doctoral dissertation). Baghdad.

Kubaisi, S. A. (2004). *Knowledge management and development organisations*. The Office of Universities, the Arab Republic of Egypt.

Kuhn, J. (2000). *Linking Continuous Improvement and Knowledge Management*. Retrieved from www.continuous-innovation.net/publications/papers-3rd/kuhn-ci.2000.pdf

Kulkarni, U. R., Ravindran, S., & Freeze, R. (2007). A knowledge management success model: Theoretical development and empirical validation. *Journal of Management Information Systems*, *23*(3), 309–347. doi:10.2753/MIS0742-1222230311

Lagrosen, S., Seyyed-Hashemi, R., & Leitner, M. (2004). Examination of the dimensions of quality in higher education. *Quality Assurance in Education*, *12*(2), 61–69. doi:10.1108/09684880410536431

Lahaie, D. (2005). The impact of corporate memory loss. What happens when a senior executive leaves? *Leadership in Health Services*, *18*(3), xxxv–xlvii. doi:10.1108/13660750510611198 PMID:16167654

Laitinen, E. K. (2002). A dynamic performance measurement system: Evidence from small Finnish technology companies. *Scandinavian Journal of Management*, *18*(1), 65–99. doi:10.1016/S0956-5221(00)00021-X

Larsson, R., & Finkelstein, S. (1999). Integrating Strategic, Organizational, and Human Resource Perspectives on Mergers and Acquisitions: A Case Survey of Synergy Realization. *Organization Science: A Journal of the Institute of Management Sciences*, *10*(1), 1–26.

Latan, H., & Ghozali, I. (2012). Partial least squares: Concept, technique and application using program SmartPLS for empirical research. In V. Pez, R. Butori, & G. C. de Kerviler (Eds.), *Because I'm worth it: The impact of given versus perceived status on preferential treatment efficiency. Academic Press.*

Lauer, T. (2001). Questions and information: Contrasting metaphors. *Information Systems Frontiers*, *3*(1), 41–48. doi:10.1023/A:1011497503953

Lee, S. K. J., & Yu, K. (2004). Corporate culture and organizational performance. *Journal of Managerial Psychology*, *19*(4), 340–359. doi:10.1108/02683940410537927

Leung, C. H. (2010). Critical factors of implementing knowledge management in school environment: A qualitative study in Hong Kong. *Research Journal of Information Technology*, *2*(2), 66–80. doi:10.3923/rjit.2010.66.80

Levine, S. (2006). High Performance Organizations: Creating a Culture of Agreement. Handbook of Business Strategy, 7(1), 375–380.

Levy, M. (2011). Knowledge retention: Minimizing organizational business loss. *Journal of Knowledge Management*, *15*(4), 582–600. doi:10.1108/13673271111151974

Lewis, W. G., Pun, K. F., & Lalla, T. R. M. (2007). Measuring employees' perception in small and medium-sized enterprises: A self-assessment scale. *International Journal of Management Science and Engineering Management*, *2*(3), 229–238.

Liebowitz, J. (2004). Bridging the knowledge and skills gap: Tapping federal retirees. *Public Personnel Management*, *33*(4), 421–448. doi:10.1177/009102600403300406

Lilleoere, A. M., & Hansen, E. H. (2011). Knowledge sharing enablers and barriers in pharmaceutical research and development. *Journal of Knowledge Management*, *15*(1), 53–70. doi:10.1108/13673271111108693

Lincoln, Y., & Guba, E. (1985). *Naturalistic inquiry*. Beverly Hills, CA: Sage.

Ling, C. W., Sandhu, M., & Jain, K. K. (2009). Knowledge sharing in an American multinational company based in Malaysia. *Journal of Workplace Learning*, *21*(2), 125–142. doi:10.1108/13665620910934825

Lin, H. F., & Lee, G. G. (2004). Perceptions of senior managers toward knowledge sharing behaviour. *Journal of Management Decision*, *42*(1), 108–125. doi:10.1108/00251740410510181

Lomas, L. (2004). Embedding quality: The challenges for higher education. *Quality Assurance in Education*, *12*(4), 157–165. doi:10.1108/09684880410561604

Lovelock, C. H., & Wirtz, J. (2010). *Service Marketing: People, Technology, Strategy* (7th ed.). Upper Saddle River, NJ: Prentice Hall.

Lusch, R. F., & Vargo, S. L. (Eds.). (2006). *The Service Dominant Logic of Marketing: Dialog, Debate, and Directions*. New York: Taylor & Francis.

Luszczynska, A., & Schwarzer, R. (2005). Social cognitive theory. In M. Conner & P. Norman (Eds.), *Predicting Health Behaviour* (2nd ed.). Buckingham, UK: Open University Press.

Lyons, K., Acsente, D., & van Waesberghe, M. (2008). Integrating knowledge management and quality management to sustain knowledge enabled excellence in performance. *Vine*, *38*(2), 241–253. doi:10.1108/03055720810889879

Mackenzie, A. (2002). *Transductions: Bodies and machines at speed*. London: Continuum.

Mackenzie, D. (2009). *Material markets: How economic agents are constructed*. Oxford, UK: Oxford University Press.

MacKinnon, D. P., & Fairchild, A. J. (2009). Current directions in mediation analysis. *Current Directions in Psychological Science*, *18*(1), 16–20. doi:10.1111/j.1467-8721.2009.01598.x PMID:20157637

Madden, T. J., Ellen, P. S., & Ajzen, I. (1992). A comparison of the theory of planned behavior and the theory of reasoned action. *Personality and Social Psychology Bulletin*, *18*(1), 3–9. doi:10.1177/0146167292181001

Maglio, P. P., Kieliszewski, C. A., & Spohrer, J. C. (Eds.). (2010). *Handbook of Service Science*. New York: Springer. doi:10.1007/978-1-4419-1628-0

Maitre, O., Kruger, F., Querry, S., Lachiche, N., & Collet, P. (2012). EASEA: Specification and execution of evolutionary algorithms on GPGPU. *Soft Computing*, *16*(2), 261–279. doi:10.1007/s00500-011-0718-z

Maktoum, A. M. b. R. (2007). *Mohammed addresses the knowledge conference*. Retrieved on February 17, 2011 from http://www.sheikhmohammed.co.ae/vgn-ext-

Malhotra, Y. (2005). Integrating knowledge management technologies in organisational business processes: Getting real enterprises to deliver real business performance. *Journal of Knowledge Management*, *9*(1), 7–28. doi:10.1108/13673270510582938

Mallak, A. L., Lyth, M. D., Olson, D. S., Ulshafer, M. S., & Sardone, J. F. (2003). Culture, the Built Environment and Healthcare Organizational Performance. *Managing Service Quality*, *13*(1), 27–38. doi:10.1108/09604520310456690

Manning, E. (2010). Always more than one: The collectivity of a life. *Body & Society*, *16*(1), 117–127. doi:10.1177/1357034X09354128

Maqsood, T., & Finegan, A. F. (2009). A knowledge management approach to innovation and learning in the construction industry. *International Journal of Managing Projects in Business*, *2*(2), 297–307. doi:10.1108/17538370910949310

Maqsood, T., Walker, D. H. T., & Finegan, A. D. (2007). Facilitating knowledge pull to deliver innovation through knowledge management: A case study. *Engineering, Construction, and Architectural Management*, *14*(1), 94–109. doi:10.1108/09699980710717007

March, J. (1991). Exploration and exploitation in organizational learning. *Organization Science*, *2*(1), 71–87. doi:10.1287/orsc.2.1.71

Marcoulides, G. A., & Heck, R. H. (1993). Organizational Culture and Performance: Proposing and Testing a Model. *Organization Science*, *4*(2), 209–225. doi:10.1287/orsc.4.2.209

Markus, M. L. (2001). Toward a theory of knowledge reuse: Types of knowledge reuse situations and factors in reuse success. *Journal of Management Information Systems*, *18*(1), 57–93. doi:10.1080/07421222.2001.11045671

Markus, M., & Robey, D. (1988). Information technology and organizational change: Causal structure in theory and research. *Management Science*, *34*(5), 583–598. doi:10.1287/mnsc.34.5.583

Marsh, S. J., & Stock, G. N. (2006). Creating Dynamic Capability: The Role of Intertemporal Integration, Knowledge Retention, and Interpretation. *Journal of Product Innovation Management*, *23*(5), 422–436. doi:10.1111/j.1540-5885.2006.00214.x

Martin, P., & Turner, B. (1986). Grounded theory and organizational research. *The Journal of Applied Behavioral Science*, *22*(2), 141–157. doi:10.1177/002188638602200207

Martins, E. C., & Meyer, H. W. J. (2012). Organizational and behavioral factors that influence knowledge retention. *Journal of Knowledge Management*, *16*(1), 77–96. doi:10.1108/13673271211198954

Mason, R. (1978). Measuring information output: A communications systems approach. *Information & Management*, *1*(4), 219–234. doi:10.1016/0378-7206(78)90028-9

Massumi, B. (2012). Technical mentality revisited: Brian Massumi on Gilbert Simondon. In Gilbert Simondon: Being and technology. Edinburgh, UK: Edinburgh University Press.

Ma, T., Wierzbicki, A. P., & Nakamori, Y. (2007). Establish a creative environment for roadmapping in academia - From the perspective of *i*-System methodology. *Journal of Systems Science and Systems Engineering, Springer*, *16*(4), 469–488. doi:10.1007/s11518-007-5047-5

Matt, D. T., Rauch, E., & Franzellin, V. (2012). Wissensarbeit in Kleinunternehmen am Beispiel des Baugewerbes. *Industrial Management (Des Plaines)*, *12*(28), 21–24.

Maykut, P., & Morehouse, P. S. (1994). *Beginning Qualitative Research. A Philosophic and Practical Guide*. London: Falmer Press.

Ma, Z., & Yu, K. H. (2010). Research paradigms of contemporary knowledge management studies: 1998-2007. *Journal of Knowledge Management*, *14*(2), 175–189. doi:10.1108/13673271011032337

McAdam, R. (2000). Knowledge management as a catalyst for innovation within organizations: A qualitative study. *Knowledge and Process Management*, *7*(4), 233–241. doi:10.1002/1099-1441(200010/12)7:4<233::AID-KPM94>3.0.CO;2-F

McAdam, R., & McCreedy, S. (1999). A critical review of knowledge management models. *The Learning Organization*, *6*(3), 91–101. doi:10.1108/09696479910270416

McDermott, R., & O'Dell, C. (2001). Overcoming cultural barriers to sharing knowledge. *Journal of Knowledge Management*, *5*(1), 76–85. doi:10.1108/13673270110384428

McKinney, E. H., & Yoos, C. J. (2010). Information about information: A taxonomy of views. *Management Information Systems Quarterly*, *34*(2), 329–344.

McNeil, C. M. (2003). Line managers: Facilitators of knowledge sharing in teams. *Employee Relations*, *25*(3), 294–307. doi:10.1108/01425450310475874

Meng, F., Huynh, V. N., Nakamori, Y., Fujimori, J., & Tatsuse, T. (2015). Development and evaluation of a social service system. *The 16th International Conference on Knowledge and Systems Science*, 155-165.

Metcalfe, M. (2006). Seeking Churchman's system for guaranteeing knowledge claims. *Systems Research and Behavioral Science*, *23*(1), 51–59. doi:10.1002/sres.676

Mezura-Montes, E., & Coello, C. A. (2006). A survey of constraint-handling techniques based on evolutionary multiobjective optimization. Workshop paper at PPSN.

Migdadi, M. (2009). Knowledge management enablers and outcomes in the small-and-medium sized enterprises. *Industrial Management & Data Systems*, *109*(6), 840–858. doi:10.1108/02635570910968072

Miles, M. B., & Huberman, A. M. (1994). *An Expanded Sourcebook. Qualitative Data Analysis* (2nd ed.). Thousand Oaks, CA: Sage.

Miles, M., & Huberman, A. (1994). *Qualitative data analysis*. Thousand Oaks, CA: Sage.

Miller, K., McAdam, R., Moffett, S., & Brennan, M. (2011). An exploratory study of retaining and maintaining knowledge in university technology transfer processes. *International Journal of Entrepreneurial Behaviour & Research*, *17*(6), 663–684. doi:10.1108/13552551111174729

Minbaeva, D., & Pedersen, T. (2010). Governing individual knowledge-sharing behaviour. *International Journal of Strategic Change Management*, *2*(2/3), 200–222. doi:10.1504/IJSCM.2010.034414

Mingers, J. (2006). Realising systems thinking: knowledge and action in management science. New York: Springer Science + Business Media.

Mingers, J. (2008). Management knowledge and knowledge management: Realism and forms of truth. *Knowledge Management Research and Practice*, *6*(1), 62–76. doi:10.1057/palgrave.kmrp.8500161

Mishra, B., & Bhaskar, A. U. (2011). Knowledge management process in two learning organisations. *Journal of Knowledge Management*, *15*(2), 344–359. doi:10.1108/13673271111119736

Mitchell, M. L., & Jolley, M. J. (2013). *Research design explained* (8th ed.). Wadsworth Cengage Learning.

Mitroff, I. (1973). Systems, inquiry, and the meanings of falsification. *Philosophy of Science*, *40*(2), 255–276. doi:10.1086/288520

Mitroff, I., & Linstone, H. A. (1993). *The unbounded mind: breaking the chains of traditional business thinking*. New York: Oxford University Press.

Mohamed, M., O'Sullivan, K. J., & Ribière, V. (2008). A paradigm shift in the Arab region knowledge evolution. *Journal of Knowledge Management*, *12*(5), 107–120. doi:10.1108/13673270810902975

Mohamed, M., Stankosky, M., & Murray, A. (2004). Applying knowledge management principles to enhance cross-functional team performance. *Journal of Knowledge Management*, *8*(3), 127–142. doi:10.1108/13673270410541097

Motik, B. (2007). On the properties of metamodeling in owl. *Journal of Logic and Computation*, *17*(4), 617–637. doi:10.1093/logcom/exm027

Muafi, M. (2009). The effects of alignment competitive strategy, culture, and role behavior on organizational performance in service firms. *International Journal of Organizational Innovation*, *2*(1), 106–134.

Murugan, M. (2010). A Study on Organizational Culture and Its Impact on the Performance of IT Employees in Chennai. The ICFAI Journal of Management Research, 8(5), 7-16.

Muya, J. N., & Wesonga, J. N. (2012). The Impact of Organizational Culture on Performance of Educational Institutions. *International Journal of Business and Social Science*, *3*(8), 211–217.

Nahapiet, J., & Ghoshal, S. (1998). Social capital, intellectual capital, and the organizational advantage. *Academy of Management Review*, *23*(2), 242–266.

Nahavandi, A., & Malekzadeh, A. R. (1993). *Organizational Culture in the Management of Mergers*. Westport, CT: Quorum Books.

Nahm, A. Y., Vonderembse, M., & Koufteros, X. A. (2004). The Impact of Organizational Culture on Time-Based Manufacturing and Performance. *Decision Sciences*, *35*(4), 579–607. doi:10.1111/j.1540-5915.2004.02660.x

Nakamori, Y. (2000). Knowledge management system toward sustainable society. *International Symposium on Knowledge and Systems Sciences: Challenges to Complexity*, 57-64.

Nakamori, Y. (2003). Systems methodology and mathematical models for knowledge management. *Journal of Systems Science and Systems Engineering*, *12*(1), 49–72. doi:10.1007/s11518-006-0120-z

Nakamori, Y. (2006). Designing, utilizing and evaluating 'technology-creating Ba' in a Japanese scientific research institution. *Systems Research and Behavioral Science*, *23*(1), 3–19. doi:10.1002/sres.705

Nakamori, Y. (2013). Knowledge reconstruction and justification for regional vitalization. *Journal of Systems Science and Systems Engineering*, *22*(4), 457–468. doi:10.1007/s11518-013-5232-7

Nakamori, Y. (Ed.). (2011). *Knowledge Science: Modeling the Knowledge Creation Process*. Boca Raton, FL: CRC Press, Taylor & Francis Group. doi:10.1201/b11054

Nakamori, Y., Wierzbicki, A. P., & Zhu, Z. C. (2011). A theory of knowledge construction systems. *Systems Research and Behavioral Science*, *28*(1), 15–39. doi:10.1002/sres.1046

Nakamori, Y., & Zhu, Z. C. (2004). Exploring a sociologist understanding for the *i*-System. *International Journal of Knowledge and Systems Science*, *1*(1), 1–8. doi:10.4018/jkss.2010010101

Nassuora, A.B., & Hassan, S. (2010). *Knowledge sharing among academics in institutions of higher learning*. Retrieved October 27, 2013 from http://www.kmice.cms.net.my/ProcKMICe/KMICe2010/Paper/PG164_173.pdf

Nathai-Balkissoon, M., & Pun, K. F. (2011). A holistic model and research agenda for organisational learning in manufacturing. *World Journal of Entrepreneurship, Management and Sustainable Development*, *7*(1), 27–38. doi:10.1108/20425961201000028

National Institute of Population and Social Security Research. (2012). *Population Projections for Japan (January 2012): 2011 to 2060*. Retrieved from http://www.ipss.go.jp/site-ad/index_english/esuikei/ppfj2012.pdf

Neely, A. (2005). Performance measurement system and relationships with performance results: Developments in the last decade and a research agenda for the next. *International Journal of Operations & Production Management*, *25*(12), 1264–1277. doi:10.1108/01443570510633648

Neely, A. D., Adams, C., & Kennerley, M. P. (2002). *Performance Prism: The Scorecard for Measuring and Managing Stakeholder Relationships*. London: Prentice Hall.

Neely, A. D., Gregory, M. J., & Platts, W. K. (2005). Performance measurement system design: A literature review and research agenda. *International Journal of Operations & Production Management*, *25*(12), 1228–1263.

Neely, A., Adams, C., & Crowe, P. (2001). The performance prism in practice. *Measuring Business Excellence*, *5*(2), 6–13. doi:10.1108/13683040110385142

Neely, A., & Jarrar, Y. (2004). Extracting value from data: The performance planning value chain. *Business Process Management Journal*, *10*(5), 506–509. doi:10.1108/14637150410559180

Newman, J. (2001). Some observations on the semantics of "information. *Information Systems Frontiers*, *3*(2), 155–167. doi:10.1023/A:1011439305567

Nguyen, N. T. (2007). *Advanced Methods for Inconsistent Knowledge Management (Advanced Information and Knowledge Processing)*. Secaucus, NJ: Springer-Verlag.

NIST. (2017). *Malcolm Baldrige National Quality Award*. National Institute of Standards and Technology, US Department of Commerce. Retrieved September 1, 2017, from http://www.nist.gov/

Nonaka, I. (1994). A dynamic theory of organizational knowledge creation. *Organization Science*, *5*(1), 14–37. doi:10.1287/orsc.5.1.14

Nonaka, I., & Takeuchi, H. (1995). *The knowledge-creating company: How Japanese companies create the dynamics of innovation*. Oxford University Press.

Nonaka, I., & Takeuchi, H. (1995). *The Knowledge-Creating Company: How Japanese Companies Create the Dynamics of Innovation*. Oxford, UK: Oxford University Press.

Nussbaum, M. C., & Sen, A. (Eds.). (1993). *The Quality of Life*. Oxford, UK: Clarendon Press. doi:10.1093/0198287976.001.0001

O'Donnell, A. (2014). The Contribution of Networking to Small Firm Marketing. *Journal of Small Business Management*, *52*(1), 164–187. doi:10.1111/jsbm.12038

OECD. (2013). Health at a Glance 2013 Report. OECD 2-13-11-21, Chapter 1.6. doi:10.1787/health_glance-2013-en

Ogiwara, N. (2009). *Knowledge Management: Case Studies for Small and Medium Enterprises*. Asian Productivity Organisation.

Okashah, A. (2008). *The impact of organizational culture at the level of functionality: an empirical study on telecommunication company "paltel" in palestine* (Thesis). The Islamic University-Ghaza. (in Arabic)

Olie, R. (1994). Shades of Culture and Institutions in International Mergers, (Periodical style). *Organization Studies*, *15*(3), 381–405. doi:10.1177/017084069401500304

Onken, M. (1999). Temporal elements of organizational culture and impact on firm performance. *Journal of Managerial Psychology*, *14*(3/4), 231–243. doi:10.1108/02683949910263756

Orlikowski, W. (1993). CASE tools as organizational change: Investigating incremental and radical changes in systems development. *Management Information Systems Quarterly*, *17*(3), 309–340. doi:10.2307/249774

Orlikowski, W. J. (2007). Sociomaterial practices: Exploring technology at work. *Organization Studies*, *28*(9), 1435–1448. doi:10.1177/0170840607081138

Orlikowski, W. J. (2010). The sociomateriality of organizational life: Considering technology in management research. *Cambridge Journal of Economics*, *34*(1), 125–141. doi:10.1093/cje/bep058

Orlikowski, W. J., & Iacono, C. S. (2001). Research commentary: Desperately seeking the 'IT' in IT research—A call to theorizing the IT artifact. *Information Systems Research*, *12*(2), 121–134. doi:10.1287/isre.12.2.121.9700

Orlikowski, W. J., & Scott, S. V. (2008). Sociomateriality: Challenging the separation of technology, work, and organization. *The Academy of Management Annals*, *2*(1), 433–474. doi:10.1080/19416520802211644

Orlikowski, W., & Baroudi, J. (1991). Studying information technology in organizations: Research approaches and assumptions. *Information Systems Research*, *2*(1), 1–28. doi:10.1287/isre.2.1.1

Orlikowski, W., & Robey, D. (1991). Information technology and the structuring of organizations. *Information Systems Research*, *2*(2), 143–169. doi:10.1287/isre.2.2.143

Ormrod, J. E. (2006). *Educational psychology: Developing learners* (5th ed.). Upper Saddle River, NJ: Pearson/Merrill Prentice Hall.

Paloti, R. G. (2010). *Knowledge sharing*. Retrieved January 17, 2011. http://eprints.rclis.org/bitstream/10760/15179/3/KnowledgeSharinFinal.pdf

Pangil, F., & Chan, J. M. (2014). The mediating effect of knowledge sharing on the relationship between trust and virtual team effectiveness. *Journal of Knowledge Management*, *18*(1), 1–24. doi:10.1108/JKM-09-2013-0341

Pankratz, G. (2005). A grouping genetic algorithm for the pickup and delivery problem with time windows. *OR Spectrum*, *27*(1).

Paquette, J., Cordeau, J., Laporte, G., & Pascoal, M. M. B. (2013). Combining multicriteria analysis and tabu search for dial-a-ride problems. *Transportation Research Part B: Methodological*, *52*, 1–16. doi:10.1016/j.trb.2013.02.007

Parise, S., Cross, R., & Davenport, T. H. (2006). Strategies for Preventing a Knowledge-Loss Crisis. *MIT Sloan Management Review*, *47*(4), 31–38.

Parragh, S.N., Doerner, K.F., & Hartl, R.F. (2008). A survey on pickup and delivery problems. *Journal fürBetriebswirtschaft*, *58*(1), 21-51.

Patel, M., & Ragsdell, G. (2011, June). To share or not to share knowledge: An ethical dilemma for UK academics? *Journal of Knowledge Management Practice*, *12*(2).

Patterson, M. G., West, M. A., Lawthom, R., & Nickell, S. (1997). *Impact of people management practices on business performance*. London: Institute of Personnel and Development. The Cromwell Press.

Paulzen, O., Doumi, M., Perc, P., & Cereijo-Roibas, A. (2002) A maturity model for quality improvement in knowledge management. In: *ACIS 2002 Proceedings* (Paper 5). Retrieved March 1, 2013, from http://aisel.aisnet.org/acis2002/5

Peet, M. (2012). Leadership transitions, tacit knowledge sharing and organizational generativity. *Journal of Knowledge Management*, *16*(1), 45–60. doi:10.1108/13673271211198936

Peirce, C. (1931-1958). The collected papers of C.S. Peirce (Vols. 1-8). Cambridge, MA: Harvard University Press.

Pettigrew, A. (1979). Studying organizational culture. *Administrative Science Quarterly*, *24*(4), 570–581. doi:10.2307/2392363

Pillania, R. K. (2009). Demystifying knowledge management. *Business Strategy Series*, *10*(2), 96–99. doi:10.1108/17515630910942223

Polanyi, M. (1962). *Personal knowledge - Towards a post critical philosophy*. Chicago: University of Chicago Press.

Pratasavitskaya, H., & Stensaker, B. R. (2010). Quality management in higher education: Towards a better understanding of an emerging field. *Quality in Higher Education*, *16*(1), 37–50. doi:10.1080/13538321003679465

Pun, K. F. (2003). *A Study of an Integrated Approach for Strategy Formulation and Performance Measurement in Manufacturing Enterprises* (Unpublished doctoral dissertation). Middlesex University School of Computing Science, London, UK.

Punch, K. F. (2013). Introduction to social research: Quantitative and qualitative approaches. *Sage (Atlanta, Ga.)*.

Pun, K. F. (2002). Development of an integrated TQM and performance measurement system for self-assessment: A method. *Total Quality Management*, *13*(6), 759–777. doi:10.1080/0954412022000010127

Pun, K. F., & Chin, K. S. (2005). On-line assessment of new product development performance: An approach. *Total Quality Management & Business Excellence*, *16*(2), 157–169. doi:10.1080/14783360500054244

Pun, K. F., Chin, K. S., & Yiu, M. Y. R. (2010). An AHP approach to assess new product development performance: An exploratory study. *International Journal of Management Science and Engineering Management*, *5*(3), 210–218.

Pun, K. F., & Lau, C. W. (2003). Integration of total quality management and performance measurements in government departments: An empirical study in Hong Kong. *International Journal of Business Performance Management*, *5*(4), 316–335. doi:10.1504/IJBPM.2003.003815

Pun, K. F., & White, A. S. (2005). A performance measurement paradigm for integrating strategy formulation: A review of systems and frameworks. *International Journal of Management Reviews*, *7*(1), 25–46. doi:10.1111/j.1468-2370.2005.00106.x

Pun, K. F., & Yiu, M. Y. R. (2017). Management of knowledge and ignorance in the context of organisational learning: A research agenda. *West Indian Journal of Engineering*, *40*(1), 24–30.

Rad, G. P., Alizadeh, N., Miandashti, N. Z., & Fami, H. S. (2011). Factors influencing knowledge sharing among personnel of agricultural extension and education organization in Iranian Ministry of Jihad-e Agriculture. *Journal of Agricultural Science and Technology*, *13*(4), 491–501.

Rahma, L., Khemakhem, M., & Semet, F. (2015). Rich vehicle routing problems: From a taxonomy to a definition. *European Journal of Operational Research*, *241*(1), 1–14. doi:10.1016/j.ejor.2014.07.048

Rahman, K. (2006). Learning from your business lectures: Using stepwise regression to understand course evaluation data. *The Journal of American Academy of Business, Cambridge*, *19*(2), 272–279.

Ramlee, A. R. (2011). *Knowledge sharing practices among researchers and officers of the national institutes of health Malaysia* (Unpublished doctoral dissertation). International Islamic University Malaysia, Kuala Lumpur, Malaysia.

Rappaport, A., Bancroft, E., & Okum, L. (2003, Winter). The Aging Workforce Raises New Talent Management Issues for Employers. *Journal of Organizational Excellence*, 55-66.

Rashid, A. Z., Sambasivan, M., & Johari, J. (2003). The Influence of Corporate Culture and Organizational Commitment on Performance. *Journal of Management Development*, *22*(8), 708–728. doi:10.1108/02621710310487873

Rastogi, P. N. (2000). Knowledge management and intellectual capital–the new virtuous reality of competitiveness. *Human Systems Management*, *19*(1), 39–48.

Rathswohl, S., & Franz, V. (2013). Auf Wissen bauen: Besonderheiten im Baugewerbe. *Wissensmanagement - Das Magazin für Führungskräfte*, *1*, 12-13.

Ratnatunga, J., Gray, N., & Balachandran, K. R. (2004). CEVITA™: The valuation and reporting of strategic capabilities. *Management Accounting Research*, *15*(1), 77–105. doi:10.1016/j.mar.2003.12.005

Reddy, A. (2011). Cultural Dimensions and Impact on Performance Management. *International Journal of Multidisciplinary Research*, *1*(6), 300–311.

Reeve, J., & Paperboy, D. (2007). Evaluating the evaluation: Understanding the utility and limitations of evaluation as a tool for organizational learning. *Health Education Journal*, *66*(2), 120–131. doi:10.1177/0017896907076750

Rennstam, J. (2012). Object-control: A study of technologically dense knowledge work. *Organization Studies*, *33*(8), 1071–1090. doi:10.1177/0170840612453527

Reychav, I., & Weisberg, J. (2009). Good for workers, good for companies: How knowledge sharing benefits individual employees. *Knowledge and Process Management*, *16*(4), 186–197. doi:10.1002/kpm.335

Ribeiro, F. L., & Ferreira, V. L. T. (2010). Using knowledge to improve preparation of construction projects. *Business Process Management Journal*, *16*(3), 361–376. doi:10.1108/14637151011049403

Ribes, D., & Bowker, G. C. (2009). 'Between meaning and machine: Learning to represent the knowledge of communities. *Information and Organization*, *19*(4), 199–217. doi:10.1016/j.infoandorg.2009.04.001

Ribes, D., Jackson, S., Geiger, S., Burton, M., & Finholt, T. (2013). Artifacts that organize: Delegation in the distributed organization. *Information and Organization*, *23*(1), 1–14. doi:10.1016/j.infoandorg.2012.08.001

Ricardo, R. (2001). *Corporate Performance Management: How to Build a Better Organization through Measurement Driven Strategies Alignment*. Butterworth Heinemann.

Riege, A. (2005). Three-dozen knowledge-sharing barriers managers should consider. *Journal of Knowledge Management*, *9*(3), 18–35. doi:10.1108/13673270510602746

Robbins, S. P., & Sanghi, S. (2007). *Organizational Behavior*. New Delhi: Pearson Education.

Rodman, H. (1980). Are Conceptual Frameworks Necessary for Theory Building? The Case of Family Sociology*. *The Sociological Quarterly, 21*(3), 429–441. doi:10.1111/j.1533-8525.1980.tb00623.x

Rose, N. (2013). The human sciences in a biological age. *Theory, Culture & Society, 30*(1), 3–34. doi:10.1177/0263276412456569

Ross, P. H., Ellipse, M. W., & Freeman, H. E. (2004). *Evaluation: A systematic approach* (7th ed.). Thousand Oaks, CA: Sage.

Rothwell, W. J. (2004, February). Capturing the Lessons of Experience: Knowledge Transfer: 12 Strategies for Succession Management. *IPMA-HR News*, 10-11.

Rousseau, D. (1991). Quantitative assessment of organizational culture. *Group & Organization Studies, 15*(4), 448–460. doi:10.1177/105960119001500408

Rousseau, D. (2004). Psychological Contracts in the Workplace: Understanding the Ties that Motivate. *The Academy of Management Executive, 18*(1), 120–127. doi:10.5465/AME.2004.12689213

Rowley, J., Seba, I., & Delbridge, R. (2012). Knowledge sharing in the Dubai Police Force. *Journal of Knowledge Management, 16*(1), 1–24.

Russell, S. J., & Norvig, P. (2009). *Artificial Intelligence: A Modern Approach* (3rd ed.). Pearson Education.

Ryu, S., Ho, S. H., & Han, I. (2003). Knowledge sharing behavior of physicians in hospitals. *Expert Systems with Applications, 25*(1), 113–122. doi:10.1016/S0957-4174(03)00011-3

Saar-Tsechansky, M., & Provost, F. (2007). Handling missing values when applying classification models. *Journal of Machine Learning Research, 8*, 1217–1250.

Sakao, T., & Lindahl, M. (2012). A value based evaluation method for Product/Service System using design information. *CIRP Annals Manufacturing Technology, 61*(1), 51–54. doi:10.1016/j.cirp.2012.03.108

Sallis, E., & Jones, G. (2002). *Knowledge Management in Education: Enhancing learning and education*. London: Kogan Page.

Sandhawalia, B. S., & Dalcher, D. (2011). Effective knowledge sharing for competitive advantage. *Journal of Knowledge Management, 15*(2).

Sandhu, M. S., Jain, K. K., & Umi Kalthom, A. (2011). Knowledge sharing among public sector employees: Evidence from Malaysia. *International Journal of Public Sector Management, 24*(3), 206–226. doi:10.1108/09513551111121347

Santos, V. R., Soares, A. L., & Carvalho, J. A. (2012). Knowledge sharing barriers in complex research and development projects: An exploratory study on the perceptions of project managers. *Knowledge and Process Management, 19*(1), 27–38. doi:10.1002/kpm.1379

Sarros, J., Gray, J., Densten, I., & Cooper, B. (2005). The Organizational Culture Profile Revisited and Revised: An Australian Perspective. *Australian Journal of Management, 30*(1), 159–182. doi:10.1177/031289620503000109

Sarvan, F. (2005). *Fundamentals of Quality Management in the University Context: The Example of Akdeniz University/Sarvan F., Anafarta N.* In The 49th EOQ Congress, Antalya, Turkey.

Sarvary, M. (1999). Knowledge management and competition in the consulting industry. *California Management Review, 41*(2), 95–107. doi:10.2307/41165988

Sasson, J. R., & Douglas, I. (2006). A conceptual integration of performance analysis, knowledge management, and technology: From concept to prototype. *Journal of Knowledge Management, 10*(6), 81–99. doi:10.1108/13673270610709233

Scardamalia, M., & Bereiter, C. (1999). Schools as knowledge building organizations. In D. Keating & C. Hertzman (Eds.), *Today's children, tomorrow's society: The developmental health and wealth of nations* (pp. 274–289). New York: Guilford.

Scardamalia, M., & Bereiter, C. (2006). Knowledge building: Theory, pedagogy, and technology. In R. K. Sawyer (Ed.), *The Cambridge handbook of the learning sciences* (pp. 97–118). Cambridge, UK: Cambridge University Press.

Schein, E. H. (1990). Organizational culture. *The American Psychologist, 43*(2), 109–119. doi:10.1037/0003-066X.45.2.109

Schein, E. H. (1992). *Organizational culture and leadership* (2nd ed.). San Francisco: Jossey-Bass Publishers.

Schwab, K. (2012). *The global competitiveness report 2012– 2013.* Geneva: World Economic Forum. Available from http://www3. weforum. org/docs/WEF_ GlobalCompetitivenessReport_2012-13. pdf

Schwandt, D. R., & Marquardt, M. (2000). *Organizational learning: From worldclass theories to global best practices.* Boca Raton, FL: CRC Press.

Seba, I., Rowley, J., & Lambert, S. (2012). Factors affecting attitudes and intentions towards knowledge sharing in the Dubai Police Force. *International Journal of Information Management, 1120*, 1–9.

Sekaran, U. (2003). *Research methods for business* (4th ed.). Hoboken, NJ: John Wiley & Sons.

Sekaran, U. (2003). *Research methods for business: A skill building approach (4th ed.).* New York: John Wiley & Sons.

Serenko, A., & Bontis, N. (2009). Global ranking of knowledge management and intellectual capital academic journals. *Journal of Knowledge Management, 13*(1), 4–15. doi:10.1108/13673270910931125

Sharma, A. (2010). *Enabling knowledge management of organizational memory for groups through shared topic maps* (Master dissertation). Iowa State University.

Shehabat, I., & Berrish, M. (2013). E-learning content enhanced by active knowledge management techniques. *International Council for Educational Media (ICEM), 2013 IEEE 63rd Annual Conference,* 1-8. doi:10.1109/CICEM.2013.6820185

Shehabat, I., & Mahdi, S. A. (2008). E-learning as a Knowledge Management Approach for Intellectual Capital Utilization. *Turkish Online Journal of Distance Education*, *9*(1), 205–216.

Sherer, M., Maddux, J. E., Mercandante, B., Prentice-Dunn, S., Jacobs, B., & Rogers, R. W. (1982). The self-efficacy scale: Contribution and validation. *Psychological Reports*, *51*(2), 663–671. doi:10.2466/pr0.1982.51.2.663

Siddique, C. M. (2012). Knowledge management initiatives in the United Arab Emirates: A baseline study. *Journal of Knowledge Management*, *16*(5).

Sim, J., Bartlam, B., & Bernard, M. (2011). The CASP-19 as a measure of quality of life in old age: Evaluation of its use in a retirement community. *Quality of Life Research: An International Journal of Quality of Life Aspects of Treatment, Care and Rehabilitation*, *20*(7), 997–1004. doi:10.1007/s11136-010-9835-x PMID:21243529

Simondon, G. (1964/1992). The genesis of the individual. In Incorporations. New York: Zone Books.

Simondon, G. (1980). *On the mode of existence of technical objects* (B. N. Mallahphy, Trans.). London: University of Western Ontario. (Original work published 1958)

Simondon, G. (2009a). The position of the problem of ontogenesis. *Parrhesia*, *7*, 4–16.

Simondon, G. (2009b). Technical mentality. *Parrhesia*, *7*, 17–21.

Simondon, G. (2011). *Two lessons on animal and man* (D. S. Burk, Trans.). Minneapolis, MN: Univocal.

Simon, H. (1997). The future of information systems. *Annals of Operations Research*, *71*, 3–14. doi:10.1023/A:1018975616482

Simonis, J. (2010). *A guide to question types: An art and a science*. Academic Press.

Simons, R. (2000). *Performance Management and Control Systems for Implementing Strategy*. Englewood Cliffs, NJ: Prentice Hall.

Singer, P., & Hurley, J. E. (2005). The Importance of Knowledge Management Today. *ALA-APA Library Worklife Home*, *2*(6), 1–3.

Skok, W., & Tahir, S. (2010). Developing a knowledge management strategy for the Arab World. *The Electronic Journal on Information Systems in Developing Countries*, *41*(7), 1–11.

Sohail, S., & Daud, S. (2009). Knowledge sharing in higher education institutions: Perspectives from Malaysia. *The Journal of Information and Knowledge Management Systems*, *39*(2), 125–142.

Soltani, E., & Lai, P. C. (2007). Approaches to quality management in the UK: Survey evidence and implications. *Benchmarking: An International Journal*, *14*(4), 429–454. doi:10.1108/14635770710761852

Soltani, E., Lai, P. C., & Mahmoudi, V. (2007). Managing change initiatives: Fantasy or Reality? The case of public sector organisations. *Total Quality Management & Business Excellence*, *18*(1–2), 153–179. doi:10.1080/14783360601053319

Spivak, V. A. (2001). *Organizational culture: theory and practice*. Sankt-Peterburg: Piter. (in Russian)

Staab, S., & Studer, R. (Eds.). (2004). *Handbook on Ontologies*. Springer.

Standards Australia. (2005). *Knowledge Management - A Guide AS 5037-2005*. Sydney: Standards Australia.

Stannack, P. (1996). Perspective on Employees Performance. *Management Research News*, *19*(4/5), 38–40. doi:10.1108/eb028456

Star, S. L. (1999). The ethnography of infrastructure. *The American Behavioral Scientist*, *43*(3), 377–391. doi:10.1177/00027649921955326

Stenmark, D. (2002). Information vs. knowledge: the role of intranets in knowledge management. *Proceedings of HICSS*. IEEE Computer Society Press. doi:10.1109/HICSS.2002.994043

Stiegler, B. (2011). *Technics and time, 3: Cinematic time and the question of malaise (S. Barker, Trans.)*. Stanford, CA: Stanford University Press.

St-Pierre, J., & Delisle, S. (2006). An expert diagnosis system for the benchmarking of SMEs' performance. *Benchmarking: An International Journal*, *13*(1/2), 106–119. doi:10.1108/14635770610644619

Strauss, A., & Corbin, J. (1998). *Basics of qualitative research: Techniques and procedures for developing grounded theory* (2nd ed.). Thousand Oaks, CA: Sage.

Stufflebeam, D. L., & Webster, W. J. (1980). An analysis of alternative approaches to evaluation. *Educational Evaluation and Policy Analysis*, *2*(3), 5–19. doi:10.3102/01623737002003005

Styhre, A. (2008). Transduction and entrepreneurship: A biophilosophical image of the entrepreneur. *Scandinavian Journal of Management*, *24*(2), 103–112. doi:10.1016/j.scaman.2008.03.004

Suddaby, R. (2006). What grounded theory is not? *Academy of Management Journal*, *49*(4), 633–642. doi:10.5465/AMJ.2006.22083020

Sujatha, R. (2007). Critical success factors for implementation of knowledge management. *Icfai Journal of Knowledge Management*, *5*(1), 61–72.

Sun, J., Nakamori, Y., Tian, J., & Xiang, J. W. (2016). Exploring academic knowledge creation models for graduate researches. *Proceedings of the 2016 IEEE International Conference on Software Quality, Reliability and Security*, 202-209. doi:10.1109/QRS-C.2016.31

Sveiby, K. E. (2002). A knowledge-based theory of the firm to guide strategy formulation. *Journal of Intellectual Capital*, *2*(4), 344–358. doi:10.1108/14691930110409651

Swanson, K. (1978). The two faces of organizational information. *Accounting, Organizations and Society*, *3*(2), 237–246. doi:10.1016/0361-3682(78)90015-6

Syed, S. A., Zaini, A., Noormala, A. I., & Zahariah, M. Z. (2009). Assessing knowledge sharing behaviour among employees in SMEs: An empirical study. *International Business Research*, *2*(2), 115–122.

Talisayon, S. D. (2008). *Knowledge Management in Asia: Experience and Lessons*. Asian Productivity Organisation.

Tan, C.N., Lye, Y.H., Ng, T.H., & Lim, Y.S. (2010). Motivational factors in influencing knowledge sharing among banks in Malaysia. *International Research Journal of Finance and Economics*, (44), 191-201.

Tan, H. C., Carrillo, P., Anumba, C., Kamara, J. M., Bouchlaghem, D., & Udeaja, C. (2006). Live capture and reuse of project knowledge in construction organisations. *Knowledge Management Research and Practice*, *4*(2), 149–161. doi:10.1057/palgrave.kmrp.8500097

Tasan, A. S., & Gen, M. (2012). A genetic algorithm based approach to vehicle routing problem with simultaneous pick-up and deliveries. *Computers & Industrial Engineering*, *62*(3), 755–761. doi:10.1016/j.cie.2011.11.025

Taticchi, P., & Balachandran, K. R. (2008). Forward performance measurement and management integrated frameworks. *International Journal of Accounting and Information Management*, *16*(2), 140–154. doi:10.1108/18347640810913807

Taticchi, P., Tonelli, F., & Cagnazzo, L. (2010). Performance measurement and management: A literature review and a research agenda. *Measuring Business Excellence*, *14*(1), 4–18. doi:10.1108/13683041011027418

Tatsuse, T., & Fujimori, J. (2014). Verification of the "Care-Will Model". In J. Fujimori (Ed.), *Research Report on Practice and Verification of the "Care-Will Model" toward Implementation of Autonomous Social Support to Male Retirees*. Research Grants of the Ministry of Health, Labor and Welfare.

Taylor, R., & Tofts, C. (2012). *Managing Complex Service Systems*. New York: Springer.

Tenenhaus, M., Vinzi, V. E., Chatelin, Y. M., & Lauro, C. (2005). PLS path modeling. *Computational Statistics & Data Analysis*, *48*(1), 159–205. doi:10.1016/j.csda.2004.03.005

Thacker, E. (2003). Data made flesh: Biotechnology and the discourse of the posthuman. *Cultural Critique*, *53*(1), 72–97. doi:10.1353/cul.2003.0029

Thiagaragan, T., Zairi, M., & Dale, B. G. (2001). A proposed model of TQM implementation based on an empirical study of Malaysian industry. *International Journal of Quality & Reliability Management, 18*(3), 289–306. doi:10.1108/02656710110383539

Tian, J., Nakamori, Y., & Wierzbicki, A. P. (2009). Knowledge management and knowledge creation in academia: A study based on surveys in a Japanese research university. *Journal of Knowledge Management, 13*(2), 76–92. doi:10.1108/13673270910942718

Tichy, N. M. (1982). Managing Change Strategically: The Technical, Political, and Cultural Keys. *Organizational Dynamics, 11*(2), 59–80. doi:10.1016/0090-2616(82)90005-5 PMID:10298937

Tikly, L., & Barrett, A. M. (2011). Social justice, capabilities and the quality of education in low income countries. *Total Quality Management & Business Excellence, 17*(3), 303–323.

Titiev, M. (1959). *Introduction to Cultural Anthropology*. New York: Henry Holt & Company.

Tohidinia, Z., & Mosakhani, M. (2010). Knowledge sharing behavior and its predictors. *Industrial Management & Data Systems, 110*(4), 611–631. doi:10.1108/02635571011039052

Treleaven, L., & Sykes, C. (2005). Loss of organizational knowledge. From supporting clients to serving head office. *Journal of Organizational Change Management, 18*(4), 353–368. doi:10.1108/09534810510607056

Tsoukas, H., & Vladimirou, E. (2001). What is organizational knowledge. *Journal of Management Studies, 38*(7), 973–993. doi:10.1111/1467-6486.00268

Tuan, N., & Huy, V. (2008). Information as system in IS: A systems thinking perspective. *International Journal of Information Technologies and Systems Approach, 1*(2), 1–19. doi:10.4018/jitsa.2008070101

Tuomi, I. (1999). Data is more than knowledge: Implications of the reversed knowledge hierarchy for knowledge management and organizational memory. *Journal of Management Information Systems, 16*(3), 107–121. doi:10.1080/07421222.1999.11518258

Turner, B. (1983). The use of grounded theory for the qualitative analysis of organizational behavior. *Journal of Management Studies, 20*(3), 333–348. doi:10.1111/j.1467-6486.1983.tb00211.x

Turner, T., Creighton, S., Nudurupati, S., & Bititci, U. (2004). Implementing a Web enabled performance measurement system. *Measuring Business Excellence, 8*(2), 40–54. doi:10.1108/13683040410539427

Ullman, E. (1997). *Close to the machine: Technophilia and its discontents*. San Francisco: City Lights Book.

van de Ven, A. H. (2007). *Engaged scholarship: A guide for organizational and social research*. London: Oxford University Press.

Van der Post, W. Z., de Coning, T. J., & Smit, E. M. (1997). An instrument to measure organizational culture. *South African Journal of Business Management, 28*(4), 147–168.

Vargo, S. L., Maglio, P. P., & Akaka, M. A. (2008). On value and value co-creation: A service systems and service logic perspective. *European Management Journal*, *26*(3), 145–152. doi:10.1016/j.emj.2008.04.003

Vayanos, P., Hamilton, B. A., & Hammoud, M. (2007). *Promoting the growth and competitiveness of the insurance sector in the Arab world*. Academic Press.

Veiga, J., Lubatkin, M., Calori, R., & Very, P. (2000). Measuring organizational culture clashes: A two-nation post-hoc analysis of a cultural compatibility index. *Human Relations*, *53*(4), 539–557. doi:10.1177/0018726700534004

Vinzi, V. E., & Russolillo, G. (2013). Partial least squares algorithms and methods. *Wiley Interdisciplinary Reviews: Computational Statistics*, *5*(1), 1–19. doi:10.1002/wics.1239

Vinzi, V. E., Trinchera, L., & Amato, S. (2010). PLS path modeling: From foundations to recent developments and open issues for model assessment and improvement. In V. E. Vinzi, W. W. Chin, J. Henseler, & H. Wang (Eds.), *Handbook of partial least squares: Concepts, methods and applications* (pp. 47–82). Berlin, Germany: Springer-Verlag. doi:10.1007/978-3-540-32827-8_3

Vogt, W. P. (2007). *Quantitative research methods for professionals*. Boston, MA: Pearson Allyn and Bacon.

von Hayek, F. (1962). The use of knowledge in society. *The American Economic Review*, *35*(4), 519–530.

von Krogh, G. (2009). Individualist and collectivist perspectives on knowledge in organizations: Implications for IS research. *The Journal of Strategic Information Systems*, *18*, 119–129. doi:10.1016/j.jsis.2009.08.001

Walter, A. (1999). Relationship promoters: Driving forces for successful customer relationships. *Industrial Marketing Management*, *28*(5), 537–551. doi:10.1016/S0019-8501(99)00081-4

Wang, C. L., & Ahmed, P. K. (2005). The knowledge value chain: A pragmatic knowledge implementation network. Handbook of Business Strategy, 321-326.

Wang, S., & Noe, R. A. (2010). Knowledge sharing: A review and directions for future research. *Human Resource Management Review*, *20*(2), 115–131. doi:10.1016/j.hrmr.2009.10.001

Wasko, M. M., & Faraj, S. (2005). Why should I share? Examining social capital and knowledge contribution in electronic networks of practice. *Management Information Systems Quarterly*, *29*(1), 35–57.

Weathington, B. L., Cunningham, C. J., & Pittenger, D. J. (2010). *Research methods for the behavioral and social sciences*. Hoboken, NJ: John Wiley & Sons, Inc.

Wei, C. C., Choy, C. S., & Yew, W. K. (2009). Is the Malaysian telecommunication industry ready for knowledge management implementation? *Journal of Knowledge Management*, *13*(1), 69–87. doi:10.1108/13673270910931170

Weick, K. (1979). *The social psychology of organizing*. Reading, MA: Addison-Wesley.

Weick, K. (1998). Improvisation as a mindset for organizational analysis. *Organization Science*, *9*(5), 543–555. doi:10.1287/orsc.9.5.543

Wetzels, M., Odekerken-Schroder, G., & Van Oppen, C. (2009). Using PLS path modeling for assessing hierarchical construct models: Guidelines and empirical illustration. *Management Information Systems Quarterly*, *33*(1), 177–196.

Whyte, G., & Classen, S. (2012). Using storytelling to elicit tacit knowledge from SMEs. *Journal of Knowledge Management*, *16*(6), 950–962. doi:10.1108/13673271211276218

Wiener, N. (1948). *Cybernetics, or control and communication in the animal machine*. New York: John Wiley.

Wierzbicki, A. P., & Nakamori, Y. (2006). *Creative Space: Models of Creative Processes for the Knowledge Civilization Age*. Berlin: Springer-Verlag. doi:10.1007/b137889

Wiley, N. (1988). The micro-macro problem in social theory. *Sociological Theory*, *6*(2), 254–261. doi:10.2307/202119

Williams, R. S. (2002). *Managing Employee Performance: Design and implementation in organizations*. London: Thomson Learning.

Witten, I. H., Frankc, E., & Hall, M. A. (2011). *Data Mining: Practical Machine Learning Tools and Techniques (3rd ed.)*. Morgan Kaufmann Series in Data Management Systems.

Wong, K. Y., & Aspinwall, E. (2005). An empirical study of the important factors for knowledge-management adoption in the SME sector. *Journal of Knowledge Management*, *9*(3), 64–82. doi:10.1108/13673270510602773

Wongrassamee, S., Simmons, J. E. L., & Gardiner, P. D. (2003). Performance measurement tools: The Balanced Scorecard and the EFQM Excellence Model. *Measuring Business Excellence*, *7*(1), 14–29. doi:10.1108/13683040310466690

Wu, W. L., Lee, Y. C., & Shu, H. S. (2013). Knowledge management in education organization: A perspective of knowledge spiral. *The International Journal of Organizational Innovation*, *5*(4), 7–13.

Wu, Y., & Zhu, W. (2012). An integrated theoretical model for determinants of knowledge sharing behaviours. *Kybernetes*, *41*(10), 1462–1482. doi:10.1108/03684921211276675

Xiao, Y. C., & Jin, Y. H. (2010). The hierarchical linear modeling of shared mental model on virtual team effectiveness. *Kybernetes*, *39*(8), 1322–1329. doi:10.1108/03684921011063619

Yamashita, Y., Nakamori, Y., & Wierzbicki, A. P. (2009). Knowledge synthesis in technology development. *International Journal of Systems Science and Systems Engineering*, *18*(2), 184–202. doi:10.1007/s11518-009-5103-4

Yin, R. (2003). *Case study research: design and methods* (3rd ed.). Thousand Oaks, CA: Sage.

Yiu, M. Y. R. (2012). *An Investigation into an Integrated Knowledge Management Approach for Industrial Organisations* (Unpublished MPhil dissertation). Faculty of Engineering, The University of the West Indies, St Augustine.

Yiu, M. Y. R., & Pun, K. F. (2014). Measures of knowledge management performance in organisations: An exploratory study based on an integrated model. *The Learning Organization*, *21*(5/6), 310–332.

Yiu, M. Y. R., & Sankat, C. K. (2008). A self-assessment model for evaluating knowledge management performance. *The Journal of the Associations of Professional Engineers of Trinidad and Tobago*, *37*(1), 39–47.

Yiu, M. Y. R., & Sankat, C. K. 2012, Measuring knowledge management performance in industrial enterprises: an integrated model. In *Proceedings of the ANQ Congress 2012*. Hong Kong University of Science and Technology.

Yiu, M. Y. R., Sankat, C. K., & Pun, K. F. (2013). In search of the knowledge management practices in organisations: A review. *West Indian Journal of Engineering*, *35*(2), 103–116.

Yun, G., Shin, D., Kim, H., & Lee, S. (2011). Knowledge-mapping model for construction project organizations. *Journal of Knowledge Management*, *15*(3), 528–548. doi:10.1108/13673271111137475

Yu, S., Kim, Y., & Kim, M. (2004). Linking organisational knowledge management drivers to knowledge management performance: An exploratory study. In *Proceedings of the 37th Hawaii International Conference on System Sciences*. Piscataway NJ: IEEE Computer Society.

Zack, M. H. (1999, Summer). Managing codified knowledge. *Sloan Management Review*, *40*(4), 45–58.

Zack, M. H. (2002). Developing a knowledge strategy. *California Management Review*, *41*(3), 125–145. doi:10.2307/41166000

Zhang, P., & Ng, F. (2013). Explaining knowledge-sharing intention in construction teams in Hong Kong. *Journal of Construction Engineering and Management*, *139*(3), 280–293. doi:10.1061/(ASCE)CO.1943-7862.0000607

Zhao, J. (2010). School knowledge management framework and strategies: The new perspective on teacher professional development. *Computers in Human Behavior*, *26*(2), 168–175. doi:10.1016/j.chb.2009.10.009

Zhou, A., Qu, B., Li, H., Zhao, S., Suganthan, P. N., & Zhang, Q. (2011). Ponnuthurai NagaratnamSuganthan, and Qingfu Zhang. Multiobjective evolutionary algorithms: A survey of the state of the art. *Swarm and Evolutionary Computation*, *1*(1), 32–49. doi:10.1016/j.swevo.2011.03.001

Zhu, Z. (2006). Nonaka meets Giddens: A critique. *Knowledge Management Research and Practice*, *4*(2), 106–115. doi:10.1057/palgrave.kmrp.8500091

Zorn, T. (2001). *Knowledge Management, E-Organization and the Technology of Management Fashion*. Retrieved from www.mngt.waikato.ac.nz/ict/mgtfashion.pdf.may

Zott, C., & Huy, N. (2007). How entrepreneurs use symbolic management to acquire resources. *Administrative Science Quarterly*, *52*(1), 70–105. doi:10.2189/asqu.52.1.70

About the Contributors

W. B. Lee is the Chair Professor of the Department of Industrial and Systems Engineering, The Hong Kong Polytechnic University. He has pioneered research and practice of knowledge management in various industrial sectors in Hong Kong which include manufacturing, trading, public utilities and health care, and has conducted research projects in knowledge elicitation and mapping, unstructured information management, and organizational learning and innovation. Professor Lee and his team have launched Asia's first on-line MSc. Program in Knowledge Management, as well as the Global MAKE (Most Admired Knowledge Enterprise) Award in Hong Kong and Mainland China and chair the annual Asian Knowledge Forum. He is currently the chief editor of the Journal of Information and Knowledge Management Systems (Emerald) and the International Journal of Knowledge and Systems Science (IGI Publishing).

Farzad Sabetzadeh has his PhD in Knowledge Management from the Hong Kong Polytechnic University. Prior to his research focus on knowledge management, he had a multidisciplinary research and work track, both in IT management and Industrial Engineering. He has also spent more than 7 years of continuous research work at the Knowledge Management Research and Innovation Center (KMIRC) at the Hong Kong Polytechnic University and engaged in various projects of knowledge management in areas like e-learning, personalized learning environment, intellectual capital reporting, big data and Massive Open Online Courses (MOOC). Currently he is working as the knowledge management specialist and research scientist in education industry. His current research is on design and deployment of the state-of-the-art web technologies for knowledge visualization and learning analytics.

* * *

Huda Alami Skaik holds Ph.D in Knowledge Sharing, M.A.in LIS, Diploma in Business, B.A. in Eng. Lang. & Liter. She currently works as Freelance Researcher, Editor, and Translator. Her research interests include knowledge management, knowledge sharing/transfer/exchange, knowledge creation, library and information science, education, and others. She has a long experience in conducting and reviewing researches and academic works, editing and proofreading, and translation.

Anas Mahmoud Bashayreh was born on the 15th of May 1982, holding Jordanian nationality. He had the Bachelor's Degree from Al-Albayt University located in Jordan in Business Administration. Then he had the Master's Degree from Universiti Utara Malaysia (UUM) located in Malaysia in Human Resource Management. Then he completed a PhD degree from Universiti Sains Islam Malaysia (USIM) majoring in Human Resource Management. His employment history working as assistant professor for one year at University of Petra, and Al-Zhra'a College located at Muscat in Oman.

Guido Bruns is an entrepreneur in the field of SME consulting and education. His main research topics are Knowledge and Strategic Management.

Carlos Catania is Associated Professor in Computing Science at National University of Cuyo . His current research focuses on in the application of Machine learning and data mining algorithms on massive data, genetic algorithms for knowledge discovery,, application of parallelization techniques on data mining and machine learning algorithms and metaheuristics-based methods for optimization.

François de Bertrand de Beuvron is an Associate Professor of Computer Science at INSA Graduate School of Science and Technology of Strasbourg (France). He is a member of the Icube (Engineering science, computer science and imaging) laboratory of the Strasbourg University. His current research focuses on hybrid, probabilistic, logic based knowledge representation systems for problem solving in engineering, extended CAD systems, and semantic image analysis.

Ingi Runar Edvardsson is a Professor at the School of Business, University of Iceland. He received his PhD in Sociology from the University of Lund, Sweden. His research and publication includes: knowledge and human resource management, outsourcing, regional universities and Nordic labour markets.

Van-Nam Huynh is an associate professor at Japan Advanced Institute of Science and Technology, Japan. He received a PhD in Mathematics (1999) from the Institute of Information Technology, Vietnam Academy of Science and Technology, and a "Habilitation à Diriger des Recherches" (2012) at Université de Technologie de Compiègne, France. He was a post-doctoral fellow (2001–2002) awarded by Inoue Foundation for Science at Japan Advanced Institute of Science and Technology

(JAIST). His research interests include decision theories, computing and reasoning with words, information fusion, kansei information processing and application, data mining and recommender systems. Currently he is a member of the Editorial Board of the International Journal of Approximate Reasoning (Elsevier) and an associate editor of the International Journal of Knowledge and Systems Science (IGI Global).

Nguyen Manh Tuan is a senior lecturer in information systems and decision sciences at School of Industrial Management, HCMC University of Technology, Viet Nam. He received his PhD in Business Administration (Management Information Systems) from HCMC University of Technology. His current research interests include pragmatism and semiotics in organizational phenomena, philosophy of inquiry and of management, systems thinking, organizational information, knowledge management, and service management.

Fei Meng received a PhD in Knowledge Science from Japan Advanced Institute of Science and Technology. Her research interest is the development of a systematic evaluation method of knowledge management activities. Especially, she has been focusing on knowledge creation and value creation of people engaged in various service systems in local communities.

Roslina Othman is a Professor at the Department of Library and Information Science, Faculty of Information and Communication Technology, International Islamic University Malaysia, Kuala Lumpur, Malaysia. Dr. Roslina has a Bachelor Degree in Business Administration – Finance, Minor in Economics from Eastern Washington University, WA, USA. She obtained a Master in Library and Information Science from the International Islamic University Malaysia, and a Ph.D in Information Retrieval from University of Wales, Aberystwyth, UK. Her areas of research interest include: Semantic-based Search Engine, Ontology, Knowledge Management; Information Retrieval System, Search Engines, Digital Library, Bibliometrics, Webometrics and Big Data Analytics.

Kit Fai Pun is presently the Professor of Industrial Engineering (IE) of the Faculty of Engineering at The University of the West Indies. He is a Registered Professional Engineer in Australia, Europe, Hong Kong, and The Republic of Trinidad and Tobago. His research interests and activities include industrial engineering, engineering management, quality systems, performance measurement, innovation, and information systems.

Issa Shehabat earned his PhD, MSc, and BSc, in Computer Systems and Networks from Kharkov State Polytechnic University - UKRAINE. He is an Assistant

Professor in MIS. His research interests in areas like Knowledge Management, E-Learning, E-Commerce, E-Government, Decision Support Systems, and Total Quality Management. He worked jointly in the development of e-learning systems under the umbrella of the UNESCO projects. Published many articles in journals and conferences, and authored one book chapter. Dr. Shehabat worked as a vice Dean and head of a department at Philadelphia University- Jordan, and as a Director of Quality Assurance at King Saud and Shaqra Universities – Saudi Arabia.

Alexander Styhre, Ph.D., is chair of organization theory and management, Dept. of Business Administration, School of Business, Economics and Law, University of Gothenburg. Styhre earned his Ph.D. from the School of Economics and Management, Lund University, in 1998. Styhre has published widely in the field of organization and his work has appeared in e.g., Journal of Management Studies, Organization Studies, and Academy of Management Review. The most recent books published are and Corporate governance, the firm and investor capitalism: Legal-political and economic views (Edward Elgar, 2016), and the making of the shareholder welfare society (Routledge, 2017). Styhre is the former Editor-in-Chief of Scandinavian Journal of Management and is on the editorial board of Academy of Management Review and Organziation Studies.

Jing Sun is a doctoral student of School of Knowledge Science, Japan Advanced Institute of Science and Technology. She has been studying knowledge management and systems sciences. Her research interest is the development of knowledge creation and validation models to support graduate students to promote research confidently.

Jing Tian is an associate professor at School of Computer Science and Technology, Wuhan University of Technology, China. She received a PhD in Knowledge Science from Japan Advanced Institute of Science and Technology in 2006. Her research interest includes knowledge management, knowledge engineering, soft engineering and system methodology.

Jianguo Wu is a master student of School of Computer Science and Technology, Wuhan University of Technology. He has been studying knowledge management and privacy protection. His research interest is knowledge management and knowledge creation in undergraduate and privacy protection by negative survey.

Man Yin Rebecca Yiu is a Researcher in the Department of Mechanical and Manufacturing Engineering at The University of the West Indies, Trinidad and Tobago. She holds an MPhil in Industrial Engineering and a Master degree in Information Systems. Her research interests are in the areas of industrial engineering, knowledge management and information systems.

Cecilia Zanni-Merk is Professor in Computer Science at INSA Rouen Normandie She is also the head of the MIND team at the LITIS laboratory (EA 4108 / FR CNRS 3638) . Her main research interests are in conceptual modelling and reasoning strategies for problem solving in Engineering.

Index

A

academics 151-153, 155, 157, 159, 161, 163, 165-166, 177-180, 225, 271

B

barriers 52, 54, 105, 151, 159-160, 178, 271
building and construction 100-101, 105-108, 113-114

E

Evolutionary Algorithms 192-193, 195, 197-198, 200-201, 204, 212, 220
excellence 225-226, 230-231, 246, 248, 257, 259, 271-272, 274, 277

G

Gilbert Simondon 38, 40-41, 46
graduate research 70-72, 77
grounded theory 5-7, 11

H

higher education 71, 151-155, 178-180, 257-259, 269, 273-275

I

individuation 38, 40-45
information systems 1-2, 4, 29, 38-40, 46, 263-264
information technology 28, 38-40, 44, 46, 153, 262, 264-265, 274
insurance 50-51, 54, 58, 62-64

K

knowledge creation model 70-77
knowledge engineering 192-193, 197, 220
knowledge loss 102, 111, 114
knowledge management 1-2, 71, 100-102, 108, 113-114, 152-154, 157-158, 224-226, 233, 235, 239, 256-257, 259-260, 262-263, 274
knowledge retention 100-103, 105-114
knowledge sharing practice 151, 160, 163-166, 168, 177

O

ontogenesis 41-42
ontologies 40, 198, 220
organizational culture 50-54, 57-58, 60-64
organizational information 1-4, 6-7, 9-28
organizational performance 50-52, 56-64

P

performance 15, 26, 50-53, 55-64, 152-154, 160, 176, 219, 224-233, 235-237, 239-242, 244-248, 256-257, 259, 262-263, 265-266, 268-270, 272-274, 278, 281
predictors 60, 151, 163, 179, 289

Q

questionnaire survey 70-72, 81, 89, 91-93, 121, 129, 135-136

Q

research ability 70-71, 79, 83, 87, 92
research environment 70-72, 79, 84, 86-87, 89-90, 92

S

semiotics 1, 3-5, 9, 13, 26-27, 29
small and medium-sized enterprises (SMEs) 101, 227
social service systems 120-121, 142
systems thinking 1, 3-5

T

Total Quality Management (TQM) 224-226, 229-231, 233, 235, 248, 256-257, 259, 268-281, 283, 286-287, 289-290
transduction 38, 40, 42-43

V

value co-creation 121, 125

www.ingramcontent.com/pod-product-compliance
Ingram Content Group UK Ltd.
Pitfield, Milton Keynes, MK11 3LW, UK
UKHW050947290926
15470UKWH00005B/1693

9 781522 556558